Wartime

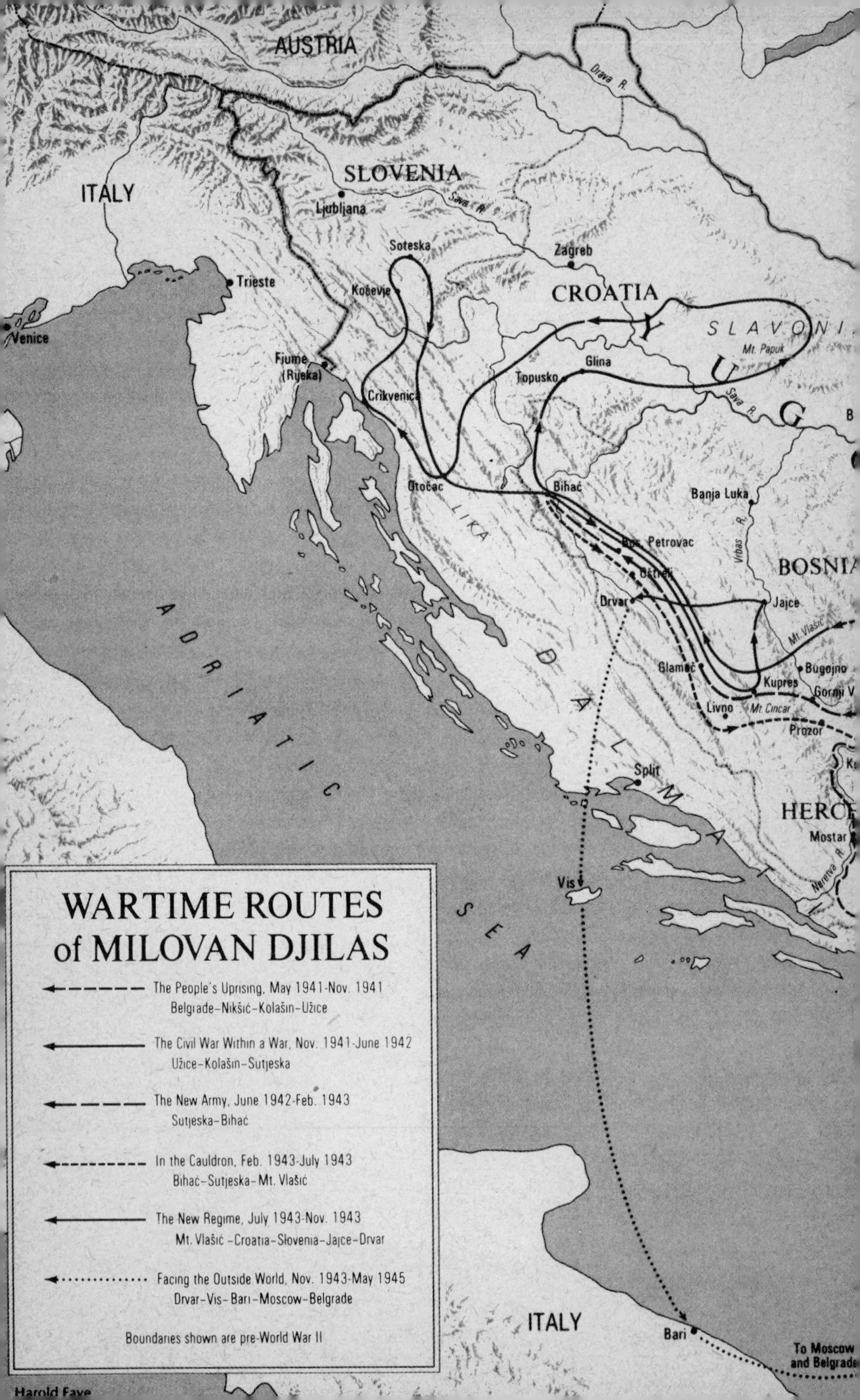
AUSTRIA
ITALY
SLOVENIA
Ljubljana
Sava R.
Drava R.
Zagreb
CROATIA
Soteska
Kočevje
Trieste
Venice
Fiume
(Rijeka)
Crikvenica
Topusko
Glina
SLAVONI
Mt. Papuk
Y
U
G
Sava R.
Otočac
LIKA
Bihać
Banja Luka
Bos. Petrovac
Vrbas R.
BOSNI
Drvar
Jajce
Mt. Vlašić
ADRIATIC
D
A
L
M
A
T
I
Glamoč
Bugojno
Kupres
Gornji V
Livno
Mt. Cincar
Prozor
Split
HERCE
Mostar
Neretva R.
Vis
SEA
WARTIME ROUTES
of MILOVAN DJILAS
The People's Uprising, May 1941-Nov. 1941
Belgrade-Nikšić-Kolašin-Užice
The Civil War Within a War, Nov. 1941-June 1942
Užice-Kolašin-Sutjeska
The New Army, June 1942-Feb. 1943
Sutjeska-Bihać
In the Cauldron, Feb. 1943-July 1943
Bihać-Sutjeska-Mt. Vlašić
The New Regime, July 1943-Nov. 1943
Mt. Vlašić -Croatia-Slovenia-Jajce-Drvar
Facing the Outside World, Nov. 1943-May 1945
Drvar-Vis-Bari-Moscow-Belgrade
Boundaries shown are pre-World War II
ITALY
Bari
To Moscow
and Belgrade
Harold Faye

HUNGARY
0
25
50
75
Miles
Drava R.
Osijek
VOJVODINA
ROMANIA
Novi Sad
Bosna R.
Danube R.
Belgrade
Sava R.
Drina R.
S
L
A
V
I
A
Valjevo
Kladanj
Bosna R.
Sarajevo
SERBIA
Užice
Morava R.
Danube
Partizanske Vode
Mt. Zlatibor
Drina R.
Čajniče
Kalinovik
Foča
Pljevlja
Nova Varoš
Niš
VINA
Tara R.
Mt. Zlatar
Sutjeska R.
Žabljak
SANDŽAK
Plužine
Mt. Durmitor
Mojkovac
Bijelo Polje
Šavnik
Piva R.
Podbišće
Kolašin
Berane
Nikšić
Andrijevica
KOSOVO
Sofia
MONTENEGRO
Dubrovnik
Mt. Čakor
BULGA
Kamenik
Murino
Danilovgrad
Bioče
Podgorica
Cetinje
Skoplje
MACEDONIA
Vardar R.
ALBANIA
GREECE

TRANSLATED BY MICHAEL B. PETROVICH 

Milovan Djilas

A HARVEST/HBJ BOOK

HARCOURT BRACE JOVANOVICH | NEW YORK AND LONDON

WARTIME

Printed in the United States of America

Library of Congress Cataloging in Publication Data

Dilas, Milovan.
Wartime.

(Harvest/HBJ book)
Includes index.
1. World War, 1939–1945—Underground movements—
Yugoslavia—Biography. 2. Dilas, Milovan.
3. Statesmen—Yugoslavia—Biography. 4. World
War, 1939–1945—Personal narratives, Serbian.
I. Title.
[D802.Y8D548 1980] 940.53'497'0924 [B] 80-16174
ISBN 0-15-694712-9

First Harvest/HBJ edition 1980

ABCDEFGHIJ

Contents

The People's Uprising 3

The Civil War within a War 91

The New Army 187

In the Cauldron 215

The New Regime 309

Facing the Outside World 367

And That Was Freedom 415

Biographical Notes 451

Index 459

Illustrations

Between pages 150 and 151

Seventh SS Prinz Eugen Division advancing toward Čemerno, May 1943

Djilas with his brother, Aleksa, in Montenegro, July 1941

Djilas and Ivo-Lola Ribar, 1942

The Partisan leadership in Montenegro, May 1942

Draža Mihailović and Dragiša Vasić, his political adviser, on Mount Zlatar, 1942

Germans near Prozor, March 1943

Tito and Djilas with the Supreme Staff, Bihać, December 1942

Germans escorting Montenegrin peasants to their execution

German, Ustashi, and Italian officers, Plevlja, May 1943

Sick and wounded Partisans in Bosnia, June 1943

Partisan units crossing the Drina, May 1943

Funeral feast for a dead Partisan, Sandžak, May 1943

The Prinz Eugen Division crossing the Sutjeska, June 1943

Partisans crossing the Piva, May 1943

WARTIME

Between pages 278 and 279

Tito, after being wounded on Mount Ozren, June 1943

A Partisan brigade marching toward Prozor, March 1943

Partisan fighters in Lika, December 1944

Ivo-Lola Ribar with F. W. Deakin and members of the British military mission attached to the Partisan army

Randolph Churchill with the leadership of the Eighth Partisan Division, June 1944

Review of the Second Proletarian Brigade and the Third Dalmatian Division, in the Bosnian mountains, 1943

Sava Kovačević, leader of the Partisan uprising in Montenegro

Croatian poet, Ivan Goran Kovačić, who was killed in 1943

The Second Proletarian Brigade crossing the Piva, May 1943

Mitra Mitrović, with Soviet officers, October 1944

Blažo Jovanović and Milovan Djilas in Montenegro, May 1945

The Partisan leadership in Vis, summer 1944

The People's Uprising

1

It is a much more difficult and formidable task to relate a historical tragedy than to take part in it. This is not just because now, some thirty years later, many once inspiring realities have turned into bitter realizations. Rather the events and persons involved did not seem all that significant at the time, and certainly were not bound up with my thoughts and my being in such measure and as painfully then as now. In setting down on paper my memories of the Yugoslav revolutionary war and of my part in it, I am of course reliving all its painful essence and insane grandeur.

I sensed, I knew the pain and suffering this account would bring, so I kept putting it off, secretly hoping that I would not have to write it. But this too has become a duty—one which seems to me today to be even greater and more significant than my participation in the revolutionary war. That experience was life itself—an unusual life, with death and in death—while this is giving an account of that life, of oneself, and of events which were historic even as they were happening, since they altered the existence, the consciousness, and the destiny of the persons and nations drawn into their vortex.

Yet that life, that war—however much we Communists prepared for it through our ideas and activities and in our whole being—inflicted on us unforeseen circumstances which ravaged our previous mode of life and turned acquired views and intellectual preconceptions upside down. Not even those who had experienced World War I and the October Revolution, nor even those Communists who had returned by various routes from the recent civil war in Spain, were spared from painful shocks. For every war, like life, is new and unexpected. Only with time do one's consciousness and conduct accept the inexorability of war and adjust to its bloody and insane devices.

The military operations which we Communists launched were motivated by our revolutionary ideology; however, for the people who joined us they were but the resumption of the war which the broken and exhausted forces of the Kingdom of Yugoslavia had already lost. Our preparations were speeded up with the beginning of the occupation. Those members of the Politburo and of the Central Committee who found themselves in Belgrade at this time had intended to strike at the forces of occupation, but only after the Germans had been spent and demoralized by the war. Some forces of the former regime had similar views: we learned of certain officers who had made off to western Serbia and hoped to re-establish the old regime with the fall of Germany. Our preparations against the occupiers were therefore preparations against the old order as well. Although the Germans and their allies had dismembered and partitioned Yugoslavia, it was to their advantage to re-establish the old state apparatus and economic life. A revolution was not feasible without a simultaneous struggle against the occupation forces.

However, not even in the declaration of the Central Committee, which Tito wrote on the very day of the German attack on the U.S.S.R., was there an explicit call to arms, but only a summons to make ready for a struggle. That was understandable: we had long had radio contact with the Comintern—that is, with Moscow—and as responsible and fateful a move as an uprising and armed struggle could hardly have been undertaken without approval from above. True, it was indicated in the conversations of the Central Committee members that armed guerrilla actions were in the offing. Nevertheless we waited for Moscow's directive, and for once Moscow did not delay.

I no longer remember how many messages reached us from Moscow on that subject, but I am sure that Moscow ordered us to begin diversionary and guerrilla actions. Tito so informed us, and it became obvious that the activity of the party had to be reoriented, and that we had to speed up the stockpiling of weapons. This brought a halt to preparations for the publication of the party newspaper *Borba* (The Struggle). The Central Committee had named me editor along with Rodoljub Čolaković, while the task of printing and distributing fell to Svetozar Vukmanović-Tempo, who had vast and unmatched experience in this area. I remember that Čolaković and I discussed the editing of *Borba* even after

June 22—in fact, up to the time the order came from Moscow to begin armed attacks. To be sure, our discussion wholly concerned the Soviet-German war, since events and new responsibilities had dissolved former habits and ways. I cite this detail concerning *Borba* as an illustration of the Central Committee's dual orientation before the outbreak of the Soviet-German war, which provided for both an armed uprising and semilegal mass propaganda.

The German attack on the U.S.S.R. changed both the mode of operation and the lives of the Communists. On that very day the police, who had been absorbed into Hitler's New Order, initiated street cordons and arrests. As part of that campaign my wife Mitra was picked up. We knew that our arrested comrades could hope for nothing but death, and that the rest of us would remain free only as outlaws and ultimately as fighters.

Even so, it would be incorrect to conclude that the Communist uprising would not have taken place without Moscow's directive. The party was undoubtedly devoted to Moscow at the time, but as confirmed revolutionaries and beyond its direct control. The attack on the U.S.S.R., and the reign of terror by the occupation forces and domestic fascists, would have led them to take up arms even without the Moscow directives. Very likely we would have begun our struggle somewhat later, but with less haste in getting organized and with a deeper, more self-reliant political outlook. From its very inception the Yugoslav revolution never quite suited the needs and purposes of the Soviet government, and the stronger it grew, the less it reflected the Soviet state's "grand designs." After all, revolutions are not the result of directives but of a specific combination of conditions in which the will and consciousness of the revolutionaries, and the talent and resoluteness of their leaders, play a decisive role.

July 4, 1941, is celebrated in Yugoslavia as the day on which the Central Committee of the Communist party launched the armed struggle against the occupation forces. The Central Committee meeting at which this significant decision was made took place that morning at the villa of the director of *Politika,* Vladislav Ribnikar. There was no deliberation: we already knew of the Comintern's directive, and our ideological, internationalist obligations and love for the U.S.S.R.—the "fortress of World Communism" and the "leading socialist country"—harmonized with our own position and aspirations. Thus the topic of discussion was how to carry out the armed struggle and diversionary action. In consultation with the other Politburo members, Tito had already decided that I was to go to Montenegro to begin the armed struggle, and that Vukmanović-Tempo was to go to Bosnia, while other assignments regarding diversionary and guerrilla operations were apportioned among the other comrades.

My older brother Aleksa showed up in Belgrade at that time. In late 1940 he had received a teaching position in Bačka. As Hungary had now taken over that part of Yugoslavia, he set out to join our parents in Gornji Milanovac, and on the way there looked me up in Belgrade. I urged him not to go to Gornji Milanovac, where he could be arrested on account of his past. I was getting ready to go to Montenegro, so we decided that he should join me, especially since he yearned for armed combat.

Several days earlier I had met a worker named Tomović and his cousin, both natives of my part of the country, who wanted to make a contact with the party. I gave them their contact and they gave me their labor cards. The artist Djordje Andrejević-Kun altered the seals and we placed my picture and my brother's over theirs. With these falsified papers my brother and I went to the Italian embassy, where they graciously gave us visas for Montenegro, which had fallen to Italy.

Had I had calluses on my hands, I just might have passed for a worker, but even with calluses my brother Aleksa could not; he wore glasses, and his complexion was fairer and smoother than mine. Nevertheless, I knew that everything would go well at the embassy, for a mass exodus, set in motion by the war and by the dismemberment of the state, was still going on; in that terrible storm everyone was headed for the greater security of his own ancestral home. What I feared were identity checks along the way.

Even so I could think of nothing I wished to do more than go to Montenegro, where, from time immemorial, ideas found consummation in sheer violence, and where one's first perceptions and first words were imbued with zeal against the aggressor. The journey was long and risky, but staying in Belgrade was even more dangerous and inglorious. As always, when confronted by a new situation, I was anxious.

Yet there was no joy in this preoccupation with my duty because Mitra had been arrested. From the start this personal misfortune beclouded the historical event and my role in it. Many sympathizers were also arrested, among them Dr. Milutin Zečević, my schoolmate from the Kolašin high school, and Slobodan Škerović, in whose apartment I had spent the previous night. I did not have a pistol and wondered inanely what I would do if a detective recognized me. Fear had seeped into every cranny of our existence. Armed combat offered an escape at least from torture, if not from death.

To this day, two figures are inseparably linked in my mind with my departure from Belgrade—Tito and Zora, Mitra's sister. The former stood for duty, the latter for personal ardor. From Tito I sought final words of advice, and through Zora I took leave of Mitra, who in my imagination was already on the scaffold.

On a damp and cloudy morning I met Zora at the soccer field—today the playing field of the *Partizan* soccer team. Upon hearing of Mitra's arrest, I had sent word to Zora, hoping that something could be done to save Mitra through their family. Now Zora came to tell me that the family no longer knew anyone in Belgrade powerful enough to get Mitra out of jail, even though she was guilty only of being a leftist and my wife. However, the police had accepted clothing and food brought for her. Zora and I walked very circumspectly through the new park, and Zora asked me if we would win in the end.

"Yes!" I answered confidently, and added, "But the victory we want so much won't be a happy one, with all the horrors and—without Mitra."

"You're right," Zora replied, "there won't be much to be happy about."

I told her I was going to Montenegro. We took leave of each other. Hunched up, she hurried away down the slope toward Topovske Šupe, while a sad emptiness and bitter determination swelled inside me like an avalanche.

The parting with Tito took place around ten o'clock that same morning, at Ribnikar's villa. At first we talked about the Eastern front. Even according to Soviet reports, the German armies were penetrating the Soviet Union as if there were no organized resistance. Nevertheless, in the spirit of the moment I expressed my confidence in the speedy victory of the Soviet Union—a confidence born of my enthusiastic acceptance of Soviet propaganda's claim that the U.S.S.R. would bring the war to the invader's soil. I was not the only one in the Central Committee to think so. Tito didn't refute my faith, but he didn't agree either. I concluded that, though he was convinced of the victory of the Soviet Union, he believed that the war could last a long time. His uncertainty set me to thinking—though not till later—that the war and my adjustment to it would reduce my romantic rapture to a verbal and stereotyped conformity.

In defining the party line in the armed struggle, Tito insisted on latitude but also on decisiveness; he was afraid of Montenegrin narrowness and sectarianism. We also spoke about the organizational forms of the struggle, but the expressions "partisan," "partisan unit," and the like were not used, nor was a name given to the command. These names were introduced after my departure from Belgrade—most probably on the Russian model, for they did not even exist among our people up to that time. As borrowed words in our language, they had a different meaning. As for the command, a military structure had not yet been developed; rather there were only military committees alongside party committees, and fighting units just in their inception.

At the July 4 meeting of the Central Committee, there was not yet any talk of the "elaboration of a plan," or partisan operations in Serbia, or of the creation of a free territory in western Serbia, as stated in some documents. There were as yet no Partisan operations, and it was only later, as a result of these operations in liberated areas, that the idea arose of any kind of stable free territory in western Serbia. All official historiographies—and ideological ones in particular—can visualize events only as something conceived much earlier in the heads of the leaders. Leaders doubtlessly conceive possible courses of action, but within the frame of events and in actual encounters. Every time leaders deduce courses of action out of their own heads and their own fond hopes, they are mercilessly punished by reality and taught a lesson—if indeed they are still in a position to learn.

There was no talk of any uprising either, but simply of diversionary and guerrilla operations. The Politburo recognized that the party wielded its greatest influence in Montenegro, and that there were considerable quantities of weapons there which had been seized and hidden away from the royal army when it capitulated. Tito said, "Be careful not to incite a general uprising. The Italians are still organized and strong, and they will break you. You should begin with minor operations." He added, "But shoot anyone—even a member of the provincial leadership—if he wavers or shows any lack of discipline!"

The entire conversation did not last more than half an hour—as though we already knew all there was to know and had said it all. Nor did the farewell with Tito reflect any special warmth, just the authority and courage which he inspired.

My brother and I had outfitted ourselves with suitable clothing and shoes. I wore the spiked boots and windbreaker of a mountain climber. To avoid a confrontation with detectives at the central station, we boarded the train in the suburbs. I assume that was July 5. Boarding the narrow-gauge train to Sarajevo, we embarked on the transformation of our own and our nation's destiny.

2

The train was not overcrowded. There were no checkpoints, no inspection of papers. But for the careworn faces, one would have thought everything normal. Passing through Gornji Milanovac, where our parents and our youngest sister and brother lived, we looked out the window with yearning and fear, hoping to catch a glimpse of them. But the train puffed on through the bushy hills and the yellowing wheat and young corn.

After stopping for a long time in one station, the train headed for the new Serbian border with Croatia, between Vardište and Višegrad—the very point that had marked the border between Serbia and Austria before 1918. On leaving the last town in Serbia I felt increasingly secure, for there was less chance of encountering police who might recognize us. Yet we were also anxious, for we did not know what to expect in the Independent State of Croatia, which had been improvised during the first days of the invasion under the aegis of the Axis Powers.

I thought that the last critical moment would come as we crossed the border, but it was as if there were no border. The conductor simply announced that the next station would be in Croatia. No one asked for our papers, even though the train stopped for quite a while. There was no commotion, no border guards or customs officers—just a gloomy mountain station. In the late afternoon the train finally started up again. The next Croatian station was not much different. The Ustashi state may have been worried about the trains from Serbia, but obviously it was not well organized.

And so, around noon the next day, we arrived safely in Sarajevo, where we were to change trains for Mostar and had to wait about two hours. We left the station for a stroll and a bite to eat. The town seemed unchanged except for the store windows, which were no longer abundant. One could still get a good meal in restaurants, though occupation money was not gladly accepted.

I knew most of our leading comrades in Sarajevo, as well as some contacts and sympathizers, but time was too short and we had to return to the station. At the corner of Marijin Dvor, we came across a poster announcing the execution in Zagreb of Prica, Keršovani, Adžija, Richtman, and others—all well-known leftist activists and distinguished Communists—in retaliation for the murder of the police agent Tiljak. I knew almost all those comrades personally; I had worked with some, and was on close terms with Prica and Keršovani.

While a prisoner at the main Belgrade police station, I had also had encounters with Tiljak. I knew that, in the first chaotic days of the occupation, Rade Končar and Pavle Pap had enticed Tiljak into some woman's apartment and strangled him. They had then stuffed him into a trunk and thrown him into a swamp near the Sava River, which dried up in the summer heat. The body was found. Tiljak was a notorious anti-Communist police agent who frequently resorted to torture. I remember a list that Ranković or Kardelj had drawn up with the names of the spies Tiljak had exposed. Yet it seemed hardly worth the risk of murdering him. For the execution of so many well-known activists—ten, if I remember correctly—this one police agent's death was obviously just a pretext.

This was the first announcement which listed executed Communists by name. The choice of victims and the announcement itself were conceived with utter ruthlessness. The aim was to strike terror and to dispel any thought of leniency. Moreover, any hope of an existence independent of Ustashi authority was being eradicated. An unfathomable, unbridgeable chasm suddenly opened within me: this was not merely a struggle for survival, but unto death.

Despite my violent agitation, I retained sufficient presence of mind to watch my words and movements. Aleksa, however, could not restrain his headstrong temperament; he clenched his fists and vented his rage in curses. This inflamed my bitterness, sharpened by vigilance. I had noticed that people glanced at the poster furtively as they passed by and then saw a pair of uniformed and armed Ustashi standing by a kiosk some thirty paces behind us. They were hiding there to observe the reactions to the poster, and were watching us with suspicion. I suggested to my brother that we walk away slowly and proceed without haste to the station. If the Ustashi stopped us and found that we were Montenegrins going home, they would let us go. The Ustashi followed us all the way to the station. The train was crammed with passengers and refugees, though it was still twenty minutes to departure time. We pushed our way into the thick of the crowd, to make it harder for the Ustashi to get at us. In the car there were only the poor, who seemed poorer and more destitute than ever. Through the thicket of sweaty limbs and heads, greasy kerchiefs, bags, and baskets, I glanced at the platform. The two Ustashi were still there, surveying the crowd.

At last the train pulled out of the station. I wanted to tell my brother that the danger was over, but around us were alien, untrustworthy people. I could only smile happily at him. He laughed bitterly and swore again. In Mostar I was still on guard, but there was no check there either. The train passed through the countryside as if there were no border between the Independent State of Croatia and Montenegro. The train was almost empty by the time we entered Montenegro in the morning heat.

At a small station near Bileća a cluster of Serbian refugees got on the train. At first they were close-mouthed and apprehensive, but once they realized there were no authorities on the train, they relaxed. They were fleeing from the Ustashi terror. A fair, robust peasant in his thirties, with bruised cheekbones and curly hair matted with dried blood, told us how the Ustashi had surrounded his village and driven everyone—men and women, young and old—to a rocky ravine, then struck them down with clubs. The peasant freed himself of his bonds at the very edge of the ravine; though he had been struck in the face, he was able to scramble into a brush-covered mound of boulders. "They are killing every Serb in sight!" he lamented. "Like cattle—a blow on the head, then down the ditch. They are mostly Turks.* Their time has come. They want to wipe out the poor Serbian people."

For me this was a new story. Later, in the course of the war, I was to hear it many times and almost always the same: a village surprised and the men all bound, murdered, and thrown into a ditch. Religious and ideological murders do not require any imagination, just efficiency: in this lies all the horror and—for the victims—"relief." Yet I was not as shocked as I should have been. I was already familiar with the Ustashi ideology—an amalgam of primitive Croatian nationalism with modern

* Reference here is to Moslems of South Slavic origin who are called Turks because of their religious identity with the Turks and their integration in the Ottoman state.

fascist totalitarianism. While in prison I had come to know many leading Ustashi. I had followed the evolution of their ideology from militant separatism to fascism and total anti-Serbianism. Reports had reached us in the Central Committee in Belgrade concerning the persecution of Serbs in Croatia; then came the first of many large droves of refugees. We knew of the circulating drumhead courts-martial and of Pavelić's "laws," which contained few articles but always decreed the death sentence. My own lack of horror reflected the atmosphere—the nature of the groups pitted against each other, the flood of propaganda, and the bloody events themselves.

Those simple people, mostly peasants, were even less horror-stricken. One could not even say that they were bitter: a misfortune had come along, terrible because it was human, but perhaps for this very reason surmountable. In the group there was also a slight, dark girl in city dress. She told us calmly that the Ustashi had assembled all the prominent Serbs in her town—merchants, priests, and officials—and two or three days later loaded them on a truck and took them away, supposedly to Mostar, but in fact to be murdered and thrown into a ravine. The Serbs who remained in the little town, mostly women and children, helplessly awaited a similar fate. This girl was fleeing to Montenegro, but she was not too happy about having escaped. Death had suddenly become commonplace, something as ubiquitous as the air and the soil.

"Well," I said turning to the injured peasant, "why don't you defend yourselves?"

"Who can defend himself?" the peasant lamented. "We didn't expect anything. We couldn't believe a government would attack people just like that. We have no weapons. We are left to ourselves like cattle. But at Nevesinje they did rise up and finish off a lot of Ustashi."

People got on and off at each station, but the telling of Ustashi massacres continued—of course, with new details. I don't remember whether my brother joined the conversation; after being followed by the Ustashi in Sarajevo and my constant warnings, he was subdued.

What was I thinking and feeling at that time? How can I be sure after all these years? Perhaps it was then that I clearly grasped for the first time the significance of my task, the extraordinary bond between my ideas and the destinies of my people—a bond not against the bourgeoisie but against the aggressors whose killing of "aliens" and of innocents was a confirmation of the "idea" if not indeed its only real aim. It was obvious to me that only we Communists could help the people who were being massacred through no fault of their own, but because of the mad ambitions of a few. This made me neither proud nor happy; that is simply the way it was.

As we approached Nikšić, Dr. Vojo Djukanović showed up. I knew him slightly from meetings with leftist groups, though I didn't know if he was a party member. In any case one could speak more openly with him, all the more because he was by temperament lively and direct. As we stared through the window at the creviced cliffs, Dr. Djukanović told me more or less the same story as the injured peasant.

"And what are you doing, the progressive people, the antifascists, the Communists?" I asked.

"We are doing all we can," he replied. "Here I've organized a public health service."

"That's fine! But are you fighting, organizing attacks?"

Dr. Djukanović looked at me with a curious amazement. "Is that the party line? Is that the directive?"

"Yes. We must begin immediately!"

After some thought he observed, "But over there in Hercegovina, there's hardly any political activity. There's almost no organization—just individuals."

"That isn't important now! We must begin with small groups. The organization and an army will be formed in the armed struggle."

We soon reached Nikšić. It was a steaming sunny day—a rare day to remember, for it is rain, fatigue, and hunger that most often remain in one's wartime memories. A bus was in the little square, but we had to wait—for some Italian chief, as someone explained.

Soon the chief arrived: a ruddy little man in a brown jacket, with a black shirt and a briefcase. He sat in the best seat, next to the driver. Someone said, "That's Serafin Mazolini, secretary to the Civil Governor of Montenegro."

It is in the nature of Montenegrins—perhaps because there are so few of them—to know and remember one another. But my brother and I were from a different region and had long lived outside Montenegro. None of the passengers recognized us. Certainly Mazolini's secretary could never have imagined who was sitting behind him.

We were in the middle of the bus. I glanced at a heavy-set, dark-haired woman who was trying to ingratiate herself with the secretary, and we exchanged looks of contempt. That, too, was Montenegro: women consorting with the forces of occupation, and occupation officials traveling about freely, mixing with the people—a people growing used to life under the occupation. I had half a mind to stop the driver at some handy spot; to drag out that official, who may have been daydreaming about launching his career; and to fill him full of holes by the road. That would even have been my duty and I would have done it, had I not had a far more significant assignment: to find the party leadership and transmit to them the line on the armed struggle. At the same time I was aware that Mazolini's secretary was a human being—maybe even an innocent one, all in all. But what about me and my people? Wasn't I a man, too, with obligations—even to murder—and with an ineradicable allegiance to my country and people, and to an ideology?

We didn't get off the bus in Podgorica (today Titograd), though it would have been easiest for me to make contact with the Montenegrin party leadership at the coffee house of Periša Vukošević. He knew me, and I considered him a member of the leadership. But Podgorica was the main city, and we might have been subjected to a special inspection. I decided that we would get off near Blažo Javanović's house at the foot of Velje Brdo, seven kilometers from Podgorica.

I knew Blažo well, though we had met only a few times in the course of some ten years. He was the most prominent member of the Communist party of Yugoslavia's Provincial Committee for Montenegro, the Bay of Kotor, and Sandžak. He was the organizational secretary, while Božo Ljumović, an older party member, was the political secretary. But Blažo was a Communist, and it was likely that his house was being watched, perhaps even surrounded. For this reason we could not approach it directly.

In those days buses stopped on request. We got off. When the bus had gone on, I went to a little stone house by the side of the road, while my brother sat in the shade. A dog barked, and out of the house ran an old woman, with her thick hair bound in a black kerchief like most Montenegrin women. When I told her that I wanted to see Blažo, she stiffened.

"This isn't his house!"

"I realize that. But I don't know if he's home or what things are like over there."

"Go and see!" the woman snapped.

"Look here, my good woman, it seems that you don't trust me. Would I be asking you about Blažo if I could go to him first?"

"There are all kinds of people about these days," said the woman. "We must be careful. Who are you?"

"I'm from Belgrade and I have something important to talk about with Blažo, so if you can find out—"

Before I even had a chance to finish, the woman ran over to a pile of boulders above the house. A swarthy head popped out from behind the boulders. It was Savo Brković, a member of the party leadership, whom I had met recently while hiding out in Montenegro. The woman was his mother. As a precaution, he had been sleeping under a lean-to at a distance from the house.

I told Savo I needed to get in touch with the Provincial Committee, especially with Blažo. With a trace of a smile he said that the leaders were already in hiding, but that Blažo was not far away. Then he guided us across the road and through a meadow, down to the muddy banks of the Zeta. Out of a copse of willows he pulled a rowboat; the three of us crowded into it and floated across the calm green waters. I felt that I was crossing a frontier, leaving torture and subterfuge behind me. Soon we found ourselves on the other shore. We were still not out of danger, but now it could find me only on the field of battle.

Blažo had settled himself in a peasant hut right across from his house, above a small pasture in front of a rocky wooded ridge. It was a convenient lookout and hiding place. We found him inside. After hearty greetings he and I went out and sat in the shade. I told him briefly of the purpose of my trip, and of the need to call a meeting of the leadership immediately to discuss our tasks and duties. There was no reason for secrecy because of my brother, even less because of Brković, since we were joined already in armed struggle. Nor did I speak with Blažo in private out of conspiratorial habit, but rather to go through channels, so that the leaders might be informed and get organized first, in order to take command. Later—after Blažo had had a talk with Savo, and sent villagers out to call a meeting of the Provincial Committee for the next day—Blažo, my brother, and I walked across the meadows at dusk. We spoke openly about everything, without anticipating the decisions of the leadership.

Not even Blažo slept in the villager's hut, though it would have been difficult for anyone to approach it unnoticed. They spread some hay and blankets for my brother and me among the boulders. We were utterly exhausted and slept soundly under the stars—just as we had done in childhood in sheds beside our cattle—as if we had already carried out the hardest part of our assignment.

3

To get to the meeting of the Provincial Committee on the afternoon of July 8, Blažo and I had to climb a rocky path for nearly an hour. My brother stayed behind with the villagers. We gathered in a forest clearing and immediately began the discussion. I was not familiar with the region and knew only that we were on the territory of the Piperi clan. In fact, four members of the leadership were Piperi—Blažo Jovanović, Božo Ljumović, Savo Brković, and Budo Tomović. Published documents state that the place of our meeting was called Stijena Piperska: it must have been so, for the surviving Piperi participants had a greater interest than I in that history.

The Provincial Committee consisted of Božo Ljumović, Blažo Jovanović, Periša Vujošević, Savo Brković, Vido Uskoković, and—as a representative of the Communist Youth—Budo Tomović. Krsto Popivoda, a member of the Central Committee, and Radoje Dakić, until recently the secretary of the Local Committee of Belgrade, were also invited to the meeting, as was the secretary of the Local Committee of Podgorica, Marko Radović. At that time Moša Pijade—for many years a member of the Central Committee—was hiding out in Montenegro. He was in prison at the time of the coup d'état of March 27, 1941, but was freed prior to the attack on Yugoslavia on April 6. He had fled to the Bay of Kotor, which may have started the rumor that he and a friend were trying to get away by ship to the West. It seems that Moša could hardly wait for my fall from power in 1954 to shower me with spiteful attacks. Most probably he was the one who then resurrected the long-forgotten tale of his flight to the West, and of course ascribed it to me. But I couldn't have carried that tale from Montenegro to Belgrade, since I returned before the Provincial Committee even learned that Moša was in Montenegro. Actually, it was Ivo-Lola Ribar who passed it on to the Central Committee. Tito was upset, Milutinović amazed, Ranković reproachful, and I derisive, but nobody took the matter too tragically. After all, we all understood that Moša was particularly vulnerable as a highly publicized Communist, and as a Jew whose physiognomy was a dead giveaway.

It should be added that, because of his long years in prison, Moša was out of touch with the new generation and the new style of action. He was beloved and treated with respect, but more as a historical figure than as an active leader. Moša's biographer Slobodan Nešović reproaches me for not having invited Moša to the meeting where a decision on the armed uprising in Montenegro was reached. I maintain that Moša had already gone or was about to go to Kolašin, a safer refuge. But even if this wasn't so, his presence wasn't essential since we were contemplating not an uprising but small guerrilla actions; he could have offered only advice. To the best of my knowledge, as long as I was in power Moša never complained about this episode, either in friendly conversation or in jest.

In any event, there were no disagreements or misunderstandings at that meeting insofar as the organization of armed operations was concerned. In Montenegro there were already "shock brigades" of from ten to thirty persons attached to the committees, and these were a good beginning. We had to determine the targets and methods of attack.

One difficulty was the extreme fragmentation of the party organization: the Local or Town Committees corresponded to the administrative units of the Kingdom of Yugoslavia. We recognized that it would be difficult to maintain ties with such a large number of Local Committees. It was therefore agreed to set up four Regional Committees, and delegates were designated to form them and to transmit instructions concerning armed actions. Savo Brković was assigned to the Kolašin region, and left with my brother the next day. Radoje Dakić, whom I knew well from party work in Belgrade, was sent to the Cetinje region, while Krsto Popivoda, a person of exceptional courage and conscientiousness, was sent to the Nikšić region. The Provincial Committee was located in the district of Podgorica, and so—if I remember correctly—a Regional Committee for Podgorica was not even formed. In due course delegates were also designated for the Bay of Kotor and Sandžak: Veljko Mićunović and

Rifat Burdžović, respectively; I knew them both from party work in Belgrade. Thus from the very start the leading cadre of Montenegro consisted for the most part of Communists trained and developed in Belgrade.

There was a great disagreement over what line to take in the struggle. We all agreed on a broad association of all who opposed the occupation. However, on the occasion of the attack on the U.S.S.R., the Provincial Committee had already issued a proclamation of its own which—contrary to the position of the Central Committee—spoke of "the imperialists of both warring camps" who were waging war for a new division of the world, and which called for strengthening "the militant alliance of workers and peasants in the struggle for the realization of Soviet power and a final settling of accounts with the capitalist system."* The proclamation was written, I believe, by Božo Ljumović. I subjected the proclamation to a criticism which was perhaps all the sharper because this was not the first "sectarian" deviation by the Montenegrin leadership. Everybody, of course, accepted the criticism, though Ljumović upheld a reservation "in principle" that Britain was, like Germany, an imperialist power, and that the war was therefore leading to the collapse of capitalism and to the establishment of Soviet power. Ljumović too was in fact a historical relic, but in the Montenegrin manner: honorable and vain, expansive and without initiative, pretentious as a theoretician. Blažo Jovanović, on the other hand, was more active and militant, and amenable to the stands taken by the Central Committee. No wonder I collaborated more closely with him, though I respected Ljumović and permitted him to remain at his post.

The comrades dispersed by early evening, and Blažo and I set out for our shelter.

In those days the Montenegrin separatists—better known under the name of "the Greens"—† were preparing to hold an assembly in Cetinje to proclaim Montenegro's independence. That act was to take place on St. Peter's Day, July 12 (O.S.)—the day when popular assemblies had met in the time of the Montenegrin prince-bishops. Montenegrin "independence" under the aegis of fascist Italy now had to be wrapped in a glorious and mythical tradition. As long as they posed as federalists, the

* *Hronologija oslobodilačke borbe naroda Jugoslavije* [Chronology of the National Liberation Struggle of Yugoslavia] (Belgrade: Vojno-istorijski Institut, 1964), p. 46. (*Note:* Hereafter this source will be cited as *Hronologija*.

† The Montenegrin separatists got the name "Greens" (Zelenaši) from their green electoral tickets for the Constituent Assembly of 1919, while the adherents of unification with Serbia got the name "Whites" (Bjelaši) from their white tickets.

Montenegrin separatists had maintained a significant influence in the environs of Cetinje and among the former Montenegrin chieftains. But the occupation reduced their influence, even though they were receiving support from the Italians. Some of their more prominent adherents rejected all collaboration with the Italians and made common cause with the Communists. Similarly the adherents of unification with Serbia, the so-called Whites, grew passive or joined the Communists as the strongest antioccupation group.

It was at that time that I met two men who were to play a significant role in the war: Peko Dapčević and Arso Jovanović.

Peko arrived from Austria, having fled there from a French detention camp where he had been held as a Spanish volunteer. In Austria he had worked in the Stayer munitions factory. He found his chance and left Austria with false papers, as other Yugoslav Spanish volunteers had done. As soon as we met, he reminded me of our meeting in Cetinje in the spring of 1936. Well dressed and tanned, he looked hardened. We talked with Peko the whole afternoon and long into the night, mostly about conditions in Germany and Austria; the next day he journeyed off to his village of Ljubotin—to continue the war begun in Spain. I liked Peko for the directness of his reasoning. We grew close from the start, and would have remained friends all our lives but for the proceedings against me in the Central Committee in 1954.

In discussions within the Provincial Committee, and particularly between Blažo Jovanović and myself, we concluded that we had to create a military command for Montenegro, the Bay of Kotor, and Sandžak. We did not formally assign the posts in that command, though in the course of the struggle I assumed the position which, given my work and role, could be called that of commander. In actual fact, everything of importance was decided by the Provincial Committee or—if that Committee could not agree—by myself and the corresponding member. We understood immediately that we needed a professional military person. Blažo proposed Arso Jovanović, a member of the Royal Yugoslav General Staff, who had escaped as a prisoner of war and was hiding in his village of Zavala, in Gornji Piperi. In other communities, too, there were officers in hiding who had met with the Communists. But Arso was nearby and, in Blažo's estimation, very suitable. This was something I had to judge for myself—especially since the two men were distantly related.

It was dark when Blažo and I started out for Arso's hiding place. We traveled over an hour up a rocky slope, and found him in a low stone hut with a fireplace and an oil lamp. With him were his mother, his sister, his wife, and two little daughters. The girls soon went to bed and we began our conversation by the fireplace, feeling slightly ill at ease before the women. We told Arso that a struggle against the Italians was in the offing and urged him to declare his stand, although he was not yet a party member. He replied without hesitation, "I agree absolutely to everything that is against the occupation."

"What about the other officers?" I asked.

"I believe that all of them will take up the struggle."

The women served us some warm milk, which was all they had. Arso then put on his officer's jacket and his boots—he was to wear those same boots and jacket throughout the war—and accompanied us outside. There we agreed that he would join us the next day at the house of Todor Milutinović, brother of the Politburo member Ivan Milutinović. Then and there began Arso's rise in the new, still unformed army. At first he was the chief of our Montenegrin command, and by the end of 1941 chief of the Supreme Staff of Yugoslavia. He would not have taken that path had the Royal Army, which he served with devotion as one of its most promising officers, been ready to wage war and die for the homeland. Arso appealed to me because of his unambiguous resolve, even though his speech and way of thinking were not Communist. He was spare, with a lean, ruddy face, dark eyes, and thin lips—the tense look that excellent students from poor families sometimes have.

Operations were planned in secret even though, in a still strongly clannish Montenegro, secrets were hard to keep; as a rule, they were entrusted only to those designated to make the actual attack. To be sure, other measures were also taken: the evacuation of exposed comrades from the towns, the acquisition of printing equipment, and so on.

We too lived conspiratorially, though we met and talked with the villagers. Some of them—the more outspoken—even inquired what the devil we were up to. We received daily reports from the neighboring villages. Despite this, our impatience grew, and with it our disappointment that we couldn't begin our attacks prior to the Greens' assembly at Cetinje.

4

Having set in motion party organizations and armed groups, we were faced with numerous unknowns and dilemmas. We knew that the fighting capacity of the Italians was not great. They had been beaten by the Greeks, and in the confused and shameful recent war even the Royal Yugoslav Army had made its only advance against them. We also knew that the Black Shirts and the Carabinieri would fight well. Above all, we knew that we would have no help from anyone, and that no one would be aware of our sacrifices and exploits for some time to come. Besides local administrations and old Yugoslav gendarmerie stations, there were also Italian military outposts. However, the local Montenegrin authorities were inactive, while the Italians were isolated amid a population which held them in contempt. You might say the authorities were there but didn't function. The entire population had become alienated from government and transformed into one large conspiracy. All political and neighborly squabbles were forgotten, along with past defeats and faintheartedness. In the memory of the people only determination and will to victory abided. There was also Mother Russia. Even without Communist propaganda, an unyielding and outraged people revived the memory of ancient ties with Russia, and the mythical images of its might and glory. But Russia was far away.

One could sense a certain excitement and, along with it, increased pressure from the occupation forces. In the first months of the occupation the Italian military authorities did not use violence, there were no arrests, and property was not touched. Indeed, goods started arriving from Italy. It looked as if Montenegro might suffer less want than before. Only the Black Shirts, who were encamped in the larger towns, presented a threat. The sole motive for attack and rebellion lay in wounded self-esteem and bitterness over an unjustly lost but unfinished war. There were but few open supporters of Italy, and still fewer agents. But how would the people hold up when Italy began to burn, imprison, and kill? How would those elements react who felt that it was "not yet time to act," or the Moslems who were inwardly bitter over their lost empire? How would the neighboring Albanians act, who had already driven the Montenegrins out of the territories which their Italian protectors had joined to "Greater Albania"? How would the Communists and peasants prevail against steel, torture, and death?

Yet wars and rebellions are not only a curse and misfortune; they also spawn hope and creativity. Wars and rebellions are a vital proving ground for leaders, ideas, and nations. Wars and rebellions are an imperative: to renounce war when it is time for war means to renounce one's own inner nature. In opting for war, we came to understand who we were. Only in armed conflict could we affirm ourselves and force the enemy to understand us and grant us recognition.

That affirmation, that self-realization—of the self and of the nation—took place on July 13, 1941. The first attacks on the strongholds of the occupation began in old Montenegro, where, inseparably, Montenegrin independence and the Serbian idea first arose. Italian garrisons were attacked and disarmed in Čevo, Virpazar, and Cuci, and in Spič near Bar. The volleys of the insurgents dispelled the invaders' illusions of Montenegrin support and the separatists' fantasies. Only the day before, on July 12, a Montenegrin assembly in Cetinje had proclaimed "the sovereign and independent state of Montenegro in the form of a constitutional monarchy," and resolved "to unite the destiny of Montenegro to the destiny of Italy"—actually, their own destiny.*

But we, the leadership, didn't learn of these battles until later. By July 15 fighting had begun around Danilovgrad, and an attack was made on a garrison in our neighborhood, next to the village of Bioče, at the confluence of the Morača and the Mala Rijeka. In the balmy night filled with scents of parched grasses and ripening figs, we listened tensely and when, finally, a volley of gunfire burst forth, we were petrified as we awaited the news. Solitary shots and brief salvos continued until sunrise, but no word came from anywhere.

Meanwhile on July 14, in our native village, my brother Aleksa had

* *Hronologija,* p. 55.

collected a group of Communist villagers, and set up an ambush on the highway in the gorge between the Tara River and Vrlostup Bluff. But the ambush was detected by a villager, the son of a hardened separatist, who stopped a truck that was going from Mojkovac to Kolašin. Somehow he conveyed to the driver and his escort that Communists were lying in wait for them, and they turned back. The ambushers knew nothing of the betrayal and waited patiently. The commandant of the garrison in Mojkovac ordered the truck to go back, this time filled with Italian soldiers. As the truck approached, my brother fired his rifle first and the rest followed. But the real job was done by a peasant, Milosav Djurašinović, with a machine gun. As he told me later, "First I mowed down one row of soldiers with one burst, and then I set my sights on another." But he did not hit the driver or the motor, and the truck sped off to Kolašin full of corpses, sprinkling the highway with blood. When I came into liberated Kolašin a few days later, I was shown helmets punctured by bullets. The Montenegrins jeered, "What a miserable empire! You can pierce their helmets with buckshot!" Whereas I was thinking: What did the brain under the helmet feel when the bullet pierced it?

Such ambushes and attacks on garrisons were but the sparks. The entire population—those with rifles and those without—rose up against the invader. Gathering at customary meeting grounds, the men came—young and old, grouped by families, villages, and clans—and set out against Italian garrisons in towns. Poorly organized but enthusiastic, they were given leadership by the Communists. Not everyone agreed to Communist leadership, but no one was strong enough to challenge it. The Communists were not only the sole organized force, but a new one, uncompromised. Their propaganda concerning the rottenness of the ruling parties had been confirmed in the recent war. No other political movement could have waged the struggle, for all were confined by regional bounds or carried away by excessive ethnic nationalism. The Communists were the only Yugoslav movement, and for that reason the potential heir of a multinational Yugoslav state. Their association with the U.S.S.R. certainly hurt them among the peasantry and the middle class. But after the invasion this association was to their advantage, for no other power save Russia was in sight. All remembered Russia's support in the past, while forgetting its betrayals; the memory of the people doesn't die, but is awakened or allowed to sleep according to the dictates of life.

Thus the people overwhelmed our leaders, going beyond our expectations and efforts. The leadership was in a dilemma: we had Tito's instructions to begin with small actions and had made our preparations accordingly, yet the people had moved ahead of us. There were other problems as well: the popular movement did not break out everywhere at the same time, and news reached us with a delay of from two to three days. We were aware that this inadequately organized, poorly equipped, and completely inexperienced army could not withstand an offensive by a modern army, not even the Italian.

Therefore we took the position—largely on my initiative—that the movement had to be scaled down to guerrilla proportions, and the people told that we were not yet ready for a general uprising. We told this to the Piperi, who had assembled under arms, all prepared to be divided into military units. We sent Brković to the Kolašin region with a directive to proceed along these lines. A similar directive was sent to Popivoda in the Nikšić region and to Dakić in the Cetinje region.

How did the people react? There was anger, desperation, even tears. And not—at least not yet—because Italian reprisals would hurt everyone, but because we had begun something which was not just ours alone but belonged to all the people, and in which everyone felt bound to take part. The peasants told us, "You don't hate the Italians any more than we do! This is our fight, too! You don't have a deed on the people and the homeland! We aren't afraid to die when our young people are dying."

I remember so well the stand taken by Todor Milutinović, a tall, bony man with a long mustache. He was perhaps the most prominent peasant of his clan, as well as sensible and articulate. He loved his brother Ivan, a high party functionary, like a father. He supported the Communists in all their activities. I had grown very close to him, and he treated me with the consideration of an older kinsman—indeed, an uncle. "Why the devil did you stir things up," he exclaimed, "if you didn't want to have the people with you? This isn't like distributing your leaflets, you know—human heads are rolling and the nation is on fire! Don't disperse the army now that it's assembled. Let's hold council and see what we can do, now that you've started the bloody dance!"

We stood by our decision, however, stating reasonable arguments—countless reasonable arguments, as is often the case when something new and unexpected is presented. But our arguments persuaded few. Moreover, the general excitement came over me as well, and with such force [illegible]at I was seized with rage.

[illegible] was afternoon. Shadows of the cliffs mingled with the teeming [illegible]s it began to disperse. We, the leaders, were left alone, saddened and dejected. Then Blažo and I left, followed by anxious looks. And as evening fell, I resolved to assume the responsibility for changing Tito's stand and our previous directives, and to uphold the people's uprising, even though it was not well prepared. To keep my head from splitting, to keep from being torn apart, I cried out, "We can't separate ourselves from the people! We must share everything with them! If they're mistaken, we shall teach them and learn ourselves! Or else we are not their sons!"

We introduced this new stand to the assembly. They accepted it, but were still confused and discouraged. Only Blažo zealously supported me. Immediately we dispatched couriers and functionaries in all directions, while our Piperi clansmen had assembled by the next morning—rather less enraptured, but still united and resolute. This confusion in the beginning of the uprising—for which I was certainly the most responsible—was later used against me, particularly after my fall from power, but in ignorance of the facts and indeed in spite of them. Pijade was especially adamant in this respect—even during the war. He never credited me with the "correction of an error." As for Tito, though he was dissatisfied with the subsequent course of the fighting in Montenegro, he never held this error against me—most certainly because he had an initial part in it; yet he didn't defend me against Pijade's slurs either, but passed them over in silence.

5

The initial confusion and faltering were quickly forgotten. Many garrisons fell, as did several small towns; some three thousand Italians were taken prisoner; and significant quantities of food and munitions were captured. Looking back, it seems that we could hardly have achieved greater success after such slim preparations, and with so disorganized and inexperienced an army. Spirits flag at the first real bloodshed, and only well-organized, wisely led, experienced fighters can stand the test and win. It was impossible to surprise the enemy after the first attacks on July 13. On July 14 the Italian command established a curfew, and on July 15 the Italian Supreme Command issued an order to Pirzio Birolli, commander of the Ninth Army, to quell the uprising.

On the evening when we decided to support the popular uprising, we also decided to create a military command separate from the Provincial Committee. Since the uprising had a broad popular character, we took the view that this aspect of the uprising had to be strengthened by the organization of a supreme military command. This was all the more necessary because a great many officers of the former Royal Army had joined the uprising and assumed duties assigned to them by the party committees, never hesitating to take on even those responsibilities that were beneath their rank. Besides a patriotic impulse, this stand expressed their desire to redeem themselves after the shame of the recent war.

We also had to designate titles for the supreme command as well as the army, and naturally this was bound up with the character of the uprising and with our position. There were proposals to call the army a workers' or a peasants' army. I was against even the expression "people's" because that term, in the language of the party, had been the usual circumlocution for Communist activity. Since all social strata were taking part in the uprising, I proposed that the designation "national army" be adopted.

It was clear to everyone that the insurgent army was not a restoration of the Royal Army, and that its new, special character had to be demonstrated by outward appearances and internal organization. Battle flags could be the national ones, Yugoslav and Montenegrin, but royal insignia were not used, though not expressly forbidden—rather, committees and military commands avoided their display. There were, however, officers who wore epaulets and royal insignia, and no one prevented them from doing so, though some Communists looked askance at this.

Overnight the Communist party emerged from semilegality to become the only real authority. But the Communists had as yet to get organized; without organized Communists, political work and influence in such a popular army were unthinkable. There was also the danger, though only a potential one, of military arbitrariness with respect to the people, and a need to bring army and people together in the matter of supplies, shelter, and so on. We were familiar, of course, with the title of political commissar in the Red Army, and knew that the title first appeared in the French Revolution; but we were unclear as to its meaning. I wished to give these functions not just a party character, but a broader one. Above all, I wished to avoid any terminological identification with the Soviet Union; ours was a revolutionary army, but a national-revolutionary one. Thus the term "liaison with the people" was devised to describe the function of the political commissar. In November 1941, when I was recalled to Užice, comrades in and around the Central Committee poked fun at this invention of mine, but when they saw that I was offended they stopped teasing me, and it was all soon forgotten.

When we discussed the creation of a broad national command, Bajo Stanišić's name came up. He was a colonel in the Royal Army who had taken refuge with his clan, the Bjelopavlići, after the occupation. I had heard of him, and I believe that Blažo Jovanović even knew him. Bajo was prominent throughout the region as commandant of the Shkodër sector in the only successful operations of the royal army in World War II. We arranged a meeting with him the very next day, in the vicinity of Arso Jovanović's house.

Bajo Stanišić didn't show up, however, until ten o'clock, by which time we had almost lost hope he would come; he pleaded the length and steepness of the way. He was in a wrinkled, shabby uniform without insignia. Rather heavy-set, he carried a thick cane, though he seemed to be only in his fifties and in such good health that he hardly needed support. He had a beard, and when we asked him why, he replied that it was the Chetniks' custom to wear a beard as long as the country was in captivity. I knew of this custom, but attached no significance to the beard or to the motivation for it; I did not expect Bajo Stanišić to be anything but an officer and patriot.

He knew that the uprising had begun, so we quickly got down to talking about our goals, and the creation of a supreme command. We proposed that he join the command. But he refused, saying, "You young people started it." He insisted that the uprising was premature, though he didn't hold it against us. "I understand you. Russia is at war. And I understand the people, who are embittered by the collapse, and feel betrayed by the Greens." Finally I proposed that he assume the post of commander; we were not interested in titles or prestige for ourselves, only the fight against the occupation. But he refused point-blank. "No, I can't do it. You young people started it—good luck to you!" I didn't urge him any longer, for I had the impression that before me was a hard-core conservative officer out of a bygone era. But others continued to urge him. Arso Jovanović was particularly insistent: he couldn't understand how one could be an officer and not take part in the fight against the occupation.

Bajo Stanišić stayed with us another two or three days, and even gave us some bits of advice. But collaboration with him ceased, without ever having really taken place. It was clear that he didn't agree with the struggle the Communists had begun. Yet there was no reason to conclude that he would go "under the wire" to the Italians.* In Užice I was chided, half in jest, for having offered Bajo Stanišić the position of commander. To this day I don't see anything wrong or unreasonable in that: it was a popular uprising and there had to be a place in it for everyone.

The next day, July 18, we set up the Provisional Supreme Command of the National Liberation Troops of Montenegro, the Bay of Kotor, and Sandžak. We called it "provisional" because we felt that only an elected legislative body could establish a permanent command. The term "supreme" was adopted to emphasize Montenegrin equality with the other Yugoslav nationalities, and also because a new Yugoslav military hier-

* The term "under the wire" referred to the Italian camp, that is, a town or garrison surrounded by barbed wire.

archy had not yet been instituted. I was in fact the commander, though I didn't use this title, but always signed "in the name of supreme command." Blažo Jovanović was the political commissar, that is, the "liaison with the people," and Arso Jovanović was the military chief.

The Supreme Command never developed its authority because the Italian armies broke through. Also, we did not give sufficient significance to military organization. However, the command was very active politically. The bulk of its orders and decisions have been lost, yet from what has survived it is clear that we recognized the Supreme Command's authority as the only authority in Montenegro—in other words, as the authority of the Communist party. We passed a decision concerning court-martials, but never got around to organizing them. Private property was protected, and looters, spies, and saboteurs were subject to the death penalty. Though in retrospect it sounds naive, it is interesting to note that the Supreme Command adopted a resolution favoring an antifascist multiparty system; the work of each antifascist party was to be defined in programs and statutes signed by three persons. Preparations for the election of new civil authorities were in progress. On July 22 the Supreme Command issued an order concerning the election of governing bodies—popular committees and political bodies in military units. Some elections of officials had already taken place, but in the spirit of the directives of the Provincial Committee. Thus on July 21, in Berane (now Ivangrad), 217 delegates elected a district people's committee of 21 members, headed by Archpriest Bojović, my former teacher of religion and ethics, who was never a Communist. It was planned that on August 5 the delegates of district and local committees would meet in Kolašin to elect a "Supreme National Liberation Administration," but the Italian offensive stopped it. It is interesting that even the military leaders were elected at meetings of the electorate, of course under the influence and with the initiative of the party, but not only from among party members.

The struggle continued. Bitter fighting took place at Berane, where the Carabinieri barricaded themselves inside the former gendarmerie station and offered resistance until our people burned it down. Pavle Djurišić, the royal officer who later became a renowned Chetnik commander, distinguished himself in this battle. Fighting was going on also in the vicinity of the Supreme Command, particularly around Danilovgrad. I went there on the morning of July 20. I took along a captured Italian noncommissioned officer who spoke French, and told him that we were sending him to the Danilovgrad garrison with a demand that they surrender, and to explain to them that we treated our prisoners well—the officer himself was living proof. He had been captured at Bioče, where the prisoners were simply disarmed and their freedom to move about limited; indeed, some got jobs working for the peasants.

Across from Danilovgrad, on the left bank of the Zeta River, I met Boško Djuričković, commander of the siege. I knew him from the student movement in Belgrade. He was respected for his quiet self-confidence and devotion to whatever assignment he undertook. Tanned, slim, and in an officer's blouse, he now assumed a martial air as well. This was in fact my first direct experience of battle, and since we were within range of random Italian firing, he cautioned me to get behind some rocks. Using a map, he explained that he was preparing to take Bralenovice Hill, which, if I remember correctly, was a key point in the defense of the town. I saw immediately that Boško understood battle operations far better than I. He was obviously conscious of this, but it didn't make him overbearing. The men were in good shape, yet Boško complained that some officers had a bad effect on morale. We agreed that he would be patient with these officers, but decisive in stamping out insubordination. It was there that I met Puniša Perović, a veteran Communist with a taste for literature, withdrawn, but popular as the doyen of the student movement. I also met Djuro Čagorović, a peasant whose powerful frame radiated a limitless courage and energy. This appealed to me, but most of all I liked his ingenuous, unassuming spirit of self-sacrifice. We were always glad to see each other—that is, until my altercation with the Central Committee in 1954, which he could not comprehend, if for no other reason than he had already become accustomed to the benefices of ceremonial office.

Around noon I returned to Piperi and the Provincial Committee. Scarcely two or three hours later a courier arrived from Danilovgrad with the report that the garrison had surrendered—even before the attack. Over a thousand Italians had been taken prisoner. Why did the Italians fight so badly? They were surprised by the uprising, militarily as well as politically. Their officers were trained, but their soldiers did not have any enthusiasm for the war. They shot dutifully from their fortifications, but once our men penetrated the fortifications or engaged them, the Italians threw down their arms.

In Danilovgrad we also captured a police official by the name of Bakić, whom I had known slightly in high school. He was a poor student, and came from a prominent family. His mother, a very aggressive woman, had somehow got him a position with the police. It was proposed that Bakić be executed. But he repented of his actions and made promises, and reputable people pleaded for him, so his life was spared amid general rejoicing. But as soon as the Italians advanced, he again joined the police force under their wing, zealously hunting down Communists and all others who opposed the occupation. Bakić's case was typical: during the uprising they would surrender without a fight, but after the Italian breakthrough they would go into the police service, becoming more cruel than they had ever been under the royal regime.

The Italian counteroffensive quickly exposed both the shakiness of the insurgents' unity and the weakness of the insurgent army. But it also showed that the Communists had become a new and independent force, capable of withstanding unheard-of trials. The Italian troops had come from Albania, gathered at Podgorica, and fanned out in three directions: Cetinje, Nikšić, and Kolašin. We had reports of Italian reinforcements, but from the Piperi cliffs we also observed, through field glasses, the yellowish dust raised by the Italian trucks along the Shkodër-Podgorica road. Anticipating a breakthrough, we issued instructions on July 23 regarding the conservation of ammunition and battle tactics: no rigid fronts or frontal attacks. Ultimately an entire Italian corps, with air and naval support, was sent into the offensive.

The first Italian move of any consequence was the dispatch of a motorized battalion in the direction of Podgorica and Cetinje. But the battalion was ambushed by Peko Dapčević in Ceklin, and a hundred or so men annihilated. It seems that the Italian command learned a lesson from that defeat: the Italians then advanced cautiously, their flanks protected, under thorough artillery cover.

The attack on the Piperi and the Bjelopavlići began at the end of July. In a surprise attack the Italians first took the mouth of the Zeta and Morača rivers, which, since it was in the plain, we could not defend. They also attacked our "rear" from the air, probably because they had learned from military manuals that "bandits" frequently hide in the woods. The key positions—the hills of Velje Brdo and Trijebač—were taken July 30. After Velje Brdo was seized, a barrage began against the Bjelopavlići. Boško Djuričković told us later that we never used our artillery, captured from the Italians in Danilovgrad, because royal officers had removed the firing pins from the breeches.

Blažo Jovanović and I were stationed with the commander of the Piperi battalion, Major Mitrović, on a craggy ridge near Trijebač. The artillery raked the rocky peak by the old Turkish watchtower on Velje Brdo with such force that the ground was spewing dark globs of earth. The peak was enveloped in thick smoke and a savage din, then all was still again. "They're about to attack," Major Mitrović observed calmly without lowering his binoculars. And indeed, accompanied by a fierce volley and the roar of bombs, Italian soldiers emerged on the peak in close ranks. None of our own men were there and I assumed that, to avoid casualties, they had withdrawn from the peak into some rocky ravines below. In tense silence we awaited a counterattack by our side, but nothing happened. To escape the artillery fire, they had abandoned the position. The peasants fled back to their homes, hoping to save their families and animals, and no force on earth could have stopped them. Blažo delivered a diatribe against incompetence and cowardice, while Major Mitrović's ruddy rotund face showed not a trace of excitement.

I had already grasped that warfare, though madness, is waged according to logical terms. I said, "Our side should have taken cover along their flank or in front of the peak, so that when the Italians—"

The major broke in calmly, "We would have been spotted. Advance units must comb the area first."

I soon realized that a guerrilla army—and that is what we were, despite our numbers and organization—was capable only of slowing down the offensive of a regular army. This means, as a rule, the necessity of evading enemy attacks; that is, one must engage the enemy only when he is in retreat or stationary. Guerrilla warfare means constant fighting rather than big battles. It is not spectacular victories and territories that count, but the annihilation of small units and the preservation of one's own vital force. Though we criticized and fumed as we watched the defeat at Velje Brdo, we knew instinctively that the Bjelopavlići could not have done much more, and that similar situations awaited us.

Indeed, that very afternoon, in spite of the lesson of Velje Brdo, we positioned the Piperi on Trijebač according to all the military rules, and ordered party members to keep up morale by word and example. The Italians repulsed our ambushes and, after preliminary artillery fire, took Trijebač. Our side did not counterattack, nor did they completely abandon their positions: small groups of Communists kept on firing in the foothills. The peasants felt weak and withdrew instinctively, but the ideologically committed and disciplined comrades continued to fight. We were on that ridge with Major Mitrović throughout the fighting. He received and dispatched couriers, steady all the while. At one point bluish smoke arose from a valley. "A mortar," the major observed, as if the Italians were throwing apples at him. The mortar shells landed some distance behind us.

The cross fire lasted until dusk, when we made our way to the village, conferring with the major along the way. We wanted him to get together as many men as possible and to take over upper Piperi, so that at least a part of the clan could be protected from the burning and looting. Major Mitrović was a regular army officer, no staff man, and unintellectual. We were in each other's company for the rest of that summer. He was withdrawn, but not a somber man. He neither flattered Communists nor hated them; he simply wished to fight against the occupation. After my recall from Montenegro in November 1941, I forgot about him in the confusion of war. But after the war, as I reminisced with Blažo Jovanović about those first days of the uprising, I remembered him. In late 1941, Blažo told me, he joined the Chetniks—on orders and without much enthusiasm—because of his oath to the King. Joining the Italians in Podgorica, he said to a kinsman, "I hope to God the first Partisan shot hits me!" In the first Chetnik attack on the Piperi, he was killed. I am sorry that I have forgotten Major Mitrović's first name. I regret that many officers experienced a similar fate. I regret this all the more because I came to know Major Mitrović as a steadfast soldier and a fine man.

Going down to the village, we came across a woman by the side of the road who was wailing over the body of her dead husband. It was dark and the armed peasants were walking past her in silence. In peaceful times they would have gathered around the body and mourned. I thought that I ought to stop and say a word or two of consolation to the woman. But that seemed senseless: the woman did not know me, and if she realized that I was the leader of this group, she might even berate me. War was war, I thought, anyone can get killed. So I passed the woman by like the rest.

During the night we held party meetings and gave out instructions on how to evacuate the villages. We met with the villagers, who were bitter and concerned, but not critical of us: they had shared in the selection of their fate. Just before dawn we got to a plateau overlooking the Piperi, where we awaited the sunrise. The Italian troop movement had begun along the Zeta Valley. In the green valley the even greener soldiers approached the stone huts, and smoke soon began to pour out of them. Blažo's house was already gutted; his mother, sister, and two children had fled to the hills. Blažo's wife Lidija was away on some party assignment. There were no people down there. They were all here on the road beside us, with their animals, children, and belongings. They passed by, as Blažo and I sat there watching them silently. Down among the houses there was no resistance, only occasional shooting—no doubt at the chickens. Along came a peasant carrying a cannon wheel on his shoulders; he sat down to rest.

"What's that for?" I asked.

"Well, it may come in handy," the peasant replied.

"How? What for?"

"Well, I don't know. Some day it just might come in handy."

"How? How? Don't you have anything better to do than drag that thing up a mountain?"

The peasant fell silent. I took the wheel, dragged it to the edge, and sent it spinning down the precipice. "There," I said, as the wheel sped downhill. "You'll find it there after the war."

The peasant said nothing. Probably he saw for himself that it was time to do something more reasonable—make war, for instance.

6

The next day I had to go to Kolašin. Arso Jovanović was already there. He sent word that, despite some successes, there were problems: the Albanians in the Moslem villages had joined the Italians, and there was trouble in the insurgent army.

I set out after midnight from a Piperi encampment, with two escorts. For a long time we tramped along rocky by-paths, and then descended into the canyon of the Morača. The dawn and the pale azure of the Morača, framed in the gray crystalline rocks of its precipitous banks, are fused in my memory. I was tired and hungry, but we had to go on. We came to a shallow place and forded the river, which was chill and inhospitable in the dawn. In the rocky hollows along the river, men and animals had taken refuge, weary from flight yet watchful. We ignored their sullen curiosity and walked further down the river toward Bioča—a crossroads that the Italians had already taken—then climbed up a scrubby cliff. At daybreak we reached the highway and made our way through the Bratonožići country. As a result of this exhausting journey, in the afternoon I felt my temperature rise. I thought I should take it, but I didn't. The journey to Kolašin—some sixty kilometers of barren road—would have been too long and strenuous even if I hadn't been weakened by my lung condition. I had to go on, so I decided not to bother with my temperature. I gave my thermometer to one of my companions as a gift, which made him very happy even though, being healthy, he hardly needed it. I continued to feel my usual early evening fevers and nightly sweating well into late fall. I think I might have died of the disease, had I not been forced into the war. That is how many of us got well.

When we had gone not even ten kilometers, we encountered an advance guard of the Kolašin units, followed by their commander, Radun Medenica. I knew Radun from the Kolašin high school; we had followed each other from school to school. He was older, but he recognized me when he heard my name. I remembered him well—as a cordial and obliging fellow, and even more as the son of Miloš Dragišin Medenica, the most renowned officer in the battle at Mojkovac in 1915. The din and blood which then shattered my native mountains had darkened my childhood and later became inspiration for my stories.

I was very happy to see Radun, and to hear that he was expecting a supply truck from Kolašin any minute. He was in the uniform of a royal officer, and his men were in good spirits and under control. He did not worry much about the Italians, but he did worry about where he ought to fight; to keep the Italians from burning their houses, the armed peasants wouldn't let him fight in "their village." "You defend your own villages and clan," they said, "and we'll defend ours."

"What can I do?" Radun asked. "They may fire on us, and then what?"

"You must explain to them."

"It's no use. We've tried everything."

"A conflict must be avoided. There's a fifth column at work here."

"Certainly. But we can't shoot at peasants. Besides, my men wouldn't do it."

Finally we decided to avoid a clash at all cost. That was the only plausible stand: the war had just begun, the peasants were frightened and disorganized, without any military organization able to impose its will by force. They were susceptible to the influences of covert opponents of the uprising, and we were in no position to force our own Communists, let alone peasants, to submit to the ultimate sacrifice. Yet the peasants' cowardly surrender of freedom and homeland for the sake of property and family stunned me and made me bitter, all the more so because it was contrary to my inherited belief in Montenegrin heroism and hostility toward the alien foe. Nor could I fit this into my dogmatic notions about the people—about the peasant and the worker ready and willing to sacrifice themselves for "their" cause, given the right conditions. In fact, I wasn't free from this idealization of the people until my fall from power, though even during the war I suspected that the people do not sacrifice for any ideal unless an organized force, or the danger of annihilation, obliges them to do so.

The truck arrived, and we were in Kolašin by noon. In addition to the local functionaries, Moša Pijade and Arso Jovanović were also there. Moša administered the local command, and the distribution and evacuation of the Italian supplies, while Arso kept in touch with commanders in the field through long telephone conversations.

My brother Aleksa was also there. He was glad to see me, and could tell me about the recent battles and the heroism or cowardice of local leaders. He had distinguished himself in the ambush at Vrlostup as well as in the taking of the Carabinieri post at Mojkovac. Under the protection of machine-gun fire, he had run up to the post without seeking cover, and tossed a bomb into it—whereupon the Carabinieri raised a white flag. In this deed my brother obviously had in mind our father, who in 1915 had burned down a Turkish guard post not far from there and destroyed its garrison.

Moša greeted me with a verbal barrage that was all the worse because he had good reasons to do so, given the contradictory directives concerning the uprising.

"For twenty-five years I've been waiting for an uprising," said Moša, "and when it comes and the people rise up, you send directives that an uprising is not necessary, but rather actions by groups!"

I explained the circumstances. He was not mollified even when he learned that this was Tito's position.

"But Tito didn't know that this was a matter of the entire people rising up!"

"Well, when I realized that the entire people had risen up, I changed the directive."

"A revolutionary must sense this in advance!" Moša insisted, aiming at my most sensitive spot.

I cannot explain to this day why Moša didn't like me. I don't think that even Moša could, except in terms of party clichés and generalizations. I won't say I never stung him with some mischievous jest or remark, but I never did him a bad turn or slighted him. Even in prison we had our differences. Moša was a leader of the more moderate "right" wing, while I belonged to the "left." Moša was shaped in factional strife and never free of factional pettiness and spite. I never took part in factions. Moša was in prison when Tito formed the party leadership, and the top leaders elected by the Fifth Conference did not include him, even though he was an old and prominent party member. But I was not to blame for this. The leadership was selected by Tito; he picked younger men who were experienced in illegal activities and of one mind with him.

Between Moša and myself there was a generation gap. On coming out of prison, he found a party of dogmatic pragmatists intolerant of time-wasting discussions and familiar relations. Though he had great merits—especially his diligence and long imprisonment—he had none of the qualities of a leader. Moreover, being old and unadaptable, he could not be a leader in a party that, though not new, had changed; one had to stand ready day and night to rush from meeting to meeting, to go into hiding, to transport and print illegal material. Despite his self-control, he must have felt it unjust that younger men took over the leadership which he deserved by virtue of his past contribution and his sufferings. He did not like Ranković either, but never showed animosity to Kardelj. As for Tito, he respected him and even liked him, if he liked anyone.

Despite the rebukes with which Moša accosted me, I consulted with him, and we decided together all that there was to decide. Anticipating an Italian advance, we decided to transfer the Italian supplies—rice, oil, macaroni, and canned goods—to my native village of Podbišće, which was some distance from the highway. That evacuation was well executed, but the people of Podbišće and of the surrounding villages—among them some party members as well—discovered the cache and picked it clean during the Italian breakthrough. Led there by spies, the Italians later took the rest.

The next day, accompanied by Arso, I rode off to the positions at Čakor. In a local command post I found Bogdan Novović, a bank clerk and friend from Belgrade. With the dignity and authority which were natural to him, and in which he delighted, he was in charge in the office of the former district administrator. I also ran into Radovan Zogović, the writer, who was a close friend of mine in Belgrade. He later told me that I said to him at the time, "The Soviet Union is at war, and we Communists shall fight to the last man." I do not remember these words, but there is no doubt that I could have spoken them. Though caught up in our own reality, we Communists were allied ideologically with the U.S.S.R. and under its influence. This may have restricted the autonomy of the uprising, but it doubtlessly inspired us with confidence.

We stopped in Murina and observed the advance of the Italians from Velika. In actual fact there were not that many Italians. They advanced slowly along the road while bands of Albanians secured their flanks, burning and looting villages on the way. Arso had a discussion with our officers, who expressed a lack of confidence and made numerous remarks to the effect that our main enemies were the Albanians and "Turks" (Moslems of Slavic origin), and that the uprising was premature and inadequately organized. But no one was yet stirring up the troops. Among the commanding officers was Major Djordje Lašić, later a commander of Montenegrin Chetniks. He was Arso's kinsman; they spoke and argued as friends. Lašić was not pleased with the state of affairs or with our prospects, and perhaps not with his subordinate role either, though he did not protest openly.

The villages of Velika and Ržanica were burning, and our one artillery piece was firing. The creativity and resourcefulness of an aroused people are inexhaustible. That gun had been left by the Royal Army and hidden away by someone. It had no firing pin, so a blacksmith forged one. Ammunition was also found, but the shells had no detonator. Expertly aimed, they landed where they were supposed to, but it was like throwing rocks at the Italians and Albanians. The artillery commander, a young red-faced officer, laughed while sweating from fatigue and the humidity; the Italians must have laughed too, but the Albanians were fearful. As for our people, they were encouraged by the sound of their own cannon.

Throngs of people, on foot and on horseback, with children in carts and with cattle, were crossing over to the left bank of the Lim, fleeing along the road. On a bluff above the road, near an inn, Ljubica Bojović joined us. With her was an officer; I thought he might be her husband, but he proved to be a relative, someone from the command. I knew Ljubica from high school and from the University in Belgrade. Well shaped, with a broad smile and free-spoken, as a young student she had been known for her frank sensuality. We were never close or particularly drawn to each other, but her bright and lively nature always radiated cordiality. At that time, in her thirties, she had a voluptuous and self-confident beauty. She was not a Communist and taunted me without inhibition—but also without anger—though we had not met in years. "So this is the mess you Communists have got us into! Oh, I know your restless nature well. But our people won't let you off that easily!"

I said, "This isn't only a Communist affair—it concerns the entire nation." I couldn't have given her a more definite response regarding our prospects, just more slogans—that victory would still be ours, etc. We parted as friends. And we have greeted each other as friends ever since, whenever we meet by chance in Belgrade—even after my fall from power.

So it was with us not long ago, though now we're a little sluggish, a little decrepit, finding joy only in memories. Desire and thoughts of love aren't blotted out by war; they intrude into the consciousness unexpectedly but are soon forgotten, discarded.

In point of fact we didn't have much of anything to recommend to the party and military leaders in Murino and Andrijevica. It was clear that, lacking heavy weapons and organization, we could not conduct frontal assaults. Through the Supreme Command we had issued directives along these lines as early as July 23. Yet whenever it was a matter of protecting the population, we could not avoid making such attacks. Moreover, we had learned from the battles in the Zeta Valley that our people's army would disintegrate as soon as it stopped making frontal attacks, that is, as soon as it stopped protecting its own villages. So we decided that we had to protect the withdrawal of the population, and then retreat into the mountains. We insisted that no one surrender his arms upon returning to his village, for we had no doubt but that the fighting had just begun, and that we would regroup and rally our forces.

In Berane the atmosphere was even more tense, although no firing was heard and the front facing the Moslems held fast. We knew of the Italian breakthroughs and did not try to hide this knowledge from small groups of prominent citizens, mostly former politicians, who demanded that resistance be offered only to the Moslems who rose up, unprovoked, against their Orthodox neighbors. The hatred between the Orthodox and the Moslems in these parts is primeval, attested by rebellions and invasions, epics and visions. I had grown up in this tradition but, like other Communists, thought a reconciliation possible. I maintained that their differences arose solely out of the malevolence of reactionary forces, and that every instance of hatred had a concrete cause or motive. But now there was no such cause or motive. In theory, the neighboring Orthodox were fighting for the Moslems, too, as the Italian occupation could not be in their interest either. Not even the anti-Communist elements among the Orthodox had any interest in provoking the Moslems. Yet the Moslems were up in arms, burning and plundering, and we Communists had no explanation other than that they were "reactionaries" manipulated by their religious leaders and party politicians. Such explanations were vague and inadequate in the eyes of peasants whose homes were being burned, and unconvincing to Orthodox Montenegrins for whom slavery under the Turks was still a living memory. Reality had demolished our wonderful dogmas and illusions; eyewitnesses of atrocities could not be refuted. Not long ago the writer Mihailo Lalić reminded me how difficult and unconvincing it was to preach brotherhood with the Moslems to a peasant who had heard his son's cries as he was being flayed alive in the traditional Turkish torture. Yet we still insisted on brotherhood and on our explanations—without, of course, ceasing to defend our villages.

Another misfortune in Berane involved four captured Italian officers, who had executed seven Communists on Jasikovac Hill when the fighting began. We did not kill the officers immediately but put them in prison; subsequently they became a source of contention between the party leadership and certain citizens, some of them decent patriots, who were stirred up against the Communists, and who demanded that the officers not be shot so as to forestall Italian reprisals. As always in politics, countless arguments were advanced—each more convincing than the one before it—to the effect that the officers should be judged by a "legal" court-martial; that they were prisoners of war and not answerable for acts committed under orders; that the Communists would retreat and abandon the population to the vengeance of the Italians; etc. The local Communists recognized their error in not shooting the officers at once, before the unity of the insurgents dissolved. Now it had become clear that this unity had been an illusion: when the enthusiasm of revolt subsided and the return of the Italians became a reality, everybody reverted to his own way of thinking and his own interests. The long smouldering struggle between the Communists and their opponents—a nationalistic and already confused peasantry on the one hand, and a democratic and patriotic but perplexed citizenry on the other—was spreading. I too was against having the officers executed on the spot; the Berane comrades could not have done it without bitter dissension in the ranks, and resignations from office. For the sake of unity, I later approved the sending of the Italian officers to the Supreme Command at Kolašin.

From Berane I went on to Bijelo Polje. Rifat Burdžović, until recently secretary of the Belgrade Committee and now secretary of the Committee of the Sandžak Region, was cheerful as always, though his situation was hardly encouraging. True, the Italians were still far away and the little town was relatively peaceful. Yet the concern was there, both among the Montenegrins who had taken part in the uprising, and among the town Moslems who had withdrawn into a secretiveness and speculation of their own. The young Moslems—largely students and Communists—who fought with the insurgent army and were especially bitterly against the Moslem peasants, constituted an exception. Meanwhile the uprising was slowly spreading to the Serbian townspeople as well; there were not sufficient forces to take Plevlje, so we agreed to dispatch a battalion of Kolašin people.

I returned to Kolašin by way of Mojkovac; hurrying through my native village, I hardly recognized it. I was torn by problems and worries, alone in the midst of the deadly vortex into which I had led my people.

7

Eighteen years before, I had lived in Kolašin as a schoolboy. The older men remembered my father, while the younger intellectuals knew me by name. I was to work with local groups and commands. However, the peasants were also beginning to turn to me for help—food, release from military service, etc. I lived and conducted my business in a little room on the second floor of a building next to the Bošković Hotel, which was later burned down. Arso worked in the same room when he was not at the front or on the telephone in the adjoining room. The local command posted a guard at the entrance to keep people from pestering us, as well as to give us an official air.

Indeed, the office began to expand. For administrative assistant we got the services of Boba Selić, a Belgrade student married to Vladeta Selić, later an official of UDBA* and a diplomat. She typed orders, received reports on the telephone, set up meetings, and took care of our meals. She never missed or forgot anything, though she was not compulsive. Dark-skinned and with thick brown hair, she had an easy pleasant manner and a big, toothy, unabashed smile. My collaboration with her lasted only a few days, but I long held friendly feelings toward her.

On the Bratonožići front the turn of events could not have been worse. The Italians were advancing slowly, but with no trouble at all. The peasants, helped by fifth columnists, were hindering the resistance of the Kolašin units, which were themselves not showing great courage or

* The initials of Uprava Državne Bezbednosti (Administration of State Security), the secret police.

discipline. I talked Arso Jovanović into going to the front to supervise the command, and the party committee into stepping up its political activity. As a result, our units began to resist the more vulnerable enemy units.

At the same time discontent broke out in Kolašin that was more open and virulent than in Berane. We found out that certain "prominent citizens" were conspiring against us. To our surprise, they themselves informed us that they had formed a committee, and invited us to a parley. Moša Pijade was in town, but in agreement with the party committee it was decided that I should go and see what it was all about. To be sure, I did not go alone, but with two or three comrades.

We entered the town freely, but as we approached the meeting place we were informed that several armed opponents of the Communists were patrolling the streets. We had some thirty armed men—enough to defend ourselves. We kept calm, but remained on guard. My brother was there, obviously spoiling for a fight, strolling through the streets with a couple of comrades to reconnoiter the enemy.

In the little tavern of the Lješnjak Hotel we met some ten men whose appearance and years gave the impression of solid prominence. Among them were Blagota Selić and Archpriest Jagoš Simonović. I remembered Blagota from my childhood, from the Kolašin market place. He used to sit in front of the little cafés with my father when he came to town on market days. Though he was near sixty he looked young, more because of his energy than his strength. He was known as a man who liked the comfortable life, but also as honest and obliging. His sons were both Communists. I was embarrassed to find him there, especially since, when I was first in town, he had come up to me several times to rekindle memories of my father, and to express his approval of the uprising and the "youth," meaning the Communists. Archpriest Simonović had been my teacher of catechism in the first three grades of high school. I did not have the most pleasant memories of him, for which—it seems to me in retrospect—my stubbornness was more to blame than his hairsplitting. Far more significant was the fact that he had been district president of the regime's Yugoslav Radical Union party, known as Jereza. Thus his presence did not surprise me at all; though Jereza disintegrated during the occupation, it bore a heavy responsibility for the collapse of Yugoslavia; its leader Stojadinović had flung Yugoslavia into the arms of the Axis, while his successor Cvetković had drawn Yugoslavia into the Tripartite Pact.

Today it may seem undignified of me that I went to that meeting, instead of receiving the "prominent citizens" in my own headquarters. But in those days we dealt with all on a broad, democratic basis, so that any such act on my part would have been taken as arrogance and an insult to older and prominent leaders. In any case I did not think that these citizens regarded my coming to them as a sign of weakness.

The first to come forward from their side was Judge Ljubo Minić, whom I had not met before but knew to be an anti-Communist. An eloquent man, full of pathos and demagogy, he was stocky and swarthy, and had a quick tongue. He wore a revolver in an open holster around his waist, though he had no reason to be afraid. He was obviously aware of the Italian advance. I do not believe that he had learned of it from the Italians, but rather from officers returning from the front. Our conversation was frank and to the point. Though a judge, Minić did not use legalistic arguments—that the state had capitulated and that we were therefore carrying out an illegal act of rebellion. Instead, he openly attacked us Communists for exposing our people to destruction for the sake of our own aims. "Where were you until your Russia went to war?" he cried, leaping up from his chair. "You are to blame for everything, and should be made to bear the responsibility—you, and not the poor people!"

His charges did not confound me; I had good arguments against them: it was a known fact that the Communists were against collaboration with the fascist powers, and for the defense of the country. And when Germany and Italy had attacked Yugoslavia, we Communists had put up the most determined resistance. We were continuing a war which need not have been lost if those in power had not been disunited and demoralized, and if fifth columnists had not gone about unhindered. We Communists were fighting not for ourselves but for the entire nation. We raised no question about who would hold power—not even in positions of command—as long as they were steadfast in the struggle against fascism. As for Russia, we Communists indeed regarded it as our guide and hope—as had our forebears, too.

The caustic exchange between Minić and myself did not last long. The meeting itself went on for just over an hour. It might have ended sooner had I not detected a certain wavering in my favor. This was especially true regarding Selić and—an agreeable surprise—Archpriest Simonović. Apparently, Selić became easily enthused, and in my exposition I had, both intentionally and spontaneously, stressed certain themes that were dear to him: the traditional Montenegrin defiance, and a self-sacrificing devotion to the homeland. As I spoke, he kept interjecting,

"That's the holy truth! No honest man would refuse to go along with that!" The Archpriest was more restrained: his face showed deep thoughtfulness and he chose his words carefully. But he too was agreeing, along with the majority. Finally Minić was left with the silent support of only two or three of the citizens.

At this I became more eloquent and open. I pressed Minić, who was grinning cynically: "What do you want? What do you propose?"

"You Communists say you're for the people," he answered. "Then you should surrender to the Italians and take upon yourselves the responsibility for all you've done."

This suggestion inspired me to a passionate exclamation, though I had no intention of fulfilling it: "We Communists would be willing to do even that, if it would benefit the people and their struggle!"

"Good for you!" exclaimed Selić, while Simonović declared that no honest person could make such an unreasonable demand. The meeting ended with my observation that we were doing, and would continue to do, everything possible to protect the people and keep them from perishing.

After the Italian entry into Kolašin, Simonović and Selić withdrew with the Communists, and took refuge with their clans. When the Partisans were forced out of Montenegro in 1942, Archpriest Simonović retreated with our units into Bosnia and shared all their trials and tribulations. He lived to see the end of the war, but was in poor health and died soon after. He received less recognition than he deserved. As for Selić, the Chetniks captured him in Rovci, took him to Kolašin, and sentenced him to death. His only guilt was that he had spoken up on behalf of the Partisans and that his children were Communists. The Chetniks were bitter because he had lent his name and his past to the Communists, and for this they decided to disgrace him. Before they led him away to be shot, they hung a cowbell around his neck and walked him through Kolašin like an animal. That bell of Blagota Selić rang shamefully in the conscience of the Chetniks—those, at least, who had not broken completely with reason and tradition.

There were suggestions—to be sure, from my brave and hot-headed brother Aleksa, among others—that we kill Minić before he left town. It wouldn't have been hard to do. But I rejected the suggestion because I didn't want trouble, and wished to avoid all such settling of accounts until our opponents themselves resorted to arms.

But Minić had his own ideas. The next day he and a hundred peasants from Rečini and surrounding villages took over Bašanje Brdo, a hill that dominated Kolašin, and at the foot of which was a garrison where we had locked up the Italian officers from Berane. He sent us a demand to turn over the officers or he would fire on us. At first we rejected the demand, and deployed the few guards we had to defend the place. Negotiations were conducted through the prominent citizens.

Though we were fewer in number, we could have defended the town, or at worst withdrawn from it. Our first thought was to kill the officers, but we rejected that solution; we were uncertain as to whether we should engage in battle with the frightened and aroused peasants, especially since the prominent citizens were against shedding the "blood of brothers." This is how I reasoned: Minić had not yet revealed himself to the peasants as a supporter of the occupation and fascism; rather, taking advantage of the violence of the Italian advance, he was frightening them with the burning and carnage, and enticing them with promises that they could save themselves. I appreciated the fact that we were not yet "ripe" for a mutual settling of scores. Besides, I adhered to the principle: Let the opponent show who he is and what he wants; let him be the first to fire. Also, the Lipov battalion, which had been sent as reinforcements to Sandžak for an attack on Plevlje, was returning in disarray, demoralized by the propaganda claim that their homes were endangered by the Italian offensive, and that their primary concern should be themselves and their own village. Minić and his cohorts made their appeals on a clan basis: why didn't the Vasojevići kill the officers, instead of sending them to us—why do we have to suffer because of them?

In view of all this we decided, in consultation with Moša Pijade and the local comrades, to avoid a clash with Minić and his peasants, and to turn the officers over to them—but on condition that they give their word of honor to return them to Berane and give them "back to the Vasojevići." I suspected that they wouldn't do this and that they needed the Italian officers so as to win recognition from the Italians. But even a false word of honor struck me as a gain in the struggle for the minds of the peasants, and in exposing Minić. So we gave them up as soon as Minić gave his word of honor. There was some grumbling and my brother was outraged, but I convinced the comrades that we must not begin a civil war. Let the fifth columnists be completely exposed, I said, and then we'll have it out with them.

That episode has bothered me ever since—even to this very day. Pijade did not hold it against me, and it soon became buried under more important events. I was not bothered by giving in to Ljubo Minić, as I regarded it as giving in to the misled peasants. Rather I was bothered because I hadn't avenged my comrades shot in Berane. Caught between revenge and clash with the peasants—caught, that is, in the beginnings of a civil war—I vacillated, then decided to give in. To be sure, Ljubo Minić greeted the Italians in triumph, and immediately became their representative and tribune. But he did not win over the peasants. They boycotted the assemblies he held in Rečine and its vicinity—within range of Italian machine guns and escorted by Italian guards. He never dared go beyond the borders of his district, not even in the Chetnik organization, and whatever success and popularity he had was due more to his violence than to any affection within the ranks of the counterrevolutionaries.

I have dwelled on Minić at length because he is the most obvious example of opposition to us in Montenegro before the inception of the civil war—that is, before the formation of a Chetnik organization, at a time when the Communists had not killed anybody, and when royal officers were participating with Communists in the same insurgent army. I had heard even then that Minić was a follower of the Serbian fascist Dimitrije Ljotić. I don't know this for a fact, but it's plausible. There were indeed fascists in Yugoslavia; the people's uprising frightened them at first, but they came out into the open with the advance of the Italian army. Minić was doubtlessly one of them. He perished at the war's end in one of those massacres for which he himself had laid the groundwork by his antinational ideological violence.

Minić's diversion caused us to bring up reinforcements from the Italian front. Luckily we did not have to use those reinforcements: Arso Jovanović arrived from the front the next day and told us that the Italians had already taken Lijeva Rijeka and put the torch to the village of Jabuka. That evening another report reached us that the Italians were already some ten kilometers from Kolašin. In great agitation, Arso Jovanović told us that the Italians were flanked by Albanians who ran like dogs through the wooded hills, shooting wildly and shouting, "Sons of bitches, Moračani,* mother cunts!" They had burst out of the woods so suddenly that Arso had barely got away.

In the course of that day all of us except the guards withdrew from Kolašin. Yet I saw to it—as I was to do later as well—that we of the Supreme Command and the party organizations left last. This was not a case of heroics, but rather of concern for what had to be done. By

* The Moračani were the largest clan in the environs of Kolašin. The Albanians certainly remembered them from the time when the army of the Pasha of Shkodër waged war on the Moračani.

nightfall some two hundred fighters, mostly Communists, arrived from the front. Strangely enough, there was no anxiety at all. Perhaps I was more worried than anyone. In the middle of the square, bathed in moonlight, in a deserted, dead little town, armed young men and women joined hands in a circle and sang as they danced a *kolo:* "Thrice the Moračani took the town of Kolašin . . ." Verse after verse of past defiance and present hopes followed, as everyone joined in the dance. Finally I got inside the circle and gave a short speech: We are leaving unconquered —to return again, as our unconquered forebears returned. . . .

On the next day—August 5, 1941—units of the Italian Venezia Division entered Kolašin.

8

In the course of the night we withdrew to my native village of Podbišće. Actually, all except some ten escorts scattered to their own villages. I stayed with Stanija Kovijanić. She was on close terms with my family, while her son was an old party member who had joined my brother in going underground in 1936. Moša Pijade was also staying there, as was Dušan Nedeljković, a professor of philosophy, along with his English mistress whose name, I believe, was Eileen. After a thorough discussion of the situation, Moša and Arso left the following day for Piperi. There were no conflicts between Moša and me, except that Moša reproached me for the confusion at the beginning of the uprising, while I explained that the error—if error it was—had been corrected without serious consequences. Anyway, it had become apparent that the people's peasant army was incapable of waging war, no matter how much it favored rebellion.

Nor was there any conflict with Professor Nedeljković, though we argued over dialectics and Hegel's influence on Marx. Nedeljković regarded Marxism as the left-wing form of Hegelianism, enriched by materialism, whereas I regarded Hegelianism, and particularly the dialectic, as an important component of Marx's original, independent doctrine. Actually, the party organization in Skopje had been in conflict with Nedeljković just before the war broke out, because of his arrogance and intolerance, and also because he had belittled the underground organization and illegal activities. Later he fled to Montenegro, where he became an active antifascist. I note this academic doctrinal discussion in Podbišće, on the day following the retreat from Kolašin, as proof that war represses, but does not kill, one's inclinations and convictions.

As for Nedeljković's English mistress, I looked upon her with a distrust which my politeness could hardly conceal. How could an Englishwoman in Montenegro—in the very midst of our armed strife—be anything but an agent of some intelligence service? I was unjust, but I was a Communist, a Communist official. Freckle-faced, wispish, and pedantic, Eileen broke up with Nedeljković, threw in her lot with our fighting men, and became a devoted fighter—only to perish in 1943, in the so-called Fifth Offensive, and remain unrecognized and forgotten.

Nedeljković did not leave with Pijade, but stayed in Mojkovac with some comrades. Later he joined the units in Sandžak and from there went on to Užice, where in November I found him serving as an editor of *Borba.* I myself stayed in the vicinity of Mojkovac to observe the further advance of the Italians, and meet with Burdžović and the Sandžak leaders.

Prior to that—on August 6, I believe—I held a meeting with the party cell of Podbišće: there had been some pilfering of our stores in the woods, and I was particularly severe because some party members were thought to be involved. The person in charge of the cache was Mališa Damjanović, my old schoolmate, a university graduate whom the war had brought back to his native village. He came from a poor but honest family and was himself unassuming and decent. But I did not spare him in my criticism. It was then that I became acquainted with his wife as well. It was her tragic fate, and Mališa's tragic role in it, that provided me with the subject for my story *Tudjinka,* which I wrote in Mitrovica Penitentiary.

Podbišće is located on the highway; therefore, to keep from being surprised by Italian motorized units, I moved over to the left bank of the Tara, to the village of Bjelojevići. My brother came with me, along with a group of comrades who had fled from Kolašin. On the right bank of the Tara my brother and I were met by Mileva Šćepanović. I hardly knew her at the time, but I was a school friend of her brother Jovo, while in their underground days she had been a friend of my brother's. She had a reputation for courage and extraordinary strength. Eventually she became a fighter in the First Proletarian Brigade and was wounded six times. Following my fall from power, Mileva let me know that she was not joining the boycott against me, and as a sign of this had her ninety-year-old father Tadiša visit me from time to time. After I got out of prison in 1966, Mileva, who was retired by then, visited me. Since then my wife Štefica and I have developed a strong friendship with her and her husband Milorad Zarubica.

What could Communism have meant then to Mileva—a capable girl, but a peasant from Bjelasica Mountain nonetheless—and why did she, like so many young Montenegrin peasants, espouse it? Communism had become linked with patriarchal justice and sacrifice. More significantly, it was the one movement that offered these bright but poor mountaineers any prospect of extricating themselves from their isolation and neglect. The ideal joined with necessity. The threat of fascism and the mythical love for Russia created the climate, and the war created the ground, for the waging of the struggle. Once the war was won, dissension and strife within Communism gradually swept away everything but the patriarchal values and personal morality. Inside herself Mileva discovered a new equilibrium and stood up for all kinds of "renegades"—whether the followers of Stalin and myself, or former "class enemies." It was from her that I heard a terrible judgment: kill him humanely but don't torture him. I noted a similar acceptance of the warriors of opposite camps by the Serbian poetess Desanka Maksimović. Mileva is a soldier of the people, whereas Desanka is a poet of tragic tenderness, but both share the same sense of justice, timeless and unconditional, universal and of the people.

We spent the night in the house of Mileva's father, Tadiša. The next morning, as we were sitting under a fruit tree, a revolver belonging to a Communist Youth committee member went off by accident, and the bullet struck the ground by my feet. Such accidents happened often among our men, as they were inexperienced in the use of weapons.

That same morning we climbed even higher into the forest, and watched the slow, cautious movements of the Italian troops below on the highway. I could also see the house I was born in, on a meadow among the hills, with no smoke rising from its chimney, no signs of life—forever sad and desolate in my memory. Several days later, when the Italians were burning Communist houses in Podbišće and were about to burn my birthplace as well, the owner dissuaded them by telling them that he had bought the house from my father just before the war.

With us was Vučić Djurašinović, a peasant from my village, with his son Milosav. Vučić had been in the ambush with my brother and riddled all those Italians in the truck with his machine gun. A stout fellow with a brown mustache, he had a reputation as a hero in former wars. Now, however, he was confused. At times he would bemoan the burning of his house and the scattering of his family, and at other times cry out that we had to sacrifice everything and not surrender. He understood that this was a different war—a war in which there would be no mercy and no rules. In the winter of 1943 Vučić was betrayed, and the Chetniks surrounded him and his two sons in a cave. On the strength of the Chetniks' word of honor that they would not be killed, they surrendered, and the Chetniks executed them just below their native village, where their graves are still to be found by the side of the road. Whenever I went that way after the war, I always recalled my childhood admiration for Vučić's heroism, his misgivings during the uprising, and the heroic but tragic youth of his sons.

That afternoon my brother, several comrades, and I set out for the highland pastures on Bjelasica Mountain. Anxious to shorten our trip, we decided to go cross-country up a steep barren slope, which exhausted us and took too much of our time. We reached Krnjače in the dark and went to my uncle Tofil's cottage. We were received by my aunt Stoja, who treated us to some dairy specialties and supplied us with sleeping mats. Full and exhausted, we were quickly lulled to sleep by the smell of new-mown hay and the sight of the clear sky above.

The war had opened up certain voids in the human consciousness which nothing could fill. A special bitterness marked my return to the virginal wilds of Biogradska Gora and its lake—a little piece of sky lost amid the timberland. We arrived the next morning at a spot where as a little boy I had herded cattle and fished for trout. I thought of it as the only place where I could have stayed forever. Yet I could be a part of it no longer, as I now belonged to an ideology and all that it imposed. On top of everything we had no fishing tackle, and during the night the damp cold penetrated our thin clothing as soon as we stirred from the fire.

The next day we were joined by Milan Kuč, Vukman Kruščić, and some other comrades from Berane. Kuč was a rawboned, tawny man, sparing of words and slow of movement, but of sound judgment. He was dressed in regulation uniform—right down to the leggings, whose virtues he stoutly defended. Vukman was altogether different: a dark-skinned, handsome man with curly hair, of boundless imagination and courage. Both distinguished themselves as commanders in the uprising—Vukman in Kolašin, and Kuč in Berane. I knew Kuč only briefly: the Chetniks killed him in the winter of 1943, as a straggler. On the other hand, I had known Vukman since childhood as a restless boy whose zest for adventure made him something of a braggart. Now he had something to brag about. He was a spontaneous and fiery orator, and a clever and capable leader. He was killed early in 1942 by the Chetniks. Some men would never show all they are capable of, were it not for war. Such were Kuč and Vukman, though they waged war all too briefly.

While at the lake with Vukman and Kuč, I realized that we must reorganize our military forces. Our popular insurgent army had in fact fallen apart, but its commanding Communist core and a good number of fighting men still held together. I concluded that we had to strengthen that core by creating small but permanent units without demobilizing the insurgent peasant masses—at least not formally; that the military committees should remain in force; that small permanent units should be formed by the refugee soldiers and distributed throughout our territory; and that we should restore political and organizational ties with the peasants who had fought with us, even if we could not count on their participation in a formal uprising.

Boba Selić was also at the lake, with her husband Vladeta and brother-in-law Vojo, secretary of the local committee of Communist Youth. Ružica Rip, a medical student and a Jew, was there as well. She had fled to Montenegro with her young man, Milutin Lakićević, a local official. The Chetniks captured Ružica in 1942, but didn't execute her by firing squad. Rather, conforming to the Montenegrin prejudice—or perhaps just their own—that a woman wasn't worth a bullet, they hanged her in the center of Kolašin. She was only a doctor's assistant—not sufficient cause for a Chetnik death sentence, had she not also been an outsider, a Jew, and the girl friend of a Partisan officer.

I remained at Biogradsko Jezero through the next day. Just before nightfall I set out with some fifteen armed escorts for Piperi, the seat of the Provincial Committee. Some comrades from Podbišće accompanied us part of the way, among them my brother Aleksa. The curse of wartime memories!—I have forgotten when and how I took leave for all time of this most beloved person. I do remember that at Sjerogošt we had to ford the Tara cautiously for fear of being surprised by Italian motorized units on the highway. It was a moonlit night and as we kept climbing, we hardly felt any fatigue until we halted, in the morning, on the dewy slopes of Sinjajevina Mountain in the village of Lučke Kolibe.

The peasants gave us plenty of milk and a little bread. They were not angry over the uprising; on the contrary, they were proud of their part in it. But they were touchy at the mere mention of new attacks on the Italians anywhere near their villages. They didn't want to see their houses burned and their families suffer. We slept around the huts, full and happy, but still on our guard. Even if party members in the village had not told us of revived anti-Communist agitation in Kolašin, the peasants would have and did in their own fashion: "Well, it would be easy to deal with the Italians, if it weren't for the spies and informers . . ."

We were still overcome with fatigue, but I was in a hurry, so we pressed on after two hours' rest. We scurried across the barren, dry heights amid rocky peaks. We hardly paused until three in the afternoon, when we stopped in a village to get some food and inquire about a dozen Italian mules grazing peacefully nearby. The peasants gave us food, but when we asked what they were going to do with the mules—captured by the insurgents in Kolašin and hid away on the mountain—a stout elderly man stepped forward, brandishing a rifle. "That is Italian government property," he said angrily, "and it should be returned to the Italian government."

"What do you mean, government property?" I retorted. "They're invaders! The mules must not be returned. We captured them and will not allow them to be returned to the enemy!"

"And what authority do you have to stop us?"

"Our own authority! The invaders have no authority in these mountains anyway."

A couple of peasants supported the angry old man, though not very resolutely, while the rest looked on with curiosity. My escorts were bristling with anger, getting a better grip on their weapons, though two or three wavered. I gave orders for the mules to be herded together and taken away, then told the escorts that we would kill them. Some two or three hundred yards from the huts, we drove the mules into a hollow and surrounded them. I explained how the mules should be hit in the shoulder, in the heart. The mules became skittish even before the firing began, as if knowing why they had been corralled together there. We aimed badly. The mules stumbled, got up again, brushed against us. This was not without its danger: bullets were hitting the ground around our feet. It all lasted longer than we had expected, but soon all was still and the hollow was covered with a pile of carcasses, blood flowing on all sides. Relieved that it was finally over, we hurried away from the hollow without glancing at each other. If I have dwelled at length on this detail, it is because it was significant—even decisive—for both me and the men who were with me. It was almost as if human beings had been involved, and not just any human beings, but my own people. Perhaps the hostile attitude of that stubborn and brazen old man, who would have defended the mules as occupation property, contributed to our feelings. The slaughter of the mules became an unbridgeable gap between him and us.

We spent the night at Bare Bojovića. There we learned of fighting between the Italians and the Durmitor insurgents at Krnovo on August 5, and of the gutting of the village of Mokro a few days later. The Italians were back in control of communications between Nikšić and Šavnik. Rested and fed, we continued the next day toward Kamenik, but not until evening did we reach the huts of the shepherd village, stuck among bare crags.

It was August 10, 1941. A provincial council meeting had been held on August 8—without my participation or that of all delegates. This upset me, not so much on account of my prestige, but because the decisions of the council were not exactly to my liking. The text of the resolution was disorganized and essentially sectarian. Blažo noted my dissatisfaction and explained that they didn't know I would be late, and that Moša had insisted that the council not be delayed. Moša was there, pleased with the enthusiasm and toughness of the comrades at the council. Among them were Mitar Bakić, Krsto Popivoda, and Arso Jovanović.

The council decided to accept Arso Jovanović into the party. I expressed my disagreement before everyone, including Arso. "This is not," I added, "because Arso does not deserve it by virtue of his conduct or character—on the contrary. But there is a certain procedure in the party. No one should be so privileged as to be accepted into the party by a council or a conference."

Nor did I accept the council's political evaluation that a segment of the people had fallen away from us. I regarded that as a superficial observation. Certainly one segment had become frightened and passive, but this segment had been with us only ostensibly, whereas our active segment was allied with us more closely than ever. Therefore, one could speak of a differentiation between segments, but not of a general falling away.

The council's resolution also linked our struggle with the U.S.S.R. much too closely and unconditionally, and needlessly stressed the change in the war with the entry of the U.S.S.R.—that is, its transformation from an imperialist war into a war of liberation. It made a distinction between guerrilla operations and the armed uprising, and accordingly between guerrilla units and "the creation of a future army of liberation." Moreover, the current situation and tasks were treated as "a new, more decisive phase . . . of the all-national liberation antifascist revolution. . . ."* The terms "guerrilla," "guerrilla units," and the like had been introduced in Montenegro, at my suggestion, in preparation for the uprising, and were applied to those armed groups, consisting largely of refugee fighters, that grew up after the uprising. I was familiar with the term "partisan" from the literature of Napoleon's 1812 invasion of Russia, and the civil war in Russia of 1918–1922. Radio Moscow had been broadcasting news of "partisan" activity and calling upon the subjected people to form "partisan" units. However, in our language the word "partisan" does not have the meaning it does in Russian, and barely existed as a borrowed term meaning "a follower," "a party man." I was against the adoption of Russian terms and preferred the international designation "guerrilla," even though guerrilla had not been domesticated either, since our native terms were either unsuitable or already adopted by other organizations hostile to us—for example, "chetnik" and "ustashi." I must have also been influenced by the volunteers in the Spanish Civil War, specifically Peko Dapčević. The comrades in Croatia and western Bosnia were also under the influence of volunteers in the Spanish Civil War. But it was not until September 12, 1942, that Tito required the Croatian comrades to substitute "partisan" for "guerrilla."

I still maintain today that although "guerrilla" was not a suitable solution, "partisan" certainly did not facilitate our becoming independent of the Soviet Union. As for the term "all-national liberation antifascist revolution," it was forced upon us, in all its awkwardness and redundancy, by the Comintern, in response to the so-called May 1941 council in Zagreb, which saw the "proletarian revolution" as the next immediate stage, thus rescinding the former division of the Yugoslav revolution into a "bourgeois" and a "proletarian" phase. The Montenegrin Communists had heard of this "correction" by the Comintern from me, and therefore automatically introduced it into the resolution.

Around that time Mitar Bakić set out for Zeta to visit his relatives and the party organization there which, because of its proximity to Podgorica, labored under difficult underground conditions. Moša Pijade went to Durmitor to help the Communists there.

Two details from a conversation with Krsto Popivoda still linger in my memory. He told me with a good laugh, "We vowed that we would

* *Zbornik dokumenata i podataka o narodno-oslobodilačkom ratu jugoslovenskih naroda* [Collection of Documents and Facts about the National Liberation Struggle of the Yugoslav peoples] (Belgrade: Vojno-istorijski Institut JNA, 1953), p. 15. [*Note:* Hereafter this source will be cited as *Zbornik dokumenata i podataka* (1953).

settle scores with the spies, and that this would be their Black Friday." Krsto also told me that the insurgents in Nikšić had joined our fighters in Hercegovina in the struggle against the Ustashi, despite the Italian offensive. That brightened my gloomy memories of the trip to Montenegro six weeks earlier, with the refugees and the account, by the fair-haired peasant with the bruised cheekbones, of the massacre of innocent Serbs.

9

All our efforts from then on sought to revive enthusiasm for the preservation of unity. No struggle of any kind, especially a war, can be successful unless it finds adequate resources. That holds true particularly for a revolutionary war, which creates its own army and organization in the course of the fighting. The talent of revolutionary leaders is measured by their organizational inventiveness and political acumen. Their philosophical views and "ultimate" aims are inherited from their predecessors. Marx and Lenin left a generous legacy, but here in Yugoslavia, in Montenegro, they served only as an inspiration and a revelation—a "creed" which had to be put into practice and confirmed. I recall that during the uprising I was obsessed by Marx's axiom that "the defensive is the death of an armed uprising." However, we were forced to be more on the defensive than on the offensive. Only the long and difficult retreats and regrouping of forces during our war could have set aside this proposition of Marx's, occasioned by the Paris Commune in 1871.

The war which we Communists conceived was not only a national war. Communists are people who are not attracted by anything, not even by national freedom, if it does not hold out to them the prospect of a specific ideologically revolutionary community. This did not deny the national role, the national idealism of Communists, but rather confirmed it—wherever the ruling classes sacrificed the national to their own class interest. However united in their ultimate aims, Communist leaders differed among themselves in their emphasis on the national role. Such differences existed also among the leaders of Yugoslav Communism during the uprising. Yet these differences were not so crucial as to cause a split: ideals gained legitimacy as the prospect of power became increasingly real. Since the Yugoslav party had entered the war as a "purged," monolithic, "Stalinist" party, the unifying elements were far stronger than the divisive ones.

Such war aims held no prospect of victory without a homogeneous and ideologized party: a special new community that identifies itself with a set of absolute social and human ideals, and thus with absolute power. In earlier uprisings and wars fought by other social groups, there was no need for such a party with such a role: the war was waged by a military apparatus consisting of rebel leaders or political parties at the head of a state apparatus. As for the Communists, it is because they are set apart and absolute in everything that they must be a party "with a difference"—a party which sustains its members, and to which the members give themselves completely. To be sure, not even I had such thoughts then. Like the rest, I recognized that without a disciplined and united Communist party there could be no successful war for national liberation, for rule by the Communists as the "finest" sons of the people.

The war improved the position of the party; in Montenegro there was hardly a village without a party cell. Communist intellectuals—students, teachers, government employees—had taken refuge in the villages with their relatives. They were on a higher plane and more vigorous than the local leaders, who were largely peasants with a partial education at best. This disparity necessitated the elevation of the newcomers to positions of leadership.

However, our most acute problem was the disorganization of the party caused by the uprising. Some Communists had become military leaders, others had made for the mountains, and still others had gone underground in the towns and villages. Many cells were divided, and committees were weakened and overburdened. Communists are only human, and many felt the pressure of families, hunger, and want. There were also the weak—more than one might have expected among Communists. There were even cases of surrender under arms. The most tragic and depressing case occurred in Crmnica, where a group of Communists dressed in parade uniforms and standing at the head of the people surrendered to the Italians. Among these men was Nikola Nikić, secretary of the Local Committee, a slender young man popular in the Belgrade student movement. The Italians sent the peasants home, and interned or executed the leaders; Nikić was executed. Every surrender, and especially the one in Crmnica, was sharply condemned by the party. Nikić has been thrust into oblivion, though it may not have been cowardice that motivated him, but "sacrificing for the people"—in that case, an inflamed and frenzied mob. In politics only those acts are "great" and "comprehensible" that advance specific aims.

And so we immediately took up the task of putting the party in order and strengthening it. It was essential to staff the committees. That, I would say, was our most significant assignment, for on the speed and thoroughness with which we accomplished it, the continuation of our fighting depended. I maintain that we carried it out successfully. Though at the time I gave the comrades no respite, nor any peace to myself, today it seems too good to be true that we reorganized the party within the space of a month and a half, and gave it the capacity to carry on the struggle ever more stubbornly and methodically.

A Communist organization, particularly under wartime conditions, is constituted in such a way that it covers basic activities and everyday affairs. In addition to the apparatus for agitation and membership, this includes a youth section, a military unit, a women's section, and so on. Each section works out its own particulars and makes assignments accordingly. The question of a political authority—the people's committees—was especially important during the uprising. "Committees of the national liberation reserves" were created—that is, committees for equipping and quartering troops. But they never had a chance to develop. We continued to hold a great amount of territory: many villages that the forces of occupation controlled by day were ours by night. It was obvious that these territories could be controlled by guerrillas, but the guerrillas did not constitute a political authority that could take charge of daily relations among citizens, and relations between the army and the population.

Also, we had to get rid of spies in our organization quickly and mercilessly. The forces of occupation inherited a spy network and a police organization from the royal regime. True, not all former spies became spies for the occupation; new ones were recruited. We hated these spies, whose activity critically affected our recovery. How could we guard our underground organization, the movements of our guerrillas, the trust of our people, as long as the eyes and ears of the enemy were in our midst? Thus the extermination of spies was often a test of our organization's strength and of the resoluteness of party members charged with this duty. Killing a spy in Montenegro was no simple matter. Spies sought protection from their clansmen, and clansmen were inclined to give it. And how does one define espionage, anyway? The delivery of secrets or subversive agitation? The guarding of roads for the invader or the spread of defeatism?

Meanwhile, in all this hurried and overzealous activity, we lived badly—worse than we had to, worse than at any other point in the war. The Provincial Committee had settled itself in a little village on Kamenik Mountain, in a couple of abandoned huts beside some ramshackle sheep pens and heaps of boulders. The Communist Youth were nearby, though their secretary, Budo Tomović, was often with us. The party leadership spent a good deal of time out in the field. At that time I ventured out only in the immediate vicinity: we often held meetings of the Provincial Committee, and meetings with members of lower committees who would come to us with problems. By day we were busy with these meetings in a nearby hut, or in the shade of a beech tree. At night we were alone with our guards in rockbound seclusion, in chilling silence.

Blažo Jovanović also rarely left the village, since I was not sufficiently acquainted with people and conditions to make decisions on my own. When we came to the village, we leaders slept in Blažo's family huts—on the floor of beaten earth strewn with straw, under coarse blankets. But we ate by ourselves so as not to deprive Blažo's children of the milk which Blažo's sister Rusa got from their scrawny cow. For a whole month—roughly the time we spent there—we ate only sugar and rice, the spoils of war. Occasionally Rusa shared with us a bowl of milk or some tough nettles and orach. We came to loathe sugar and rice; and there wasn't much one could do with them—except add a little more sugar or a little more rice. After supper, in the company of our couriers we would sing around the fire, usually beginning with revolutionary songs and ending up with folk songs of love.

I too sang, and needed to: singing helped me suppress my bitter emptiness at Mitra's absence. Rifat Burdžović told me that he had heard from a student, who had made his way from Belgrade to Sandžak, that Mitra had been executed. Before falling asleep and again on awakening, I had to shake off the vision of Mitra, with her intelligent bright eyes and small breasts, in front of the firing squad, facing the rifle barrels as she sorrowed over our severed love.

It was there, on awaking in the middle of the night, that I got the idea of amending and supplementing the resolution of the provincial council, and thus to explain to the party membership changes brought on by the end of the uprising. I walked out of the hut and into the stony silence of the village, under a canopy of blue. I paused beside a dilapidated sheep pen, my thoughts unraveling easily and logically as if nothing else existed on earth. The next day I presented my thought to Blažo. He agreed straightaway, and I retired into the shade to begin writing. The following day we discussed the letter at a meeting of the Provincial Committee. This is how the "Letter from Comrade Veljko" came about, a letter I addressed to the membership as a delegate of the Central Committee and signed with my party pseudonym of Veljko.

As I reread this letter thirty years later, it seems shorter and less effective than I remembered it. Of course, as a historian it is not proper for me to evaluate my own writing. Still, I do remember that at that time, coming in the midst of feverish efforts to put the party in order, the Letter was inspiring. That was the general consensus, even Pijade's. The letter reproached the council "for not taking into account the fact that guerrilla actions did not exclude other forms of resistance, such as an armed uprising within local limits. Which form one chooses depends on the specific situation, the mood of the masses, and other factors. One should similarly observe the relation between regular and guerrilla warfare."

With respect to the contradictory positions taken at the beginning of the uprising, the Letter said, "To be sure, those directives, taken by themselves, brought about grave errors in the field, but, looking at the course of events as a whole, they saved the party from total isolation from the masses, and from total disorganization of the liberation movement and even of the party itself."* The Letter described the division which occurred in the ranks of the insurgents: the Greater Serbian groups and former officers secretly hoped for the restoration of the old army and gendarmerie, and stirred the people up against the Moslems and Croats. The Letter warned, however, that "it would be just as wrong for the Communists to identify these groups with a fifth column."† The Letter instructed: "Communists must show elasticity, agility, and political acumen in winning over these groups and persons, even if only as uncertain allies, or in neutralizing them."‡

This Letter also used the term "liberation antifascist revolution," and introduced still another for the first time: "the people's revolution." That might have had a sectarian ring to comrades outside Montenegro. But Montenegrins are partial to unanimity even when they aren't Communists: it had to be made clear that the struggle they were waging was in fact that same revolutionary future about which they had dreamed. At

* *Zbornik dokumenata i podataka* (1953), p. 21.
† *Ibid.*, p. 25.
‡ *Ibid.*

the same time the Letter warned against sectarianism as the chief danger. It treated committees of national liberation as the embryos of "popular government," and stressed that these committees ought to be elected, not appointed. Soon after this came the creation of people's committees. The Letter also indicated the need to frame "a platform for the struggle of the Montenegrin people," that is, a program to unite all those favoring the armed struggle against the occupation.

It seems to me even today that the Letter saw the revolution and the new government as a rallying of all who fought against the occupation, regardless of class or viewpoint. In this respect—all terminological differences aside—the Letter was neither narrower nor broader, unless by its frankness, than the prevailing line of the Central Committee.

The situation within the Party was rapidly improving, as the forces of occupation settled down in the towns. In late August or early September we descended to the village of Radovče, on a broad mountain plain above Piperi. Whenever we made our way down the rocky ridges to Piperi, we could never resist pausing on the edge of the Piperi Crags to view those rare plains from whose mists the hills arise as if from the sea, and to vent our resentment at having been driven from those lush lowlands. At the village of Radovče, which stood beside a meadow, we moved into some vacant huts. We were very comfortable there, and gained in status even in our own eyes. The air was dry and mild, and though we ate only a little better than in Kamenik, I noticed that my lungs were getting better.

Our political course was growing firm and steady. To be sure, there was a bit of yawing and tacking, but no stopping and no going back. Fresh ideas occurred to us. The Communist Youth League was getting in shape. The young Communists accepted the war as a natural obligation, and the Provincial Committee of the youth took up our course more devotedly and more readily than the party leadership itself. I held a series of meetings with the Provincial Committee of the Communist Youth, and became particularly close to its secretary, Budo Tomović. From these meetings and my discussions with Tomović, I got the idea of creating an organization around SKOJ* which could rally all young people willing to fight in whatever way against the occupation. As a Communist party for the youth, SKOJ itself was too narrow an organization. This idea did not put an end to existing Communist Youth organizations; rather it made them the core of a broader, nonideological

* Abbreviation for *Savez Komunističke omladine Jugoslavije:* the Communist Youth League of Yugoslavia.

organization. I also devised the name of a new mass organization: the People's Youth. Later the young people in Bosnia and Croatia were organized on a similar basis under different names. Experience in war is more readily transmissible than any other. I cannot gauge to what degree the Montenegrin People's Youth served as a model for the rest. But I remember that when we met in Užice in November, the secretary of the Central Committee of SKOJ, Ivo-Lola Ribar, was enthusiastic over what we had done in Montenegro. Tito and the other party leaders agreed. The name "People's Youth" was adopted after the war for the Yugoslav political youth organization.

The consolidation of SKOJ and the expansion of the People's Youth would not have progressed so enthusiastically and intelligently, had Budo Tomović not been the secretary of its Provincial Committee. Stocky, goggle-eyed, with golden flashes in his hair and eyes, he was a good talker and a careful implementer of his own words. He was from Podgorica, a town with humor and imagination. Budo had a lively sense of humor himself. A recent but confirmed Communist and not a careerist, he was, in my judgment, the most gifted and brilliant member of the party leadership, in which he participated as the representative of the Communist Youth. The other leaders had neither Budo's breadth nor his creativity. Though I worked well and closely with the others, only with Budo Tomović did I reach an identity of views without effort.

The Communist organization cannot act or survive without its press. Influencing the people is secondary; the essential thing is to link and direct the party membership. Life constantly requires explanations in accord with the basic dogma, or else the membership will become disoriented and waver in its faith in the dogma. The illegal press, though often operating on the lowest technical and literary level, wields special magic, an almost mystical power over Communists: it wraps them in their own invincible world and reveals to them how unquenchable and indomitable their faith is. The party press was particularly important under conditions of war and revolution. In Yugoslavia hundreds of little party and military newspapers appeared which did more, perhaps, to satisfy a yearning for Communist fellowship than to inform. I felt a need for a press that would uplift and direct the party consciousness, and introduce party organization into guerrilla warfare. Therefore we inaugurated the *Narodna Borba* (The People's Struggle), an organ of the Provincial Committee which I edited and mostly wrote. The first issue appeared September 15, mimeographed, but in a fair number of copies. The technical side was managed by the painter Božović.

Narodna Borba was a continuation of an ongoing propaganda activity. Yet from the very first issue it noted certain changes—above all, the crumbling of the insurgents' "national unity." For example, in the *Bulletin* of the Supreme Staff of the National Liberation Army of Montenegro, the Bay of Kotor, and Sandžak, No. 2, of July 20, 1941, there was a report concerning "attempts by certain elements linked with the enemy to destroy the unity of the people. Thus in the district of Lješani our units carried out a flank attack against the enemy, because of the activities of said elements among the population of the Lješani district."* And on July 27 the Supreme Staff polemicized as follows:

> Unfortunately, our enemy have found accomplices among dishonest people and cowards who serve their ends by agitating against the uprising, sowing doubt and mistrust. . . . Some do it openly, others conceal it by saying, that they favor an uprising but think it premature. . . . We ask: Has any people ever liberated itself too soon? The enemies of the people say that the uprising is led by the youth, that they have no confidence in them. . . . The uprising is not led by the youth but by the entire people, although the youth have shown that they are ready to lay down their lives for the freedom of their people, seeking neither glory nor reward. . . . The people have elected officers in whom they have confidence. And the officers have shown that they are with the people and are dedicated to winning the ultimate victory.†

The very first issue of *Narodna Borba* named three "antiparty elements"—Milovan Andjelić, Slobodan Marušić, and Aleksa Pavićević, while the article "The Fifth Column at Work" reported in unsparing detail: "In addition to a handful of well-known Montenegrin separatists (Sekula Drljević, Dušan Vučinić, and others), there have also appeared certain outspoken lackeys of the enemy, fifth columnists who have tried to smash the struggle of the Montenegrin people. . . . In the district of Kolašin, Ljubo Minić and Novo Medenica, both judges, have attempted to organize this shameful work. . . ."‡

The passivity of the Italians and their quest for supporters made our reorganization easier. They made one sortie, consisting of a whole battalion, into Piperi country. But the separatists in the town administration had warned us that Italian units would be passing through the villages,

***Zbornik dokumenata i podataka o narodno-oslobodilačkom ratu jugoslovenskih naroda* [Collection of Documents and Facts about the National Liberation Struggle of the Yugoslav peoples], Vol. III, Book 1 (Belgrade: Vojno-istorijski Institut JNA, 1950), pp. 17–18. [*Note:* Hereafter this source will be cited as *Zbornik dokumenata* (1950).

†*Ibid*, p. 23.

‡*Ibid.*, p. 39.

and we were asked to lie low so as to avoid burning and destruction. We looked on, assembled and ready, as the Italians moved past. But we did not attack, because the peasants had begged us not to, and because we ourselves were not sure of our strength. The Italians obviously wanted peace, and we were preparing for more vigorous resistance.

In those days I had two personal experiences which I remember well—a dinner with the Vučinić family, and supper with Major Popović. The Vučinić father had fled to Radovče with his wife and children. They lived in the city, but kept their farm for food supplies. I was on friendly terms with the parents, as well as with their two sons and daughter, who, though still a young girl, had joined the insurgent army. One day Mrs. Vučinić invited Blažo Jovanović and me to dinner in their rather spacious cabin, and treated us to an unforgettable and opulent meal, just as in peacetime.

Major Popović was staying with his family at Živa, overlooking Radovče. He was our neighbor and, though not particularly active in the uprising, he often came to chat. The collapse of the army and the state dismayed him, but did not turn him to the Communists. Swarthy, rather small and puny, he took meticulous care of his uniform and showered us with unanswerable questions. One day he invited me, alone, to supper. Blažo Jovanović had no doubts about him (his sister was a party member), nor did Arso Jovanović, who knew him to be an honorable officer. So I went to Živa. We talked and talked, had supper, and he proposed that I spend the night in a shed next to the little house, where, he said, no one would disturb me. I demurred, making all sorts of excuses, in which I was supported by his Communist sister. So he went there with me. We crowded into a narrow bed and resumed our conversation. Finally he dozed off, or pretended to. I remained awake with my pistol under my head. Whenever he turned over, I did, too. And so we greeted the morning. His sister, who was faithful both to him and to me, brought us breakfast. Later Major Popović was among the first to run to Podgorica, offer himself to the Italians and Chetniks. After the war, when I reminded Blažo Jovanović of my supper at the Major's, he remarked, "Who knows, maybe he was up to something." Maybe not. But my suspicious vigil was proof enough for me at the time, of a change in human behavior.

10

On September 20 a mission from the royal government-in-exile arrived by a British submarine on the Montenegrin coast. They were rowed on to a deserted beach near Petrovac. All three members of the Yugoslav mission, two air force majors and a non-commissioned officer, their radio operator, were Montenegrins. However, this was not the reason the Montenegrin coast was chosen: it was in Montenegro that the mission could feel most secure and most easily make contacts—through relatives, if in no other way. In the mission also was a British representative, Captain D. T. Hudson, who of course, kept his government informed as to what was going on.

The mission did not know what it would encounter. The royal government and the British had heard about the uprising in Montenegro. They had also established a link with Colonel Draža Mihailović, who, with a group of officers, had taken refuge in the mountains of western Serbia. The government-in-exile knew that the Communist movement was strong in Montenegro, and that the Communists, and not its own followers, had initiated the fighting. The exile government had been cautioning that "it was not yet time" for an armed struggle. However, the government could not know the actual strength and role of the Communists in the resistance. At that time Draža could not have had a clear idea of the situation except in Serbia and eastern Bosnia.

The mission immediately sought contact with our guerrilla units and party organization. It could not in any case have gotten about without our knowledge, because we not only controlled the villages but were on close terms with the peasants. There was nothing for the mission to do but establish ties with us or risk falling into Italian hands. The party leadership was informed of its arrival two or three days after it had landed. We issued an order that the mission be escorted to us at Radovče. Our local groups did exactly that even before receiving our order, so the mission was in Radovče only five or six days after landing. Orders were that they travel only by night and take cover in the forests by day, though our own couriers and even our officials constantly crossed the roads by day. They arrived with a group of guerrillas, and with horses carrying the British radio and other equipment.

We received them pleasantly but in no great spirit of trust. The British were Soviet allies and therefore our own as well, but we had inherited an ideological intolerance of "English imperialism," and were particularly mistrustful of the Intelligence Service. The Communist press had frequently attacked that secret service, while Communists trained in Moscow warned against its refined disregard for the kind of methods it used. The Intelligence Service had been often mentioned in the so-called Moscow trials: we believed—or forced ourselves to believe—that many leaders of the Russian Revolution had been led astray by the clever English. In our own party the legend had been spread that Gorkić, Tito's predecessor as secretary of the Central Committee, had been a British agent: I heard this from Tito himself, though I don't think he made it up, but rather was himself duped by NKVD fabrications. My generation lived in the era of British Intelligence and Cheka, just as today's generation lives in the era of the CIA and KGB. The British press also contributed much to a romantic, mythical notion of the omnipotence of British Intelligence. All of which gave us reason for extreme caution: we saw in Lawrence of Arabia not an idealistic hero, but the perfidious, arrogant champion of an empire.

Captain Hudson did nothing to confirm our fears and our prejudices. On the contrary, he hid neither his considerable knowledge of the Serbian language nor his mission, which was to send back information on the enemy forces. He was particularly interested in three well-known people: Miljan Radonjić of Montenegro, Gani-beg of Kosovo, and the leader of the leftist Agrarians, Dragoljub Jovanović. A radio transmitter had apparently been left with Jovanović which had made no contact. In conversation with him I remarked that British Intelligence was perhaps the best in the world, to which he replied, "No, the Soviet Union has the best intelligence service because it is helped by Communist parties all over the world."

I didn't let this pass. "The Communist parties have solidarity with the Soviet Union for ideological reasons," I said, "but they are not agencies of the Soviet intelligence service."

Hudson replied, "I know, but these parties are a reservoir of political information such as no other state has."

We were also suspicious of Hudson's declaration that the British government would help all who fought the Axis Powers. We gave him only general data about our forces and organization. Restrained, even cold—just as we imagined Englishmen to be—and very sparing in humor, he was more like a civil servant than an adventurer. But it was wartime, and Captain Hudson represented an ally of our protectors, so we were prepared to help him, and with him carried on open, friendly conversations, just like sovereign powers.

We were more open with the Yugoslav officers because they were our compatriots—and not very distinguished or impressive. We spoke with them courteously, but in a somewhat critical tone. But "our" officers—royal officers who, like Arso Jovanović, were active in the uprising and had joined guerrilla commands—displayed a fiery bitterness, reproaching the royal government for having fled the country. "They too could have done what we and the Communists are doing!" They were even more bitter over the capitulation: "The government forced the surrender on the army, which didn't have to surrender, and then they fled the country!"

The royal officers did not defend their government very strongly, but did insist on the need to rally the "national forces" around what had been started with the coup of March 27, 1941. They were not too interested in the Italian garrisons but rather in the domestic situation, especially in "nationalists."

We talked with them at length. For me there was practically nothing new in their views, except perhaps for their highly political orientation, unusual even for royal officers. They remained conservatives, as if nothing had been changed by the occupation and the uprising, and maintained it was the Croats who sold out. We did not deny the Croatian people's disinterest in the war (except for Dalmatians, and the Communists and their sympathizers), but pointed to the confused and defeatist conduct of the Serbian General Staff and ruling circles. For those groups, the term "worker" was identical with Communist, and "Communist" was identical with someone antinational and anti-Serbian. But all these conversations ended on a polite note: we did not wish, and they did not dare, to exacerbate our differences.

To be sure, there were differences in character among them. Ostojić was hard and reserved, and Lalatović open and obliging—a sign that the royal government was no longer in full control. The royal mission also included a member who was partial to us, the radio operator, Sergeant Veljko Dragićević; finding in Montenegro a kinsman and even his wife, he offered to stay with us. We had a time convincing him that this was awkward, that the officers would think we had brainwashed him. Dragićević did eventually join the Partisans; he worked as a radio operator, and as communications chief in the Supreme Staff, until he and his wife lost their lives during the German raid on Drvar on May 25, 1943.

Major Ostojić asked me at one point why we called ourselves "guerrillas" instead of "Chetniks." I replied that the Chetniks had once been national liberation troops, but that after World War I they had been transformed into a chauvinist organization which terrorized the non-Serbian population. On top of that, the Chetnik leader Kosta Pećanac had issued a call to the Chetniks to collaborate with the Germans. Ostojić made no reply to this.

With the help of our officers, Hudson set out to measure the plain of Radovče, to see if aircraft could land there. It was concluded that, with a little leveling of the ground, it would be possible, but there the matter stood.

The peasants greeted the mission with hope and joy, though they reproached the officers for the unheroic and mindless conduct of the royal government and its army. The peasants rejoiced in the unity of the royal and the Communist representatives, although neither the mission officers nor the Communists had any such illusions. We Communists were amenable to a reasonable and mutually beneficial agreement, but it never occurred to us to lay down our arms. In the world we lived in—and especially the one we desired—we knew that our value was equal to our armed strength. The royal officers could not accept this: the ideas to which they adhered left no room for Communists—even unarmed ones.

The mission stayed with us for five or six days. They made it clear from the start that they intended to go to Serbia, to Colonel Draža Mihailović. We knew precious little about Mihailović. We did know that in Serbia both Chetnik (royalist) and Partisan units were active. One of our groups, which included Rifat Burdžović, my brother Aleksa, and Komnen Cerović, had already penetrated to the Zlatibor Range, made contact with the local Partisans, and sent us a report. Just as we had anticipated, the Partisans were more militant than the Chetniks, but we could not understand why the Chetniks were more numerous in Serbia. Wanting to establish closer contact with Partisan Serbia, and particularly with the Central Committee and Tito, we decided to send to Serbia, along with the royal mission, a delegation of our own which included Arso Jovanović and Mitar Bakić.

We gave the mission horses for the transport of their radio transmitter and baggage, and the two groups set out together by way of free territory for liberated Užice. The Germans had evacuated Užice on September 21, and on September 24 the Partisans had driven out the Chetniks of Boža "Javorski,"—wildcat units not under the command of Draža Mihailović. Bakić and Jovanović stayed with the Supreme Staff, while Majors Ostojić and Lalatović and Captain Hudson continued on their way to Ravna Gora, the headquarters of Colonel Mihailović.

Thus contact was established between the Chetniks in the Serbian country and the royal government-in-exile. Up to that point there had been only an unreliable link between them. There also had existed a certain tolerance, and even co-operation, between the Chetniks and the Partisans. It was then that the conflicts began which were to become ever more bloody and merciless. How great a role Majors Ostojić and Lalatović played in this is hard to say. Judging by their subsequent activities with the Chetniks, as fresh representatives of the émigrés they instigated a confrontation with the Partisans. As for Captain Hudson, the Partisan leadership did not regard him as an instigator of conflict, though we noted that he was not kindly inclined to us.

Whatever the case, as the mission left I remember someone proposing that we set up an ambush and kill them. I took that as a joke. Never—not even in the most bitter bloodshed with the Chetniks—was I sorry that we had received the royal officers graciously and saw them off even more graciously.

While the royal mission was still with us, we were planning a major operation against the Italians. Though we didn't involve the mission in this operation we didn't hide it from them either. We planned the operation with extreme care, and if I speak of it in detail it's largely because it signaled the revival of the uprising in Montenegro.

The leadership, and I in particular, felt distressed because fighting had slackened off since the July uprising, though we knew it had to slacken: the party had to reorganize, and the insurgents' enthusiasm had subsided. It was evident that the struggle could be continued only by a strong organization able to overcome its own feelings of guilt about enemy reprisals against the people, and able to oppose all vacillation and crush all resistance to the struggle whatsoever.

We listened regularly to both Soviet and British radio broadcasts and published their news. To be sure, we understood clearly by then that what was happening on the Soviet front was not a matter of "tactical retreats" or of deliberate enticement of the Germans into the interior. But we never doubted ultimate Soviet and Allied victory. Arso Jovanović would say, "Over there on the Eastern front—that's the real war, where whole divisions burn up like matches." And when a peasant heard that Kiev had fallen, he asked me, "Is Kiev Kievo?" In times past, Montenegrins used to call Kiev "Kievo," and when I confirmed that this was the same city, he cried out, "That's no joke! An enemy in their house is just like one in our own! Kievo is a holy city in the middle of Russia, just a bit smaller than Moscow!" But far more important for us were Soviet news reports of fighting in other parts of Yugoslavia. They aroused our shame because we were lagging behind, evoked fear about our responsibility to the Central Committee, and intensified our will to fight.

The proposed operation was carefully planned: we had learned from experience that spontaneity in war produces significant results only if organized and led. We gathered all the necessary information about the enemy—the Italians, and the village guards as well whom they had organized, with the help of fifth columnists, under the pompous name of *"milizia volontaria anticomunista."* We chose only young fighting men and officers who had already distinguished themselves. Blažo Jovanović was designated leader of the operation, but the immediate command was in the hands of Radovan Vukanović and Radomir Babić, old Communists and reserve officers, and later distinguished leaders of major units.

As a precaution, the Italians always moved in several large motorized columns. As a further precaution, the trucks maintained a fair distance between each other—about 200 meters. Having decided to ambush such a column, we picked a stretch of road above the gorge of Mala Rijeka, in Bratonožići, in a deserted place called Jelin Dub. Our units made their way there during the night and posted themselves on a rocky ridge alongside the road. October 18 dawned, and the scouts signaled that trucks were approaching. They blocked the road with boulders. A village guard came along and saw the road was blocked. The first truck was approaching and the guard ran to meet it. He tried to explain to the Italians in Serbian, gesticulating like a madman, that there were Communists in the vicinity. The rest of the trucks jammed up behind the first truck, just as we had foreseen. And as the column of forty-three trucks bunched together, our side opened fire. The fighting was brief. The Italians had no choice but to die or surrender. The soldiers threw down their weapons, and the officers followed suit.

As a result there were not many losses—about fifteen dead and wounded Italians, and one dead and one wounded on our side. The booty was enormous: over a million lire, a lot of food supplies—even fine wines for the officers' mess—but less munitions than we had hoped. As their officers looked on in amazement, the Italian soldiers helped our men set fire to the trucks and hurl them into the gorge some four to five hundred yards below. About seventy Italians were taken prisoner—not the 150 of official histories. Victories are greater in official histories than on the battlefield. Loaded down with the booty, we then set out across the rough rocky trails into the canyon of the Morača, and again up the rocky slope of over 1,500 yards to Kamenik Mountain. Blažo Jovanović took a shortcut to Radovče and arrived the same day. We were overjoyed to hear the news: we perceived the operation as a turning point, and took pride in our recovered glory and prestige.

The Italian command reacted on the following day—much sooner than we expected. Pirzio Biroli was an energetic commandant and a cruel man. He appeared on the scene of the raid and executed some ten peasants from the neighboring village of Bioče, who had been enrolled in the pro-Italian guards. A second Italian unit descended upon Kuči, where they encountered organized resistance and were beaten back after two or three days.

We were informed quickly and accurately of the Italian actions. Bitter over the shooting of the peasants, we concluded that the people would receive this news with better grace if we took revenge on the prisoners. We were of two minds over whether to shoot all the Italians or just the officers. The former opinion prevailed, though I favored the latter. There had to be an end, after all, to our releasing prisoners while the Italians carried out reprisals against our population. Thus primed, we set out for Kamenik, heading straight for the prisoners: Italian artillery raking our villages only intensified our determination. We found the prisoners sitting comfortably with our men on a wooded ridge. We immediately ordered that they be herded into a nearby hollow.

Among the prisoners taken in action was a Montenegrin who was carrying a payroll for the employees of the pro-Italian administration, and a pistol with an Italian license. We were more bitter against him than against the Italians, which was only natural. Without hesitating we ordered the couriers to shoot him, and they did so forthwith. The shots drowned out the entreaties of the frantic little man, as he fell over the rock he had been sitting on.

In the party there was still indecision and reticence about the killing of spies and collaborators. Killing is a function of war and revolution. Or could it be the other way around? Excesses were always recognized—excesses that "did harm" to the movement. But those who want to wage wars and revolutions must be prepared to kill people, to kill their compatriots—even their friends and relatives. In a country with living clan traditions everyone is of a glorious lineage and related to someone important; so, too, this little civil servant was the relative of a party functionary. That party functionary was angry: he knew his relative to be an honest man. Probably he really was honest, in a bourgeois and human sense—that is, outside the scope of our own ideological and revolutionary requirements. And our victim never understood how "natural" and "logical" for us his death was. Should one feel remorse? To be sure. But for what? For bad ideas and worse realities? Or for oneself—for one's devotion to bad ideas, for one's inability to come to terms with bad realities?

We began to deliberate among ourselves on how best to execute the Italians. Then something happened that not one of us had foreseen: the Italian soldiers realized instinctively what was going on. The prisoners, who had established a camaraderie with our men through cigarettes, food, and the carrying of loads, clasped our men about their knees and began to beg and weep. Our own men begged us and wept: "We can't kill them. They're ordinary soldiers, not Black Shirts! Comrade Djido, how can we shoot them when we broke bread with them? They aren't to blame—the fascists forced them into the war!"

And while all this was going on, the Italian officers were below in the hollow, shocked and dumbfounded, but dignified.

I looked at my comrades, but they turned their eyes away in confusion. "All right," I said, "let's discuss this again, calmly." But there was hardly any discussion. We immediately agreed to let the soldiers go. As for the officers, we were close to deciding to execute them in retaliation, though we were in no hurry, when someone—I think I was the one—proposed that we offer the officers in exchange for our arrested comrades, whom the Italians might also execute in retaliation.

Among the captured officers was a reserve major who spoke French, so that I too could communicate with him, and not just Božo Ljumović, the committee secretary, who spoke Italian. The major was not too happy writing a letter noting the names of the comrades we sought in exchange,

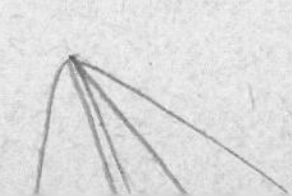

and the time and place for the exchange. We also threatened to shoot all prisoners in the future, unless the Italian commands stopped burning our villages and interning and killing our civilians.

The letter was carried by the captured soldiers, who were overjoyed to be freed. But before they left, our men exchanged their threadbare clothing and shoes with the Italians: they had no regrets whatever about letting the Italians pass through our villages and enter their garrison in rags. The peasants will laugh at them, they said, and their commanders will be ashamed. Why not?

In a couple of days we had a favorable reply from the Italian command. We watched through binoculars as the exchange was carried out at the appointed spot—one for one, just as we demanded. Among the released comrades were Stefan Mitrović, Drago Vučinić, and Dr. Martinović, the last of whom perished during the Fifth Offensive.

The raid at Jelin Dub and the repulse of the Italian punitive expedition against Kuči revived the fighting spirit of our people. But that was not all: the organization and escalation of the fighting made it possible for the leftist separatists—those who had collaborated with the Communists until the outbreak of the war—to have a say in the leadership and come out openly with their idea of "conciliation." They were led by Novica Radović, whose reputation was based on his many years of imprisonment rather than on his political acumen. He did not favor an unconditional collaboration with fascist Italy, nor was he intolerant of the Serbs of Serbia, as was the separatist leader Sekula Drljević. Radović's group attributed the uprising in part to the Italians' inability to understand the inborn Montenegrin love of independence. The uprising had separated the Communists from the "left" separatists, but not the Communist "Montenegrin national question" from the separatists, though we were also making advances at the time among the supporters of unification with Serbia.

Soon we got an invitation to hold talks with Novica Radović and Petar Plamenac, with a postscript that the Italians would also attend. We decided to accept, but not in order to seek any agreement except—insofar as the Italians were willing—over prisoner exchanges and the observance of the rules of warfare. Blažo Jovanović, Peko Dapčević, and Djuro Čegarović made up our delegation, and we dressed them as well as we could. Nothing came of the negotiations, though the Italians offered amnesty to all except those who had killed Italian soldiers, and probably would not have insisted on that exception either. So as not to offend "Montenegrin military honor," they proposed to have our weapons collected in the townships. We demanded the withdrawal of the Italians from Montenegro, the Bay of Kotor, and Sandžak—which the Italians had no thought of doing, and which would have meant the end of separatists of both the left and the right. Nevertheless these talks revealed that the Italians were no longer depending on the separatists—whether of the left or right—but were trying to find some other way out of an unpredictable and hopeless war.

Soon Arso Jovanović returned from Serbia, enthusiastic over the success and organization of the Serbian Partisans, and particularly over the work of the Central Committee and the Supreme Staff in Užice: "A real army, a real government!" he exclaimed. He also brought news that my wife Mitra had escaped from prison and was in liberated territory. This did not really excite me, rather it filled me with peace and joy in my work.

In the spirit of the decisions by the Central Committee and the Supreme Staff, we immediately organized a Central Staff, substituted the term "partisan" for "guerrilla," and introduced the clenched-fist salute used in Spain. Instructions to this effect were written up by Arso Jovanović, of course with the participation of the leadership. We even put together and published the first issue of the *Bulletin* of the Central Staff.

Meanwhile I went to visit the Cetinje leadership with Peko Dapčević and Bajo Sekulić, both of whom we were transferring from Cetinje to the Central Staff—the former in a military, and the latter in a political capacity. We got to just above Spuž by day, even though the Italians held the road, but got caught in a pouring rain and so were drenched to the skin when we finally arrived at our base on Stavor Mountain. The committee lived in dreadful want: in a lean-to beneath a cliff, miserably fed, without beds. The saddest-looking was Dika Marinović, a committee member whom I had known as a student in Belgrade: she was as quiet and patient in privation as she was in her political work. The only thing we had enough of was fire: we dried out as best we could and then went to sleep.

I awoke before daybreak with the idea that we were entering a new and more difficult phase, but better prepared and linked with the fighting in Serbia. I wrote a leaflet for the Provincial Committee about the uprising in Serbia, and called on Montenegrins to show solidarity with our brothers in Serbia. Everyone liked the leaflet, so I sent it by courier to the Provincial Committee to be reproduced. On our way back we waited for nightfall in Zagarač, then crossed the road and the plain. At one point an Italian searchlight from Trijebac shone across the bleak yet enchanting panorama of rocks and bushes. We crouched amid the boulders until the glaring beam passed on, then resumed our journey.

The new enthusiasm—my major contribution to the uprising in Montenegro, whether one accepted that uprising or not—brought me misfortune and suffering. On the evening of November 3 Blažo Jovanović informed me that my brother Aleksa had been killed the day before, while serving as acting commander of the Komovi Detachment. Struggling against my shock, I replied in the spirit of inherited sacrifice and pathos, "You can't have a wedding feast without meat!"

The report was incomplete, but the fact of his death was undeniable. The Partisans of the Komovi Detachment had been ordered to destroy the wooden bridge over the Ribarevina—a focal point on the road between the Italian garrisons in the Lim Valley. The bridge was guarded by local peasants—the *milizia*. They had dug themselves into trenches, and the Partisans encountered stronger resistance than expected. Still, the Partisans pressed on and crushed it. My brother led the attack against the trenches on the knoll in front of the bridge. The burning and demolition of the bridge, though poorly carried out for lack of gasoline, was a success.

Feeling good, the Partisans had then set out along the road to Mojkovac. At Slijepač-Most, however, where my brother had served as a teacher ten years earlier, the militiamen recognized him and opened fire from their houses. Hit by two bullets, my brother cried out and fell. The Partisans scattered without retrieving him or his weapons. One of the men later told me that they had heard the clattering of the barrel of my brother's gun: even in his death throes he was guarding his weapons. The next day the militiamen loaded him onto a horse as a trophy—he was a well-known Communist—and took him to Bijelo Polje, where his corpse was exhibited all day in the market place. Our sister's husband obtained permission from the Italians to bury him, and he was laid to rest, without a funeral, in a wild meadow outside the cemetery.

After the initial shock I resumed my regular activities. At night memories of my brother wakened, as did my pangs at his death: I could have kept him beside me or given him some lesser assignment behind the lines, had that not constituted nepotism—an "un-Communist" favoritism. However, whenever I blamed myself for bringing my brother and my family to Communism, I easily put the thought aside: there is no ideal without a sacrifice. The ideal is more important than any sacrifice, especially an individual one. At a meeting the next day, we analyzed the raid in which my brother had died. Savo Brković, who had been sent there as co-ordinator and supervisor, was back. After the meeting the comrades dispersed to their assignments. The Staff and leadership settled down on Gostilje, and I walked out to the promontory overlooking the Zeta Valley. It was cloudy, and the wind howled amid the crags. I sat on a rock and sought peace in the storm: it was as if my brother Aleksa were calling to me, as if he were merging with me, my childhood, and my unpredictable life, along with all its passions and transgressions.

The next day Ivan Milutinović, a member of the Politburo, arrived at the headquarters on Gostilje. He bore a letter from Tito, with a seal and Tito's signature: I was dismissed for my "errors," Milutinović replacing me as delegate of the Central Committee and Supreme Staff. He presented the stand of the Central Committee to the Montenegrin leadership. The letter announced: ". . . We shall present these errors in greater detail in a letter which is to follow soon, along with a short resolution which you are to communicate to the party organizations." The same letter annulled the resolutions of the Provisional Committee's council, the directives of the Provisional Committee, and my letter to the party organizations.

The letter closed as follows:

> And now a few words about your most serious errors.
>
> 1) It was not incorrect to start an uprising, but it was incorrect to start that uprising without political operations from below. It was incorrect for you to separate the Partisan struggle from the people's uprising. It was incorrect that right from the start you created armies and fronts, instead of adopting the Partisan method of warfare. Your basic error is not a premature popular uprising, but your military strategy. Your frontal struggle forced you into a frontal withdrawal; it was pointless to expect that you would be able to put up strong resistance to a much stronger enemy by a frontal defense. Furthermore, it was incorrect for you to dismiss the armed masses instead of breaking up large units into smaller Partisan detachments and carrying on a continuous Partisan war. It was incorrect to call the National Liberation Struggle an antifascist revolution. Here are but a few examples; as for the rest, Milutin will tell you orally.
>
> We have asked you once before to send us immediately 2,500 to 3,000 fighting men, for the success of the struggle in Montenegro and the feeding of the Montenegrin people depend on the struggle in this territory. Send us 100 to 200 well-armed men in short intervals. In our first letter we indicated the route they are to take.
>
> This letter also serves as Comrade Milutin's authorization.
>
> (Seal)
>
> Tito*

* *Zbornik dokumenata,* Vol. III, Book 1 (1950). p. 68.

It immediately occurred to me that this letter not only was written in haste and under pressure, but was disorganized and in poor form: point "1)" is cited, but not followed by any other points. I recognized the hand of Tito's secretary Davorjanka Paunović—"Zdenka"—who did her work in an unsystematic and careless way, resentful that she had to conceal her intimate relationship with the secretary general and supreme commander under the guise of an "ordinary" clerk's job. I called Milutinović's attention to this aspect of the letter. And although he, too, was a stickler, he replied, "Don't latch on to trifles! Zdenka just didn't do it right."

On November 5, the day Milutinović arrived, we held a session to review the "errors" presented in Tito's letter. The leaders were unanimous in refuting Tito's criticism—of course, while recognizing my mistakes and inability to cope with the situation. Tito's criticism was based on a formal comparison of the armed struggle in Serbia with that in Montenegro—the former having developed out of the gradual strengthening of Partisan detachments, and the latter out of a people's uprising. For the most part, I kept silent at the meeting because I was the "guiltiest." Of course, Tito had in fact directed me not to create an uprising, but to set in motion actions by small compact units. In the meantime Moscow and the Central Committee of Yugoslavia had adopted a stand in favor of an uprising and not simply guerrilla actions; in his letter Tito tried to reconcile the two by approving the uprising while stressing the error that "this uprising" had not been politically well prepared from below.

Especially unfounded was the reproach for calling the National Liberation Struggle an "antifascist revolution." This term did not differ in content from "National Liberation Struggle," and both were adopted from Soviet directives and propaganda. True, the expression "revolution," even though preceded by the adjective "antifascist," could be criticized as limiting the uprising, but it could also be defended as being aimed against fascism. That term was used only in Montenegro, whose Party membership was burdened with leftist and dogmatic phraseology. These people had to be convinced that the struggle against fascism was that very revolution for which they yearned, but in the only possible form—that is, the form which life offers. That this is true, and that there were no "leftist" deviations while I was in charge in Montenegro, is evident from the documents of the time. Moša Pijade has persistently charged me with "opportunism" in the uprising, on account of the directive limiting us to guerrilla actions. But at the Fifth Congress of the Party in 1948, Tito, on observing my resentment over Moša's gibes, explained to me behind the scenes how he himself understood those errors. "With you it was not a case of opportunism, but of failure to cope with the situation."

Indeed, in retrospect, though the criticism in Tito's letter was based on developments in Serbia and on ignorance of the events in Montenegro, I must admit there was some truth to it. We in Montenegro were too slow in undertaking the formation of armed units capable of operating outside their own territory. There were examples of such hesitation in the creation of permanent military units in other places as well, wherever the uprising was more or less spontaneous and massive, as for example in Bosnia. Our own circumstances had impelled us to mobilize special groups of experienced raiders who afterward melted back into the peasant mass. This was partly due to my own view that the people's uprising must rely on a people's army. I also had in mind to strengthen the Communists, not so much through political power, but rather through the people and as a movement of the people. There did not seem to be a difference of opinion between me and Tito. Tito was a staunch enemy of "sectarianism" and of a dogmatic approach to the masses. But in his view a movement, no matter how broad, was of no value unless based on firmly knit, disciplined, and indoctrinated units. Milutinović told me, "The Old Man* looks on our Partisan detachments as the future Red Army." No matter how broad and popular, the revolution had to have its own army, even if a small one. A people's army evolves gradually, as the authority of the revolution gains ground, with the development of a revolutionary government.

Such was my understanding of the letter. Before long, circumstances led me to adopt these views and approaches myself. But I learned something else from that letter which was perhaps more important for me, and of which I was not fully aware at the time; namely, that theoretical deductions must be one's own, arrived at independently, or there is no theory, no real thought. It was not until after the clash with the Soviet leadership in 1948 that the Yugoslav party leadership spoke up for itself—in good part, because of my own initiative. It was this turnabout, this frankness about our own past, that finally weaned us away from Moscow as the source and interpreter of revolution. Yet Communists are not able to cast off one dogma or mythology without adopting another: the revolution which had been accomplished in covert fashion quickly acquired, after 1948, a dogmatic mythology of its own.

Apart from these "errors," I had a feeling that Tito's letter masked

* Tito was commonly referred to as *Stari* (Old Man).

some hidden motives. The first, though not the most important, reflected Mitar Bakić's dissatisfaction with the situation in Montenegro and the role assigned him. I too was of the opinion that he was not carrying out duties equal to his capabilities. But the Montenegrin leadership was disciplined if not exactly qualified, and there was no reason to disturb the existing state of affairs. I kept Bakić in mind, but could promote him only by stages. We had used him as an instructor, and had him transferred to the Staff. It was by his choice and mine that he went to Serbia with Arso Jovanović and the royal mission: I held that, as a party activist, he would be in a better position to give a fair account of the situation in Montenegro. Bakić certainly did just that. Yet with all his many good qualities, he had his faults like everyone else. While working with Tito in Zagreb, Bakić had spent a lot of time in the company of the Central Committee members, and they would all recount, half in jest and half seriously, the mistakes and shortcomings of underlings. Bakić knew how to be witty about this sort of thing. I don't think he spoke badly of me—we were close and he respected me—but he certainly didn't hold back his thoughts on Montenegrin Communism. Thus the Central Committee could have gotten a one-sided picture of the grotesque fantasies of the Montenegrin Communists. I don't believe that Professor Dušan Nedeljković, who had also gone to Serbia at that time, could have contributed to this negative image of Montenegro, for he had no standing with the leadership. However, I don't doubt that, when given a chance, he spoke disparagingly about the state of affairs in Montenegro.

The party officials in Sandžak—the first to establish contact with Serbia—could also have contributed to this one-sided view; they wanted to separate from the Montenegrin leadership, and looked to the movement in Serbia as their model.

But all of this is of secondary importance. The main reason—and one that had some basis in fact—was Tito's suspicion that I was growing too independent. This was reflected at the end of Tito's letter: "We have asked you once before to send us immediately 2,500 to 3,000 fighting men. . . ." I had resisted this request—I don't recall under what guise, but probably by asking for more time. I also remember that at the end of my own letter, sensing that I might be recalled, I had asked to remain in my native region: my emotional ties with Montenegro were all the greater in that the party was still reorganizing, and the fighting beginning anew. Tito had reason to reproach me for dilatoriness in sending information to the Central Committee, and this might well have engendered his doubts over my independent tendencies. The journey to Belgrade, where the Central Committee remained after I had left, was difficult and dangerous. After the uprising, there was nothing for us to report except disorganization and demoralization: I wanted to get us back in shape first. True, we did send a report by way of Sandžak, but the Central Committee didn't receive it until after Užice had been liberated.

That my suspicions regarding the "nonpolitical" reasons for my dismissal were well founded became apparent as soon as I arrived in Užice: my wife Mitra teased me for failing to become "prince" of Montenegro because the Central Committee had sent me to never-never land.

Perhaps all these explanations would be unnecessary, were it not that to this day partial, twisted, and even contradictory interpretations still appear. Thus Edvard Kardelj—who not only knows why I was recalled from Montenegro, but even shared in the decision—writes today, "During the War of National Liberation we had to recall him from Montenegro because of his personal responsibility for excessively repressive measures which resulted in unnecessary bloodshed and greatly impaired the reputation not only of the Communist party, but of the National Liberation movement in general."* Something of the sort was also written by Vladimir Dedijer, whose garrulous "conscience of the historian" has always strained to attain that of a court councilor.

The Chetniks were the first to launch the myth of my responsibility for reprisals, probably not for propaganda purposes alone, but as the honest result of misinformation: they knew that I was the leader at the time of the uprising, but not that I had been recalled, so that during the reprisals I was no longer in Montenegro. Thus exile propaganda claimed that the "Dogs' Graveyard" in Kolašin—where executed Chetnik sympathizers had been buried along with dead dogs—was my work, whereas in actual fact Kolašin was in Italian hands at the time I left Montenegro in early November 1941, and again in March 1942 when I returned. The "Dogs' Graveyard" incident took place in the interim (December 1941–January 1942), after the Italians withdrew from Kolašin and before the Chetniks retook it. Walter Roberts was also taken in by these fairy tales in his book *Tito, Mihailović and the Allies, 1941–1945*.† He makes no mention of the Dogs' Graveyard, but, rather, states that the uprising itself was accompanied by reprisals. And so it goes: people who play a public role, even against their will, must count on having untruths spread about them—especially when their role strikes their opponents as dangerous.

* *Borba,* Belgrade, May 1–2, 1974, p. 9.

† Walter R. Roberts, *Tito, Mihailović and the Allies, 1941–1945* (New Brunswick, New Jersey: Rutgers University Press, 1973).

While the party was being reorganized and the uprising revived, twenty men were executed in Montenegro, most of them spies. The names of all were announced in the Communist publications of the time. The list ended with "To be continued." I believe that was on Peko Dapčević's suggestion, and the rest of us agreed to it. Today it sounds like a cruel mockery. At the time we meant it as a warning and as a statement of our determination. Among those executed were several prominent men who were not collaborators with the occupation, but who had undoubtedly come out against the Communists and the armed struggle—either as the leaders of reactionary parties (Member of Parliament Nikola Jovanović), or as Communist renegades (Milovan Andjelić, Slobodan Marušić). Practically all the executions were carried out as ambush attacks; we had no courts or adequate military organization yet. This was done to avoid blood feuds—exactly as Montenegro's rulers had done, before they established their rule and state. In any event, my role in the "reprisals"—both in Montenegro and in Yugoslavia—is described in these pages.

Would the civil war have flared up had I not been pulled out of Montenegro? And how great was Ivan Milutinović's role in the reprisals? The July 13 uprising bore the seeds of future conflict: it gave the Communists power—a power not yet translated into laws and institutions, but extending over most of the population and territory of Montenegro. Just as profascist groups could not accept that power in its inception, so the adherents of the old regime could not accept it in its consolidation. Having acquired that power in the struggle against the occupation, the Communists treated any denial of that power as aid to the enemy. This was natural, since this denial came from persons who had long persecuted Communists: revolution too is an extension of politics by other means.

In actual fact, the war which the ruling classes had lost without serious resistance was simply being continued. At first the Communists relied on the elemental resentment of the people—particularly the Serbians—at the unjustified defeatism of the former regime. The armed struggle against the occupation strengthened the Communists. Only through this struggle could they win or, for that matter, survive. A hitherto suppressed patriotic feeling awakened in them: power is also the fatherland! Their opponents—those who were against the invader—took a completely opposite stand: they favored waiting for the invaders to be weakened. They were also anxious to prevent, at any cost, the creation of a Communist army and Communist power. That stand was, of course, further complicated by the existence of several occupation territories, the creation of a vassal Croatia, as well as by the influence of the Allies. But the essence remains: each strove for his own power, and no one strove for political tolerance. All the active political forces faced the same mortal danger: no one could predict whose would be the victory, though we Communists were confident that we were the most vital, the most idealistic force. That conviction carried over to our potential supporters—those who opposed the occupation and were dissatisfied with the old regime. They accepted the "truth" of the revolutionary movement not because they believed in its totality, but because it offered a way out of the mortal crisis of the nation.

The role of the leaders was undoubtedly more important than might appear from my account. Yet not even by reducing "history" to the role of leaders can one ascribe the reprisals in Montenegro to Ivan Milutinović. He was a mild man who reacted to events after they happened; he did not foresee or create them. He was totally ethical in his personal and political life. Though open-minded and amenable to co-operation with non-Communists, he had no broad intellectual qualities. Everyone who knew him liked him, but few feared him. In the swift and stormy turn of events following my departure from Montenegro, Milutinović let both men and events slip from his control. The reign of terror in Montenegro, and the even more terrible one in Hercegovina, took place largely because of willful local leaders and doctrinaire reactions. Milutinović's responsibility was not so much for what he did as for what he did not do: order should have governed zeal, and common sense, retribution. Yes, *should have!* It is easy to wage flawless battles after they have been lost.

I was never on close terms with Milutinović, though we never had any disagreements either. This is why now, so many years after his death, I am pleased to find confirmation of his decency in the published documents—especially regarding my recall from Montenegro. In a letter of November 6, 1941, Milutinović wrote to Tito in Užice:

> From the oral reports [at a meeting of the leadership—M. Dj.] submitted to me and from what I saw in the field at the time of my arrival, a completely different picture presents itself, and there is really a state of affairs different from the one gained by reading the resolution and the letter of Comrade Veljko [my party pseudonym—M. Dj.].
>
> The letter and the resolution were examined and discussed by all the party organizations, and it was on the basis of these documents that a consolidation was carried out, regardless of the fact that there were errors in both and that both were rejected by the Central Committee as erroneous.

The erroneous names for the staff organizations, and the subsequent election of leaders—that is, the commanding personnel of the units, troops, battalions, and brigades—can easily be corrected, and a staff was in the process of being formed even before I arrived . . . in fact, it was simply a matter of changing the names, but the composition of the staff is the same.

At the same time they established a correct relationship between the staff and the party, as well as the party's activities in the rear. At first there were some irregularities, as will be seen from Comrade Veljko's oral report.

. . . You ought to wait for Comrade Veljko's arrival. . . . Only after his report will many things become clear. . . . Your resolution ought to take note of the successes and condemn the errors committed during the period prior to the council [the party council which met after the uprising—M. Dj.], successes achieved after the council, and the good work among the broad popular masses.

The question of the youth, as you have seen, has been correctly treated by them. . . .

They received your letter and my explanations without any reservations. . . ."*

Had I known of Milutinović's letter to the Central Committee, my suspicions regarding the unstated reasons for my recall would not have driven me into a state of despondency. Had Tito shown me that letter in Užice, I might not have taken such pains during the following months to be so determined and devoted. I do not believe that there was anything sinister in the "concealment" of that letter, and yet—when the criticism of the Central Committee was shown to be unfounded, everything was hushed up so as not to stir up "bad blood."

And so, broken and battered, but still with a firm grip on myself, I set out for Užice with a group of comrades on November 11, 1941. My depression was so pervasive and irrational that it crossed my mind that I might be shot. Why all these overpowering suspicions? Did they come from Tito's directive to me that even members of the Montenegrin leadership should be shot if they wavered? Or from suppressed doubts about the credibility of the "Moscow trials"? Or from the futility of resisting injustice and arbitrariness in our own ranks? The vile weather seemed to reflect my own dark mood. We journeyed through rain and snow along muddy mountain slopes from early morning on, and sometimes all through the night. Anxiety and impatience goaded me on, while I in turn goaded on my fellow travelers and guides.

At dusk the following day we reached the home of my aunt Draguna in Bakovići, near Kolašin, where I had lived during my first two years of high school. The Italians were in Kolašin itself, but the road and the

* *Zbornik dokumenata*, Vol. III, Book 1 (1950), pp. 118–19.

damaged bridges were in the hands of the Partisans. We dined and my neighborhood chums gathered, including Mihailo Ćetković, with whom I used to walk to school and play. We sat around and talked late into the night, but we felt an uneasiness, mostly because of fifth-column activities in the villages. My aunt and friends urged us to spend the night with them, but I pushed on to my native village of Podbišće. I never saw either Aunt Draguna or Mihailo again; she died and he was killed.

Just before dawn we arrived at a safe refuge—the house of Stanija Kovijanić, whose son Vuko was a Communist, and a schoolmate of mine from elementary school. I stayed with Stanija, who by then had rebuilt her burned-down house. The neighbors put up my companions, who went to sleep right away. I, however, couldn't fall asleep. I felt anxious, and was filled with new impressions and excited about being in my native region again. I sat with my fellow villagers and talked late into the night.

The talk continued into the next day, for we had to spend the daylight hours in Podbišće. I suddenly found myself in the midst of hatred and feuding of which few besides Montenegrins are capable. A couple of days before, the Communists had killed Milovan Andjelić, a person of ascetic character who had been a renowned Communist before the period of "Bolshevization." He came from a large prominent family. He had given up his government service, and with his brothers had built up a sizable estate. Meanwhile he had turned against the party, and during the Italian breakthrough after the uprising had urged his fellow villagers to surrender their arms. As far as I remember, the Provincial Committee did not order Andjelić's killing; it may have been Milutinović, or some member of the Provincial Committee, or simply the Regional Committee. I don't say I wouldn't have approved of that killing, though it would have troubled me: he had been my teacher in elementary school. He had in fact already been attacked in the press of the Provincial Committee for his open and reckless stand against the party and the armed struggle. Revolution and a totalitarian idea do not know or recognize anything but themselves.

That killing became intertwined with the death of my brother Aleksa. My home village of Podbišće was Communist; it rejoiced at Andjelić's death and mourned Aleksa's. They said of Andjelić, "That son of a bitch got what he deserved. They should have gotten his brothers, too! The youngest, Djukan, is worse than he was! A troublemaker and traitor!" But of my brother they said, "That grey falcon! If only he had died at the hands of men! But the scum got him . . . ! They'll pay for it, though; we'll find them even if they hide in Mussolini's behind!" There was also one touching incident: Mileva Šćepanović presented me with a sweater which she had not had a chance to give to my brother.

Hatreds flared which we high up in the party didn't really comprehend in all their elemental and destructive dimensions, even though we had inflamed them and directed them through our ideas and organizations. Yet there was no choice. It was as if life itself was metamorphosized into an idea, and all outsiders were consigned to hell.

The very next day we heard of a clash between the Chetniks and Partisans in Serbia, even though just two weeks earlier, on October 27, an agreement had been reached between Tito and Draža Mihailović. Two days later we obtained more detailed news of that clash from Rifat Burdžović and Voja Leković in the village of Radojnja, to which the party leadership of Sandžak had moved for reasons of safety, and to be nearer the party center. We were glad for two reasons—that the Partisans had beaten the Chetniks, and that a clash had occurred in the first place. Our bitterness over the old reactionary regime had grown too great for us to look upon its restoration with equanimity. Yet in contrast with Burdžović's revolutionary romanticism, I expressed the fear that this might complicate and impede our further struggle.

It was in joy and apprehension, with personal doubts and anxieties, that I arrived the next day in Užice—fresh from a people's uprising which treachery and irreconcilable antagonism had transformed into organized and raging feuds.

The Civil War within a War

1

It was finally on Zlatibor Mountain, at Borova Glava, that we got to a road, and managed to order a truck from Užice. We arrived in Užice in early evening. The local command found quarters for my companions, while I made my way to the comfortable little Palace Hotel, in which the editors of *Borba* were staying, and my wife Mitra as well.

My meeting with Mitra turned out to be inexplicably, incomprehensibly official—as if she had not been in prison, living "on borrowed time," nor I so desolate over her. She had changed. Not so much outwardly, though she had dyed her hair to escape recognition by the police on leaving Belgrade. She looked rather like a hairdresser or a nightclub singer. However, what struck me the most was Mitra's inner transformation: hardened, self-confident, sarcastically vivacious. She recounted, in proud sorrow, how she had made her way in a cart across Mačva while that province was caught up in an uprising and filled with the acrid smoke of villages burned by the Germans. She injected into her account many details of Partisan life, as well as Serbian jokes. She did not spare even me. Worried and unhappy as I was, I could not respond with anything but the suspicion, kept to myself, that this hard strength of hers was not just a reaction to her long expectation of death, but perhaps a cooling toward me.

I moved into Mitra's hotel room rather than into the bank where Tito, Kardelj, Ranković, and some other comrades slept. Mitra told me that I was to present myself to Tito at headquarters, which was located in the former National Bank and had a shelter dug into a hillside next to it.

To be sure, first I spruced myself up, and even got a short leather coat which a local shop was making for functionaries. Tito received me solicitously, even tenderly. He did not reproach me for anything, nor did he show much interest in the situation in Montenegro. He told me that I was to take over the editing of *Borba* and other propaganda activities, so that Kardelj could attend to organizational questions and governmental duties. I expected Tito to call a meeting of the Central Committee to consider my "errors" in Montenegro, but no such meeting was scheduled, much less held.

Headquarters seemed orderly and organized—especially in comparison with our offices in Montenegro. The men were not in strict uniform but well dressed, nearly all wearing garrison caps—the Serbian military caps worn by the Royal Yugoslav Army. The only exception was Tito. He had a Soviet *pilotka* made for himself, a pilot's cap which was later dubbed a *titovka,* and which was to become the regulation cap in the army of the new Yugoslavia. Officers' insignia had already been introduced in the army units, but at Supreme Staff no one wore them until 1943, when military ranks were inaugurated. Everybody had red stars of thick fabric sewn on his cap, while Tito wore an enameled Soviet star with a hammer and sickle which came from who knows where and had been presented to him as a gift.

Even so, this was more of a guerrilla party headquarters than a military one: staff positions were still in their infancy. The duties of the chief of staff were being carried out by a former royal second lieutenant —a diligent and obliging man in all respects. Tito and he engrossed themselves in wall maps (at a 1:100,000 scale), though I maintained that maps of a larger scale would have been more suited to our kind of warfare. Tito was not yet at home with all those hypsometrical lines. As for me, privately I was not at all pleased that he was the supreme commander: I was of the opinion that the function of party leader should be separate from that of military commander, at least insofar as operational direction was concerned. This was, in fact, the reason I had not been formally named commander in Montenegro. Yet no one could say Tito was not doing the job with vigilance and industry, despite his inadequate knowledge and rashness.

The food at headquarters was meager, almost ascetic. The leaders' attitude was that they had to deny themselves as much as they denied their men: all food supplies had to be purchased, and the peasants were not happy about selling, though we did not lack for money as a considerable sum had been seized as booty in Užice. Everything was clean and served on plates. Only the top echelons of the staff and Central Committee ate at headquarters. We would sit at a long table and wait until Tito appeared from the staff room. The Central Committee members sat around him. This was dictated more by the necessity of exchanging views and by the deference of juniors than by established order. Hierarchical precedence grew with the revolution, and apparently became as important and necessary to it as fervor and egalitarianism.

Having just come from Montenegro, where populist egalitarianism prevailed, I found this hierarchical and official decorum shocking. It all seemed somehow premature. Besides, the leading men in the Central Committee—Tito, Žujović, and Ranković—were followed about by pretty young secretaries who were obviously more intimate with them than their duties required. Only Kardelj was an exception. Later I discovered that Ranković's secretary did not have intimate relations with him. There was something ugly about all of this, though perhaps not because these comrades had new mates: no one is sentenced to live out his whole life with the same wife or husband. Rather it was because among the troops the strictest puritanism was practiced in every respect. When I spoke to Kardelj about it, his response was, "What can be done about it? Men at war live under pressure, and after all, this is a private matter." Mitra observed cynically, "It goes with power. In Serbia, a minister without a mistress is unthinkable."

At the same time, I marveled at the organization and learned from it. Without doubt a real government was evolving which gradually encompassed many regions and directed them in war and in all its aims. The territory was small and the pretensions premature, but the National Liberation Committee of Serbia was the germ of a government, and would have quietly become a broader one had not the fortunes of war been so fickle.

Though still largely spontaneous, the hierarchical order and organization could not overcome the air of impermanence which marked the "Republic of Užice" and which expressed itself in the ever more fearful and reserved attitude of the townspeople toward the new authorities. Certainly the bloody falling out with the Chetniks contributed to this. So did the shooting of all adult males in Kragujevac and Kraljevo, which the Germans carried out with the help of Nedić's people and Ljotić's fascist movement. It was believed at the time that some 5,000 were executed in Kragujevac, and 1,700 in Kraljevo. These figures grew with time—by the thousands in both places—though the actual figures have never been confirmed. *Borba* wrote of this but we spoke of it reluctantly, feeling the deathly horror that had seized Serbia.

What "reason" did the Germans have for undertaking measures which were then so unthinkable? Was it retaliation, the killing of one hundred Serbs for every dead German, which they proclaimed during the very first days of the occupation? Was it the destruction of Serbia's vital centers—towns known for their national consciousness and Communist influence? At Supreme Headquarters plans were being made to take Kragujevac; in fact, the party even slowed down the exodus of members so as to use them from within. Did not the Germans guess this, especially since Kraljevo had already been attacked, though unsuccessfully? Certainly the Nazis found sufficient justification for crushing the will to resist in a people whom Hitler considered the most politically creative and thus the most dangerous in the Balkans.

In politics nothing happens by chance, or remains without consequences. The tragedy gave to Nedić "convincing proof" that the Serbs would be biologically exterminated if they were not submissive and loyal, and to the Chetniks "proof" that the Partisans were prematurely provoking the Germans and thus causing the decimation of Serbs and the destruction of Serbian culture. As for the Communists, they were given the needed stimulus to call the population to armed struggle as the only salvation. It there was treason, and I hold that there was, it justified itself with biological survival. Counterrevolution finds justification in legality and "realism," while revolution finds it in inevitability and idealism. The massacres in Kragujevac and Kraljevo were, by their calculated total horror, beyond the comprehension of both Germany's collaborators and its opponents, and could be resisted only by a movement to which its enemies offered death as the only alternative.

Quite apart from these horrors, war had the strange power of changing the features of persons, objects, and places overnight. Užice was filled with want and tension, misery and decay. Shops were empty, market places bare. Only the tailor shops and bakeries were in full swing, as they were working for the army. The streets and parks were deserted; there were a few people—mostly older people—here and there, all shabbily dressed, though the war had just begun.

Ranković was in Čačak, negotiating with the Chetniks, when I arrived, so I spoke with him on the telephone at headquarters when he called to give Tito the basic points of the agreement. Not only was there no "censorious" reservation in his voice, but he was genuinely happy to speak with me. As for Žujović, I visited him in the hospital at Sevojno, where he was recuperating from a serious wound received in a battle with the Chetniks around Požega. But in those days my closest relations were with Kardelj and Lola Ribar.

Even had I not known of Tito's distaste for theorizing, I would not have been able to discuss with him—busy as he was—certain issues that troubled me; namely, was a revolution taking place in Yugoslavia, and had the fighting in Serbia assumed such proportions that it could be called a national uprising, as party propaganda was already doing? The question of revolution was crucial to me, because of Tito's criticism of my views in his letter. I could discuss such issues with Lola Ribar to my heart's content but could get no explanations from him, as more often than not he sought explanations from me. I turned to Kardelj, all the more eagerly since he was virtually the party's official theoretician, and because I had had similar discussions with him before the war.

Kardelj, too, believed that a revolution had already begun in Yugoslavia: in a socially differentiated and politically divided Serbia, this was even more apparent than in Montenegro. To be sure, it was not a "pure" proletarian revolution: the uprising against the occupation made the revolution a national one, while the bankruptcy of the old regime, and its collaboration with the enemy, propelled the Communist party to the fore as the leader. Yet if we employed the term "revolution," the reactionaries and profascists would depict the armed struggle against the occupation as the Communists' struggle for their own and Soviet power. Thus it was tactically more opportune—all the more so since the Comintern and the Soviet leadership so believed—not to flaunt revolutionary phrases. The term "National Liberation Struggle" was more attractive and accurate. This explanation gave me nothing essentially new, and for this very reason was all the more comforting. Clearly I had done nothing in Montenegro to restrict the struggle, I had only made the dogmatic consciousness of the party membership correspond to the reality of the uprising.

In any event, despite the terminology, the armed struggle and the Užice experience revealed the tendencies of the uprising unequivocably. In Yugoslavia all political parties had dissolved, with the exception of the Communist party and the Communist Youth (SKOJ). There were individuals in the democratic parties, including prominent ones, who favored the uprising and collaboration with the Communists—but no group did so unreservedly. Roughly speaking, Serbia was divided into three groups. The first was for local collaboration with the occupation forces. Its most influential representative was the fascist Dimitrije Ljotić. The former royal minister of war, General Milan Nedić, had placed himself at its head by becoming premier at the outbreak of armed conflict. A second group was for awaiting "favorable" conditions for an armed struggle. At its head was Colonel Draža Mihailović. The core of this group were royal officers who had fled to the forests; they were surrounded by many adherents of the old order. The third group was for an unconditional armed struggle against the occupation forces and the gradual creation of a new political and social order. This group was led by the Communists.

In the course of the war the influence of these currents in Serbia varied. However, since political authority in the villages had been reduced to the barest administrative apparatus, it was easy for the Communists to demolish it and quickly gain support. Mihailović's Chetniks were the first to take to the woods. Moreover, they had one or two encounters—defensive ones—with the Germans. But because of reprisals and the exile government's policy of waiting, they came to oppose an armed struggle—except against the Croatian fascist Ustashi. But the growing strength of the Communists and their military operations forced the Chetniks to take a stand. Since they were against the occupation in principle, they also joined in the struggle, and here and there co-ordinated their operations with the Communists. By appealing to legitimacy, they even achieved a numerical preponderance. Yet what kind of preponderance was it? Most of it was on paper, or represented men who were loafing in the villages. In joint attacks on larger towns they would retreat at critical moments and expose the Communist flank, not out of treachery but from lack of organization and militancy.

Užice was a particular illustration of the initial discord between Partisans and Chetniks. Chetniks entered the town and began to drink and celebrate, never lifting a finger to mobilize for war; some even menaced the townspeople. The Communists arrived two days later and immediately began to organize a government, the youth, the workers, the economy, and of course the manufacture of ammunition and repair of rifles. After several days the Chetniks were forced out without a struggle. There were cases also where the opposite happened—where Chetniks, relying on village traditionalism, took over towns from the Partisans.

In battle and in negotiations, deception and trickery are necessary. But revolution and counterrevolution cannot hide from each other. The Communists had no reason at the time to clash with the Chetniks. (Moscow did not want such a clash because of relations with Great Britain.) But the Chetniks could not look on calmly at the subversion of their order and the termination of their authority even before the war had ended. There can be no revolution without leaders capable of combining reality with utopia. But this in turn requires a belief in inevitability and an adversary who looks to the past for that reality and ideal.

The day after the signing of the Tito-Mihailović agreement—a modest one—the Chetniks killed a Partisan commander. Some ten days later the Chetniks moved on Užice and committed violations on a vast scale. Yet in Užice, at Supreme Headquarters, no one claimed that the Chetniks had consented to that agreement with the Partisans just to conceal their insidious designs. Indeed, they were inclined to ascribe the murder to the local Chetniks of Požega and not to Draža Mihailović. But everyone now believed that after the agreement something had changed among the Chetnik leadership—that Draža's staff had probably received a directive from the government-in-exile, if not from the British, to attack the Communists. The Chetniks prepared that attack secretly, focusing the main thrust on Užice, the seat of the Communist leadership. But the Chetnik preparations were so careless and inadequate that the Partisans found out about them. Tito rallied the troops in time and defeated the Chetniks with relative ease—though not without losses—and took the greater part of their territory away from them. With his own headquarters in danger, Draža sought new negotiations. Ranković signed a second agreement in Čačak—one that was more concrete and substantial, but which also remained a dead letter.

The party leadership did not find the conflict pleasant, much less desirable, despite the rejoicing at the victories. Tito showed courage, but also anxiety. Even at that time Tito had a low opinion of the Chetniks, because of their lack of organization and discipline, and the primitivism in Draža's own staff. It seemed that one group of Chetniks was even preparing to attack Tito's negotiating team. Yet Tito offered Mihailović the post of commander of the joint forces in Serbia, and perhaps in Bosnia; I don't believe all Yugoslavia was included, because the Chetniks were a distinctly Serbian nationalist movement. Tito later recalled one funny detail: Draža offered Tito some "Šumadija tea," which Tito gladly accepted, not knowing that it was hot brandy, and then choked on inhaling the alcoholic fumes. Nor was Kardelj happy over the conflict, though he saw in the Chetnik attack a "logical" and "inevitable" culmination of the policies of the reactionary and Greater Serbian forces. Ranković, as usual, did not give it much thought, but was bitter against the Chetniks for having betrayed the national traditions of Serbia. He referred to the Chetniks as *"četojke"*—an ambiguous term that smacks strongly of the derisive taunts against effeminacy in which the jargon of Serbia abounds. Žujović appeared to be the most militant regarding the Chetniks, not because they had wounded him in battle but because he saw in them the ruination of Šumadija, his own defiant region in Serbia.

Everybody was bitter over the Chetnik massacre of some thirty Partisan followers, mostly girls and wounded. The girls had just managed to slip out of Belgrade; ignorant of the Chetnik-Partisan conflict, they had made their way to the nearest unoccupied territory, which was Chetnik. They were killed not far from Draža's headquarters, on the orders of Draža's staff. In the minds of police agents and officers of the Kingdom of Yugoslavia, in which Communists had always been jailed and beaten, the war did not change a thing: it was "natural" and "lawful" to settle accounts with the Communists, now using the bayonet and bullet. Among those girls was Olga Jojić, a sickly but diligent worker, a zealous party member and Mitra's close friend. While in a sanatorium for tuberculosis, Olga had fallen in love with Vladimir Bakarić, member of the Central Committee of Croatia, and after the war one of the most prominent leaders of the Croatian Communists. From time to time Olga had gone to Zagreb to see Bakarić. Whenever I ran into him in our underground work, half in jest I would transmit to him the greetings of his girl friend in Belgrade. At the time I was also told that, thanks to his cleverness and good fortune, my younger brother Milivoje, acting political commissar of the Takovo Battalion, had escaped the Chetnik bullets at Požega while in charge of a train to Užice. If war could leave unscarred the memory of its participants, hatred and revenge would not become the most profound passions. There was even greater bitterness over the Chetniks' surrender to the Germans at Valjevo of about a hundred Partisans captured through trickery. It seemed as if the realities of war most convincingly confirmed the Communist doctrine that class and ideological hatred overrode national solidarity; this only fortified the Communists in their intolerance and rejection of the old regime.

Ranković's agreement with the Chetniks called for a joint commission for the investigation of crimes and conflicts. The commission met, but it was too late: the Chetniks had already decided on a fight, and the Communists on revenge. The civil war had begun, though everyone denied it, each placing the blame on the other side. At the time the Communists were relatively considerate in their treatment of the Chetniks: they released captured peasants and didn't shoot captive officers. That doesn't mean there was no killing—mostly covert—of Draža's followers. For though Draža was against collaboration with the Germans, he attacked neither the state apparatus which the Germans controlled, nor Nedić's State Guard or Ljotić's Volunteer Corps. Among those whom the Communists killed as fascists and servants of the occupation there were bound to be men whom Draža's followers regarded as "nationalists" and "good Serbs." In a divided nation the traitor is simply the person who betrays one's own side. There is immeasurably more bad blood in a fratricidal war than in any war between countries.

Much later, on a trip from Slovenia to Jajce in 1943, Kardelj let slip the remark that grave "sectarian errors" had been made in Serbia in 1941. I can only guess as to what he was referring to, for he never spoke of it again and I found it awkward to ask, because the person most responsible for these "errors" had to be Tito; Kardelj might have feared that I was setting up a rift between them. By "sectarian errors" in Serbia, I believe that Kardelj meant the narrowness of our propaganda and perhaps also of our organization, with its exclusively Communist commanding cadres and insignia, etc.—the red star, the clenched-fist salute. Yet later he himself adopted similar forms in Slovenia. Or was it that he defined every failure, or every stand that differed from the official one, as an "error" and "deviation"? In any event, those remarks of his did not seem to me to have been just a thoughtless reaction—so frequent with politicians—to some momentary impression or idea.

In retrospect, I think that one of these "errors" might well have been that parade in Užice on the anniversary of the October Revolution—with Tito and the leadership on the reviewing stand, in obvious imitation of the parades on Red Square in Moscow. Eyewitnesses say that this instilled enthusiasm in the troops: the revolution was undergoing a growing radicalization as it went deeper into the ranks. Yet it is also undeniable that such manifestations narrowed the struggle against the occupation and revealed the Communists as a force by themselves, a force with exclusive aims and methods. "Errors"—that is, internal divisions—are a threat to revolution: the Yugoslav revolution did not contain such "errors," at least not substantial ones, and thus was able to correct its real errors—illusions about its own strength.

For me, Tito's view at that time of the peasants was unusual, even shocking; perhaps there was a connection between it and Kardelj's observation about "errors." In a conversation in Tito's study, when I brought up the peasants' rallying to the fifth columnists during the Italian punitive expeditions, Tito remarked, "The peasant goes with whoever is the strongest." Later, when the villages became the chief source and suppliers of the Partisan army, we changed our minds about the peasantry to such a degree that we thought the Marxist theory of revolution ought to be expanded, to give greater stress to the role of the peasantry. Tito's remark in Užice reflected the initial preponderance of workers and students in Partisan units, and the submission of the peasant to the strongest side in a fratricidal struggle. Perhaps, too, this view of Tito's emanated from his experience in the Soviet Union, where—contrary to the fine theory of the peasant as a working man, and the peasantry as a natural ally of the proletariat—the peasants are subordinated to the power of the party.

As for the propaganda definition of the struggle in Serbia as a popular uprising, Kardelj saw it as the initial phase of the uprising by the broad masses. With this in mind, it was only natural to see those initial conflicts as a popular uprising, particularly insofar as they expressed the vital interests of the Serbian people. I accepted that explanation, but clung stubbornly to the view that it would be more useful to prepare for the uprising than to assume it already existed. However, discussions of this issue were cut off by events, and the idea stuck that the uprising began first in Serbia. Thus in 1945 it was decided that July 7, 1941—the day on which, in the village of Bela Crkva, Žikica Jovanović fired two revolver shots at gendarmes and not at the Germans—should be observed as the anniversary of the national uprising in Serbia. When I remarked to Ranković at the time that we should have picked a more significant event to mark the beginning of the uprising, he replied, "Well, you know, the Supreme Staff was located in Serbia, so it was more convenient to declare that the uprising began there."

Nevertheless it was beyond doubt that the Partisan units in Serbia, though relatively few, were the largest: around 2,500 men in Valjevo, around 1,600 in Užice. Composed largely of Communists and their sympathizers, they penetrated all Serbia and created an intolerable situation for the Germans, undermined the economy, and broke up Nedić's administration. They were a model of organization, sacrifice, and self-denial. Yet the people possessed a certain sense more infallible than the analyses of leaders: they gladly greeted both Partisans and Chetniks in Serbia, but didn't flock into their embattled ranks; rather they were horrified at their bloody discord, sensed black days to come, and took note of the mindless and hardly tapped strength of the invader.

2

In Užice I immediately took up the task to which I was assigned. I even managed to write an editorial for *Borba* on vigilance—inspired by a call from Radio Moscow for vigilance in the U.S.S.R., and by the treacherous atmosphere following the Chetnik-Partisan conflict. Yet in my vigilance there was no premonition of the explosion which was to take place in our munitions factory, in the underground vaults of the bank, around three o'clock on the afternoon of November 21, 1941.

I was with Mitra in the examination room of the Circuit Court. Ranković had asked me to complete the interrogation of a prisoner who had managed Horizon, the party bookstore in Paris before the war, and was now an agent provocateur of the royal secret police. Prison matters were generally handled by Slobodan Penezić, under Ranković's supervision; that was how their association and work in the secret police began. I was involved in that investigation because I was somewhat familiar with Horizon, and we hoped to uncover something important about the Paris émigrés. I must say that I uncovered nothing at all, and completed my investigation in the course of a single afternoon. The prison was well run, on the whole, and did not differ essentially, as far as I could tell, from the prisons of postwar years. I would say the same about the method of examination. Torture was applied selectively, in special cases, and executions were carried out secretly at night. Characteristic of prison conditions in Užice was an anecdote which Penezić told me. They had arrested a certain well-known village politician, and when Penezić upbraided him for his hostile speeches against the Partisans even though they had brought him freedom, he replied, "God save me from Partisan freedom and English cunning!" They let the peasant go, Penezić said, because he had done nothing besides criticize the Communists. At the end of the war a person could be executed for such "hostile propaganda," and in the postwar years sentenced to many years in prison.

And so, while Mitra and I were compiling our report on the prisoner, we heard an explosion—a big one followed by a smaller one, but no drone of an airplane. We wanted to go on with our work, but a guard rushed in and whispered to me that an explosion had occurred in the bank building where the Supreme Staff was housed. We gathered together our papers, returned the prisoner, and hurried outside. About a hundred yards from the court, next to a little park, we met Tito, Zdenka, and all the rest—grimy and agitated. "A raid on the munitions factory!" Tito said, while Zdenka, all pale, her face twitching, couldn't utter a word.

No one knew exactly what had happened. I made my way to headquarters. Ranković was already there. The bank's vaults, in which the Partisans had set up a munitions factory, had several wings. The right wing was used for the factory, the left wing as an air-raid shelter. The entrance from the shelter into the corridor connecting the two wings was boarded up. The entrances to the shelter were a few yards from the bank. The explosion shattered the bank's windows and brought down a few ceilings. Inside the bank several people were hurt, none seriously.

No one understood how the disaster had occurred. Grenades were bursting inside the vault, while smoke poured out. The engineers proposed that the entrances be boarded up to keep air out and thus smother the fire. That decision was approved—reluctantly, because someone might still be alive inside. The closing off of the tunnel was carried out sloppily. Tito returned to headquarters, where we stayed with him, anxious and disheartened, until late at night, when the closing of the tunnel was completed.

The Germans had begun their offensive in Šumadija and were converging on Užice. It seemed probable that German agents had caused the explosion in the munitions factory. But the actual cause of the explosion, and the number of victims, were never determined. It was estimated that there were about a hundred victims, most of them workers. Chetnik officers have boasted that it was their doing, but postwar investigations, so Ranković has told me, supported no such claim. The explosions might also have resulted from carelessness, of course. The factory had all the virtues and all the faults of Partisan improvisation, being useful but unreliable; there were catwalks, machines crowded next to boxes of powder, flickering light bulbs, frayed electric lines. It was not difficult to assume that a raid had taken place, in view of the poor state of affairs and bad security.

The next day the tunnels were opened up. The bodies were carried out—an operation which Ranković and I helped organize. In the corridor which connected the wings of the vault, there still were occasional explosions. One had to run past that entrance and into the depths where the workers had perished. Several bodies, among them party members, were found huddled together with handkerchiefs and rags over their mouths: their exit cut off, all had suffocated.

We decided to get the factory going again, though most of the raw material had been destroyed. However, the German 342nd and 113th Divisions were already advancing on Užice. On November 29 a hasty evacuation from Užice was carried out along the road to Zlatibor Mountain, including the wounded, printing presses, boxes of silver coins, food, and clothing. The Central Committee remained in town. Idle and worried, we hung around Tito and bothered him. The sky was overcast. Tito told us that he had heard the Germans over the telephone: the wire was uncut. After lunch he ordered me and Lola Ribar to go to the front to raise morale.

Tito was more familiar with the area where operations had just been completed than where they were now in progress: there, in northwestern Serbia, he had fought as an Austro-Hungarian against Serbia. He had told me the story before the war. "There was," he said, "a Serbian howitzer which hit us every time. And we'd recognize it every time it fired, and we called it 'Saint Nicholas.' " Ranković told me that on their way from Belgrade to join the Partisans, Tito reminisced about the villages where he had fought in 1914. Ranković cautioned him not to speak of it, because the Serbs were sensitive on the subject. This was later hushed up, especially in Dedijer's otherwise extensive biography of Tito.

Glad to be of some service at last, Lola and I got into a car and set out for Valjevo. We were making our way up the serpentine road of Crnokose, when we heard an artillery piece fired above us on the hill. We continued to drive up, and as we rounded a bend I saw a group of Partisans descending. I told the driver to stop, got out, and called to the Partisans. But they didn't respond, as if the enemy were close. Lola also got out. At that point a truck came to a stop next to us. In back there were several soldiers, three or four of them wounded. They told us that German tanks had broken through. "Where are the Germans?" the two of us asked. But they rushed off without an answer.

As Lola and I deliberated whether to turn back or go ahead, firing was heard from above. A grenade raised a small cloud of dust on the road down which the truck was speeding away. That meant that a German motorized unit had appeared on Crnokose Mountain, up which we were driving. We ordered the driver to turn around. Several more grenades were flung from above at the receding truck, and we heard the low hum of a motor. I instructed the driver to get down to the plain with the greatest possible speed. We were not fired upon as we did so. In Karan we parked behind the tavern, from where we could observe Crnokose Mountain. On the peak we saw the armored cars of a German advance unit. The Valjevo Partisan detachment—our largest and most reliable force, and the one to which the two of us were going—had been forced back, and there were in fact no troops in the fifteen or so kilometers between Užice and the Germans.

In the tavern there were a fair number of peasants who watched us with reserve, even hostility. I had learned from experience that people never like retreating armies. Moreover, as Lola explained, "The majority are for the Chetniks anyway." While we were waiting, Koča Popović arrived. He had been commander in Posavina, but Tito had called him back to rally the troops for the defense of Užice. Koča complained wryly, "But I don't have anyone to rally! One small unit with rifles and two or three submachine guns: that ought to hold the Germans for an hour or two, before they spread out their flanks—on condition that the road is mined."

The day was ending. We wanted to go back to Užice, but the German vehicles were making their way down Crnokose, turning off their headlights at a level stretch below the peak. We counted fifty vehicles. The Germans occupied a key position and prepared to bed down for the night. The darkness was growing, and the hostility of the peasants became more evident. We took off for Užice. At a crossroads we encountered our supply units pulling back from Požega, but they knew nothing except that Požega had been taken by the Germans. We stopped on a hill where Koča and the troops were planning the defense of Užice. We left Koča and his camp.

Reaching Užice, we immediately proceeded to headquarters to report to Tito. Tito was abashed—as if he were to blame for everything. He told us that the road was to be mined during the night, and that reinforcements would be sent from Užice. We went to bed. Mitra had been evacuated that evening, and I was beset by insomnia—as always after an intense experience or before grave decisions.

Kardelj, Lola Ribar, and other leaders drove out in the early morning to Zlatibor Mountain, while Ranković and I stayed with Tito. German planes circled over Užice and occasionally strafed us. Around noon Tito ordered Ranković and me to take the Požega unit—a crack guard assigned to headquarters—to Zlatibor; he would follow by car. The town saw us off in silence. No one opened a window to wave good-by. Not even those who liked us and felt sorry for us dared to mourn our departure. We got out one by one, scurrying through the little streets, our eyes on the planes above. When we got to the hill, we started to climb for dear life. Surprisingly, the planes did not bother us. Even so, we clung to the orchards and the brush.

From the top of the hill we looked back at Užice. On one side of the town the Partisans were spilling out in panic, though no firing was heard. The German penetration seemed too soon to me: the mines were supposed to hold them up. Later we learned that the mines either had not been placed at all, or had been taken out to let our two tanks through, whereupon the Germans had surprised the sappers. Anyway we had to hurry off. I hated to see the enemy enter a town which I had grown to love amid the fortunes and misfortunes of war.

It was around two in the afternoon when we stopped on a plateau in the woods by the road, and Tito's limousine arrived from the opposite direction, from Čajetina. The driver told us that he had left Tito and Captain Hudson in a gorge, where the road was to be blown up to stop the rapid advance of the Germans. Tito and Hudson had stopped to observe the fighting around Užice, and Tito had sent the driver to Čajetina with a message. We did not like the idea of Tito being alone; after a consultation with Ranković, I got into the car and drove off to find him.

We had hardly driven two or three kilometers, when I heard the planes again. Knowing that the car would be their target, I ordered the driver to stop so we could find some shelter. They were not Stukas but light two-engine bombers—probably Dorniers, even more maneuverable than Stukas. There were six or eight of them, flying low; two pounced on us as we were getting out of the car, and peppered the road with machine-gun fire. I scurried forward some ten yards and lay down in the ditch. The driver did the same, having gone in the opposite direction. In front of us, bombs were dropping. I took stock of our position: a clearing, and a clump of oaks on a little hill about a hundred yards to the left. Machine-gun fire cut off the road, the bullets pinging off the splintering rocks above my head. I made another attempt to run to the woods, but a second plane dove at me as if to slice me with its wing; I dropped down again. Two or three times I tried to get away, but each time the planes pinned me down with bursts of fire. My humiliation at my impotence was greater than my horror.

The area of the attack was some ten kilometers in diameter, and as the planes came one after another, they attacked everything in sight. Our limousine was one of the more attractive targets. It lasted—I looked at my watch—all of twenty minutes. Then they dropped three more bombs. I heard them thump on the ground about a dozen yards from me and calmly awaited the thunder. Dirt and sand splattered my coat as the whole area roared, the earth and clouded sky caught in the same vortex. No respite, not a moment to think, yet only presence of mind could save us. I got up and ran another dozen yards, then spotted a water conduit across the road. I shouted to the driver to run over to me, but he didn't hear. I crawled into the conduit, where I was safe at least from the machine-gun fire. But I couldn't see anything, so I got out again. I began to maneuver: when a plane came by, I jumped into the hole, and as soon as it left, I got out of it. A half hour had gone by when I realized that the planes were pinning us down until a motorized unit could arrive. Then suddenly I heard a different, deeper sound of machine guns, not from planes. I ran to the car. There could no longer be any doubt that the Germans had broken through and were nearby, maybe just around the bend. Already their bullets whistled overhead. I shouted to the driver to start the car. He shouted back, while running away, "It's damaged!"

"Turn the car around or I'll kill you!" I yelled, slipping off my rifle. The driver was prepared only to chauffeur officials and big shots, not to take to the forest, therefore he had chosen to desert. But at my threat he came back and turned the car around in no time. The only damage to the car consisted of two bullet holes through the roof. The car started off. The planes were still flying overhead but now I no longer cared. I was overcome with shame: I was fleeing in a car like the officials of the old regime. I had to get to Ranković, we had to find Tito. On the way we encountered a roadblock—an overturned cart, the horses all shot up. I ordered the driver to proceed to Čajetina and tell the local command of the German breakthrough at Zlatibor. The driver listened attentively and agreed to go. We parted when we reached the little woods where I had left Ranković with his unit.

But there was no sign of Ranković or the unit. I looked around: to the left there was a clearing—they could not be there. To the right there was a rocky ravine: it seemed unlikely that they had descended that breakneck slope, though as it later turned out, that was exactly what they'd done. Just then I caught sight of five or six Partisans crossing a crest by the woods at a run. It occurred to me that they might be the rear guard of Ranković's unit, so I followed cautiously.

When I had arrived at the hillside slope and was still about a hundred yards from the top, I heard heavy machine-gun fire behind me and the whistling of bullets overhead. I turned around. At a curve in the road, about four hundred yards away, stood two armored cars, their machine guns firing away. In a split second I realized they were shooting at me—there was no one else around. I ran off as fast as I could, the bullets raising puffs of dust around me. It seemed that my legs were paralyzed, that I couldn't run as fast as I should. I remembered stories from my childhood of men who perished because their legs gave way—and their minds along with their legs. Probably I too was going to die, and nobody would even know where. Suddenly my rifle became heavy and awkward; I wanted to get rid of it, and I might have done so had I not felt shame and a sense of responsibility. I decided to run in zigzags—first a short distance, then a long one. I must have been running very fast, for I was panting hard when I leaped over the crest of the hill with a sense of defiant joy.

I descended the other side at a lively pace. The road curved around the hills, so I had no reason now to fear the Germans. I came across the discarded barrel of a machine gun, and I lifted it onto my shoulder. As I came down into a dale, at the end of the path I found a dozen tired and dejected Partisans. "Who threw away a machine gun?" I asked. Nobody answered, but I noticed a weaponless lad dressed partly in peasant clothing. "You threw it away," I said to him. "Here, take this, and don't let it happen again." These Partisans were from various units. No one knew anything about Ranković or his troops. They had no idea who I was, but by my appearance and bearing I must have impressed them, and I organized them into a squad. Soldiers stop being soldiers as soon as they lose their command.

Even before the retreat from Užice, Kraljeve Vode (today Partizanske Vode) had been selected as the new seat of the Central Committee, that is, the Supreme Staff. It was getting dark, and no one knew the way, not even a young man from Zlatibor who thought that, with the fall of Užice, the war was over and he was on his way home. He did lead us to his village. I deployed the men for combat and went to the nearest house. I knocked. Inside someone moved, and the door finally opened. The man of the house refused to act as a guide, he said, because he had already retired for the night and was not feeling well. All my helpless anger against the Germans, which had been building up the entire day, boiled over, to say nothing of the wrath I felt for this peasant who should have been out fighting for his country, instead of bedding down with his wife. "You get ready," I shouted, "or I'll kill you even if you have nine lives!" I certainly would have acted on this threat had not the peasant immediately agreed, begging only that his neighbor join him. Later we chatted with the peasants agreeably and offered them a smoke along the way.

Kraljeve Vode had two or three little hotels and a few villas, the largest belonging to a lawyer named Pavlović. It was not difficult to guess that it was there that the members of the Central Committee had gathered. On the road leading to one of the villas, there were supply trucks, the wounded, and refugees. There I stumbled on Vladimir Dedijer, who was threatening someone with a big pistol. He was glad to see me and explained, "Imagine, this man is spreading the rumor that Užice has fallen!" I called him to one side and whispered to him, not without cynicism, "The Germans have reached Zlatibor. They're already in Mačkat!"

It was around midnight, but Kardelj, Žujović, Filip Kljajić, and Father Vlada Zečević were still up. I brought them the news of the breakthrough. They urged me to get some sleep. I went upstairs and flopped on the bed. I couldn't sleep, but just lying there rested me almost as much. About two in the morning I heard a commotion. I ran downstairs. Tito was there in dirty boots, crestfallen and dejected. He told us of his experiences. He, Captain Hudson, and an escort were barely out of sight of Užice when the planes began to attack. They crouched behind a wall as the planes strafed all around them, but apparently they were not discovered. After a while the escort got up and spotted a German infantry unit. In fact, they heard orders shouted in German. Then they scrambled down to a brook and escaped by following a ravine.

It was clear to me that the Germans would continue their advance in the morning. Tito agreed to a further general retreat. There were no organized auxiliary units to slow down the German advance. Tito, Kardelj, and Žujović drove off, and the rest followed. There was as yet no news of Ranković. I hastily organized the loading of twenty chests of silver onto a truck, then drove off to find a suitable place for their safekeeping. I entrusted to Marijan Stilinović, a long-time Communist, the transportation of our printing press, paper, and the still unbound copies of Stalin's *History of the CPSU* (*b*). We had no luck at all with that book, though we treated it as if it were the Holy Scriptures. Some devilish thing would always prevent us from printing it.

Some fifteen kilometers from Kraljeve Vode there was a breach in the road. The Central Committee members got out of their cars and made their way on foot to the village of Draglica; some went on to Sandžak that same day. As for me, I found a good spot next to a stream and buried the silver. It was hard work, digging in the cold, and carrying the chests that weighed close to a hundred pounds. Snow was falling steadily, promising to cover our tracks. Indeed, the chests remained concealed until 1943, when a flood uncovered them and the peasants stole the silver.

Stilinović had little luck with the printing press and the *History of the CPSU* (*b*); his truck was requisitioned for the wounded, and the printing press and paper were unloaded in a cottage by the road. At dawn I came across Stilinović, sitting in despair before the cottage. I picked him and his group up in my truck.

At Borova Glava, where the road was cut off, I noticed boxes of supplies in the ditches waiting to be transported from there by horses and men. Tito's limousine was also there, and not overturned. That was the very spot where the truck had picked us up so recently when I had come from Montenegro. I made my way along a ridge, behind which were the village of Draglica and a tavern with some sheds. The tavern and the sheds were jammed with Partisans. It looked more like a refugee camp than an army. In a cottage nearby I found Tito with Zdenka and a guard. He was tired and unhappy, concerned about the wounded and the units that were defending Užice. We took a walk and talked about our problems and responsibilities. Tito agreed that we should try to get the printing press out by evening, and gave orders that the supplies be retrieved from the ditch beside the road. The day was cloudy and it was snowing a little, so the planes let us alone. I lay down, but got up after a couple of hours to supervise and expedite the removal of supplies. The party members, mostly workers and students from Užice, submitted to discipline and got out all there was. The night found us assembled and busy.

That evening Tito proceeded on foot to Sandžak. I picked out about twenty comrades, among them Dedijer; around ten o'clock that night we boarded a truck, and with our headlights off drove back to get the printing press left by the roadside. Dedijer remarked that I was taking along too many "responsible comrades," to which I replied, "I'm preserving our cadres!" We did not know where the Germans were, but met some peasants on the way who told us that two German automobiles had been there and had gone back. Our printing press was still some five or six dangerous kilometers away. The road, which stood out dark against the barren white crags and gray against the lonely black pines, led us all too slowly toward our goal, and all too quickly into an ever vaster uncertainty. We knew that we might encounter a German vanguard, but placed our hope in the night and in the military conventionality of the enemy—the indispensable and sure allies of guerrillas. Two automatic rifles on the roof, and some twenty rifles and submachine guns, were ready to answer any German volley.

But everything ended well for us. The Germans were encamped in Kraljeve Vode. Some firing was heard from there in the morning, and we had a terrible presentiment that they had killed the gravely wounded whom we hadn't managed to evacuate. Reaching our destination, we hastily loaded the printing press, the newsprint, and the *History of the CPSU (b)* and returned to Borova Glava. We unloaded our treasures, and sent the truck crashing down into a ravine.

At two or three in the morning of December 1, I finally settled down on the floor among the soldiers, in the little house where Tito had stayed the day before, and fell into a sound and peaceful sleep. But I was up with the dawn. I had to awaken the men, and get on with the evacuation from the ravine over trails and footpaths to Sandžak. Exhausted, the soldiers got up from their fires and kettles and began to pull out. A stocky worker from Užice, a reliable and cheerful fellow, took a group to the ravine below the road, while two newly formed companies moved out along the hillsides in order to protect the road to Draglica.

Around eight o'clock I was in the tavern having breakfast, when firing was heard not far away. We ran out: it was coming from the other side of the hill, where the road broke off. People were fleeing past the houses and out onto the road and the field, while those two companies were streaming down the hillsides in panic: though inspired by the same idea, units whose men don't know one another aren't capable of skirmishing. Revolutionary guerrilla units develop their own sense of the possible, which commanders must understand if they wish their commands to be carried out. Assembling a group of men and assigning reliable leaders to them doesn't mean one has a battle-ready unit, particularly when retreating from defeat.

I ran up to the ridge above the inn to see what was going on. Vlado

Dedijer and Mirko Tomić were with me. The Germans had reached the break in the road in a couple of trucks and armored cars. They were at a distance of some five to six hundred yards, and we could see everything. A German infantry formation was moving into the ravine below, while bullets whizzed all around us. The men who had been retrieving the supplies couldn't get out: they all died—for the printing press and the *History of the CPSU (b)*. The guards we had posted on the road had either fled or were hiding in the houses.

We had no choice but to get away ourselves. The night before, after his departure, Tito's horse had been brought to us. It was a magnificent animal, and I was ashamed to abandon the horse of the supreme commander. I stopped in front of the stable to lead the horses out. Fleeing along with Tomić, Dedijer cried out to me, "Run, Djido, or the Germans will kill you!" But I didn't run. Dedijer felt badly about this—as if he had deserted me. He told me after the war, when he was at odds with the Central Committee, that Tito said I had complained about him. If I said anything, it was in jest and well intentioned: Dedijer acted more rationally than I, and one certainly cannot accuse Mirko Tomić of cowardice—a man who lost his life in Belgrade in a shoot-out with police agents.

In front of the inn were two cases of silver: I told the innkeeper's wife what was in the cases, and while I was saddling and bridling the horses, she raised a trapdoor under a pile of manure in one corner of the stable, and threw the cases into a pit. Since the territory was then occupied by Nedić's men, I was to send troops later at night to recover the cases. We had a patriarchal, fetishistic attitude toward silver. We always dragged several cases around with us, in the belief that we might need it in some critical situation, even though no one valued silver currency much any longer.

I brought out the horses. Down the ridge, disheveled and breathless, ran "Vera Blondie"—Dara Pavlović, a robust girl of proud courage—shouting to me, "Djido, my godbrother, don't leave me!" This "godbrother" was pure invention. (She had escaped from the ravine, but would be taken prisoner later in Serbia and shot with her sister—though not until she had danced a Cossack dance just before, out of spite.) She didn't know how to ride a horse, so I helped her mount one, placed her feet in the stirrups, and told her to hang on tight to the pommel of the saddle. I tied her horse to mine and set off at a trot. The horses were skittish, but Vera Blondie held on tight.

Below the crest we caught up with Dedijer and Tomić. Dedijer was limping, for his leg wound hadn't quite healed. We began to chatter joyfully, convinced that we had escaped the German onslaught. But as we reached the top of the ridge, a machine gun peppered us from a hillock above the inn, and a plane droned in the sky overhead. And so it went into the afternoon: the Germans chasing us from one hill to another. I observed them through my binoculars: young men, collars unbuttoned, agile. The German higher ranks may have suffered from conventionality, but the lower ranks had initiative and resourcefulness. No wonder they disregarded our sporadic firing; they would have advanced boldly even if it had been more organized. We had to rally our scattered forces and push ahead, gasping for breath before the pursuing Germans. It seemed to me that every German was chasing me personally, and no doubt the others had the same feeling: the war and the enemy assume their true likeness when they become one's personal fate.

By evening they had driven us across the Uvca River. This was the boundary of the Italian zone, but we couldn't rely on the Germans caring much about that. I rushed into the village of Radojnja and hastily gathered some fifteen party members with submachine guns to guard the bridge over the Uvca, while German machine guns kept on firing, and the fugitives and soldiers scattered over the hill overlooking Radojnja. But in the end the Germans did stop. Having cleared their own territory, they left it to their Italian allies to contend with the remnants.

Tito and the Central Committee members were quartered in several houses on the hill overlooking Radojnja, where I went to see them that same night. Tito was glad that I had rescued his horse, but passed over this quickly and told us in great excitement, "I received a report—Ranković is rallying the troops!" As it turned out this was not quite so: Ranković had gathered together some three hundred soldiers and had sent many of them to Radojnja. But Tito grasped feverishly at any good news.

It was decided that some units must return immediately to Serbia: the Serbs pressed impatiently for this decision, being homesick and disenchanted by the poverty and bareness of the Sandžak mountains. Captain Hudson also wished to go back with the Serbs—so as to rejoin Draža Mihailović. We were correct in our dealings with Hudson, though there was talk that he was partial to Draža and that he was misinforming his government: how else could one explain that constant blaring by the BBC, and by Radio Moscow as well, of misinformation and partiality? I whispered to Tito that, while inspecting the evacuated supplies and food in Radojnja, I had given orders that Hudson's radio transmitter be concealed, and told Hudson that we were still looking for it. Tito remarked, "That doesn't make sense—after all, he's an ally." The next day I returned the "recovered" transmitter to Hudson.

We subsequently learned that on November 30 the Germans had killed our wounded in Kraljeve Vode. Dedijer could not have known this at the time, though he noted it in his diary under the correct date. It was estimated that there were some ninety wounded, with an undisclosed number of medical staff. The wounded had scattered over the bare hillsides, and the Germans hunted them down, searching every ravine and crevice. This held the Germans up, and that is why it was not until December 1 that they drove us back toward Sandžak. The leaders were bitter and depressed over this, all the more because, just the day before, we had released 250 captured Germans. Even so, we never regarded Tito's decision to release the captured Germans as wrong. What could we have done with so many prisoners in those bare mountains? We recalled our "educational failures" with German prisoners: after nearly two months of political education work with them, directed by a Comintern expert, when asked whether, now that they knew about Nazism, they would kill the young and the old, they all answered, "Befehl ist Befehl" (Orders are orders). The massacre of the wounded was a decisive turning point in our dealings with the Germans: thereafter the Partisans gave the Germans measure for measure and killed their prisoners, except in special cases; nor could we in the leadership come up with any reason to oppose this.

Though we suspected it, we did not know at the time that Željko Djurić, secretary of the Regional Committee for Užice, had been killed, along with Dragojlo Dudić, president of the National Liberation Committee for Serbia. We were puzzled by the disappearance of the chief of the Supreme Staff; it was suggested that perhaps he had gone over to the Chetniks, though just before the end of the war I heard that the Germans had captured and shot him. We heard about the losses suffered by the Workers' Battalion of Užice, but we did not know that it had perished almost to a man. War has a way of quickly pushing misfortunes and defeats aside by generating new and greater disasters. As for the events that cannot be forgotten, they are passed over in silence until they become a legend.

It was past midnight when I left Tito and the leadership and went down to Radojnja. December 1—the Day of Yugoslav Unification, a national holiday—went by without our remembering it, though we were waging a pitiless war so as to inherit that very state. I spent the next day organizing the return to Serbia of the Posavina Regiment. There were only a hundred in the regiment, yet they were well disciplined, and exceptionally homogeneous and unique. They reminded one of the photographs of their fathers in the war of 1914—well outfitted and groomed, calm and determined. They drove their vehicles up to the roadblock, then loaded everything onto their big sorrel horses. The Posavina Regiment was accompanied by Father Vlado Zečević and Captain Hudson, who was later to join Mihailović and the Chetniks. Vera Blondie also left. The Posavina men got through to their homeland and perished there.

All this reveals that our leadership did not comprehend the change that was taking place in Serbia. For this offensive the Germans had picked a corps that had been slated for the Eastern front, and they had added to it their own occupation troops and Nedić's. At the time of the offensive, Draža Mihailović met with Wehrmacht representatives and promised that his forces would not work with the Partisans, as long as the latter fought against the Germans. We did not know this at the time, but knew that Draža had disbanded his army, and that he and a group of followers had gone underground. Nedić established his authority wherever the Germans went. But the organization of his authority did not proceed quickly or easily, and the Chetniks were recalled as soon as the offensive against the Partisans had started. In the villages Nedić's authority became intertwined with Draža's. Village guards were established, prison camps filled; lists of executed "Communists" and all who favored the Partisans were published daily in the towns and in pro-German newspapers. Villages where authority was weak—where the people stood together despite the pro-German police apparatus—were wiped out in a massive reign of terror. Without the co-operation of Nedić and Mihailović, the German forces would have been inadequate and ineffective beyond their lines of communication. The peasants would not have waited for the Germans to come, there would not have been massive espionage against the Partisans, and even more massive persecution of them. Yet our men, and even our leaders, still had a picture of the Serbia of the early days of the uprising, when in every village small groups of Partisans were met with open arms, held meetings, and attracted volunteers.

The area in which Tito and Kardelj were staying was not safe. On one side of the hill stood the village of Radojnja, which had no organized units. On the other side stood Nova Varoš and the Italians. They moved to the village of Drenovo, to get closer to the liberated territories in Montenegro and Bosnia. I remained in Radojnja to join the leading

Serbian comrades in rallying units which were retreating from various parts of Serbia. On December 3 Ranković arrived, calm and collected as though nothing had happened.

We were quite bogged down with "incidental" problems at this time—orders for food, the transfer of units, the issuing of passes—and Ranković even ferreted out a spy. Most pressing of all, however, was the blocking of the way from Nova Varoš. We whipped two regiments into shape by filling in two which had somehow preserved their organization during the retreat. It was in the early afternoon, and a winter rain was drizzling down. The regiments were mustered in front of the schoolhouse. "Give them a speech," Ranković said to me. "You know you're better at it." The soldiers had had some lunch but were exhausted, and in tattered clothing and footwear. I gave them a speech, short and sensible. They listened quietly and with understanding. Then I explained their assignment to them. "See that hill there? You have to take it and prevent the Italians from attacking. We don't have any place to retreat to. The only free territory for us is the one we're standing on, and holding by the force of our arms!"

The commanders issued their orders, the regiments were on the move, singing, "Little raindrops, don't you fall on me. . . ." Of all the armies that ever fought on our soil, I would give first place to the men of Serbia because of their patience, common sense, and ability to adapt to all conditions. They are true warriors; a people who regard the army as their own, and war as something natural. Many times I heard them complain, but never saw them in a panic, nor did they ever lose their sense of humor.

But our two regiments never had a chance to show what they could do: the Italians withdrew from Nova Varoš. Their fear was greater than our unpreparedness: local Partisans had captured an Italian patrol along with a major. And there was reason for them to be afraid: a considerable force of ours was on its way from Serbia, and another was breaking through from Montenegro. With the Italian withdrawal we gained a new base which, however poor and cramped, temporarily served our needs. Bits of liberated territory were linked together, simplifying our communications. So when Ranković, Lola Ribar, and I set out on December 7 for Drenovo to see Tito, we journeyed by way of Nova Varoš and not across footpaths at night.

On entering Nova Varoš we asked if we could have lunch with a regiment which was just being served. Three soldiers offered me their rations, while Ribar and Ranković were holding the horses and talking to the commander and the political commissar. Beside the soldiers stood a middle-aged, solidly built man in civilian clothes whom I recognized as Mika, a Jew, the "stocking king" from Belgrade; I believe that his surname was Albahari. He organized the lunch and supervised its distribution. He recognized us by our leather coats, our belts, and horses, and shouted to the cook, "Lay it on for the comrades from headquarters!" As we accepted the food, I cautioned Mika with mock severity, "Remember, we are not a bourgeois army!" Ribar and Ranković chuckled. After that, Mika's words became a favorite saying—whenever we got, or failed to get, anything special. In actual fact, we ate out of the common pot—that is to say, very badly, since only operational units received food of any quality. It was only when we stopped in some little town that we reverted to a more agreeable diet. The single exception was Tito, thanks more to our solicitude than his wishes: he ate better. He even had a cook, and we took along a cow to give him milk.

The three of us stayed in Nova Varoš longer than intended, so that we arrived in Drenovo in the middle of the night.

We knew of the failure of the Montenegrin attack on Plevlja, but we were not aware of the extent of the failure or its effect on the prestige of the leadership, especially in connection with the spread of the Chetnik movement in Montenegro. Tito was more critical and caustic over Plevlja than we were, but not nearly as much as in his postwar declarations. His criticism grew as the negative consequences of the battle became apparent, and particularly after Arso Jovanović, the commander at Plevlja, went over to Stalin's side in 1948. At the time of my dismissal in Montenegro, Milutinović had told me that the Supreme Staff intended to link up the liberated territory of Serbia and Montenegro by driving out the Italian garrisons in the lower course of the Lim. Plevlja was a key point in that junction. My impression was that the attack on Plevlja, as Milutinović described it, was prepared with Tito's knowledge. But it is also true that, in a letter, Tito cautioned Milutinović that our army was not ready to attack fortified places, and that Plevlja should be besieged first. I also cautioned Milutinović, but my words did not carry much weight. Nevertheless, Tito was not categorical in his disapproval of the operation, nor was he realistic or consistent in his views. Thus in his order of December 13, twelve days after the failure at Plevlja, he ordered that the Montenegrins liberate Prijepolje, although it was a fortified town. The immediate danger, and the need of a secure crossing into Bosnia, were likely uppermost in his mind.

Plevlja was attacked on December 1 by some six thousand Montenegrins without heavy weapons, and by units with little military experience. The men of Lovćen penetrated the town and held out for a full day, but some units scattered before the first shots. The Italian commandant was courageous and stubborn. Our losses were considerable: some three hundred dead, and two to three times as many wounded. But this would not have been nearly so disastrous, had it not given rise to Chetnik charges of the "inexpert" and "imprudent" handling by the Communists. In Montenegro the Chetnik movement would have spread anyway, but Plevlja brought out hitherto latent counterrevolutionary forces. On October 15, 1941, Draža Mihailović had named Major Djordje Lašić Chetnik commandant of Montenegro. Blažo Jovanović conducted negotiations with Lašić, vainly offering him positions and collaboration if only he would desist from forming separate units: those who cannot live together can hardly fight together.

As time went on I grew close to Arso Jovanović, even closer than when we commanded together in Montenegro. I spoke with him about Plevlja many times, though he found it disagreeable: his view of the defeat came down to what I have already presented. Yet no one criticized Arso Jovanović for Plevlja until after 1948, when he sided with Stalin. In fact, Tito had met Arso at the time of the Plevlja battle, liked him, and made him chief of the Supreme Staff, which post he held to the end of the war. Thus Tito's more recent claims that he wanted to try Arso by court-martial, which did not formally exist at the time, arose from a subsequent re-evaluation of the past, which is typical Communist pragmatism. The only person who proposed an investigation of the defeat at Plevlja was Moša Pijade—in a letter to Tito from Durmitor Mountain. But Pijade was also generally dissatisfied with his subordinate role in Montenegro. And though Tito liked and respected Pijade, he didn't consider him capable of more independent and responsible assignments. To this day I hold that the chief culprits in Plevlja were the party leaders of Montenegro, especially Milutinović, in that they overestimated the capacity of their predominantly "popular" irregular army. As far as Arso Jovanović is concerned, I believe that he was a brave and educated staff officer and a good friend, as well as a talented military leader, though perhaps better suited to regular warfare: as co-ordinator of operations in the Littoral and Istria in 1945, with relatively well-equipped and complete units, he was both imaginative and effective. The same was true in his organization of the Supreme Staff and staff services during the transitions to frontal warfare in 1944–1945.

With Tito in Drenovo we found Bajo Sekulić and a regiment of

Montenegrins who had fought in the battle of Plevlja. There was some fear that the Montenegrins would want to return to their homeland, and Kardelj, Sekulić, and I gave them speeches: we told them they could return to Montenegro, but none volunteered to do so.

Late that day—December 7, 1941—we held a Central Committee meeting. That meeting in Drenovo has scarcely been noted in semiofficial memoirs and publications, though it is among the more significant events in party history. But what in fact was the Central Committee? The Plenum of the Central Committee, which was elected at the Fifth Conference in 1940, did not meet until after Stalin's denunciation of the Yugoslav leadership in 1948, and even then it was incomplete because several of its members had been killed. The Politburo of the Central Committee, on the other hand, met frequently before the war, but less so during the war. We were all together—at least the majority—and so current questions were decided as we went along, while meetings on important questions were also attended by Plenum members who happened to be with us or in the vicinity (Pijade, Žujović). The Central Committee was, in fact, the group around Tito, the most prominent among whom were Kardelj, Ranković, and myself, largely because in time we became close, having been with Tito since he had become leader of the party.

This meeting was held under extraordinary conditions—in the corner of a room, around a little table which one could reach only by stepping over the bodies of sleeping soldiers. It was stifling, and dark, except for a gas lamp. On the agenda was an appraisal of the political situation and the designation of further assignments. Without any reports, in brief dialogues in which Kardelj held forth at greatest length, we agreed that the armed struggle against the occupation had developed into a class war between the workers and the bourgeoisie. This development was the result of our domestic conditions as well as of international changes: the Red Army's counteroffensive at Moscow, Britain's passivity toward Hitler, and tensions between the West and the U.S.S.R. We would, of course, continue our armed struggle against the occupation forces, not only because we thereby affirmed ourselves as a patriotic force, but because this was an integral part of a worldwide class struggle led by the Soviet Union. I too favored this appraisal: I saw no better one, though I found it difficult and awkward.

Our new position in turn dictated the ideological and organizational strengthening of the party, dependence on the working class, and its alliance with the poor peasantry. Tito presented the view that we should merge the units from Serbia into a brigade, which would not be tied to any specific territory but fight wherever needed. Tito then unexpectedly announced that he considered it necessary to resign as party secretary, and have Kardelj assume that post. "The party should not shoulder the responsibility for all the failures," he said, then added as if taking an oath, "I will continue at the head of the party and devote myself entirely to its advancement."

I barely had a chance to interject, "But that doesn't make sense," when Ribar and Ranković expressed much the same thought. Kardelj took the floor, also rejecting Tito's self-abnegation, but less with emotion than reason: in the given situation, Tito's act might be interpreted as an admission of an incorrect policy. We calmed down and, discovering within ourselves the inescapable shadow of Moscow, elaborated on our reasons: Moscow would not understand Tito's resignation and would conclude that there was disintegration within the party and the revolutionary movement.

Tito was clearly pleased with our reactions. Yet one should not conclude that he was simply "testing" us. No, he was guided by a feeling of responsibility—especially for the failure in Serbia, though he had no reason to feel responsible for Plevlja. Tito was to threaten to resign once more—as premier, in the spring of 1948, when we received those critical letters from Molotov and Stalin. On that occasion I could tell by his sly smile that he was indeed testing us; all of us exclaimed that it was out of the question—all but Žujović, who remained silent because he had already opted for Stalin.

In the course of the meeting, Tito collected himself as if nothing had happened. He is a person of quick, keen reactions. A rapid return to reason and recovery following depression was always one of his essential qualities. Was it duty or the advent of some new misfortune that prompted this? Or is it a quality of all born rebels and rulers?

Late that same night Ranković and I returned to Radojnja to reorganize the thinned-out and dismembered Serbian units. Five or six days later the Italians surprised Tito at Drenovo. Our side might not have come out so well, had the enemy known who was involved. Tito was bitter and disturbed: the local security unit had retreated without firing a shot at the Italians, who were no more than 250 to 300 yards away. Under fire, Tito rushed to the nearby woods and the Italians looted the house Tito had been staying in, and killed the host's daughter-in-law; she had her baby at her breast.

That was when our feud with the neighboring village of Rutoši began. The Rutošani killed several Partisans who had just come out of Serbia. Why was Rutoši a Chetnik village and Radojnja a Partisan village, even though they were neighbors? The people of Radojnja regarded the people of Rutoši as backward and bad. Besides, Radojnja had two full-fledged Communists—Voja Leković and Momir Bošković, whose households were the bastions of the village and of the regional party leadership. Rutoši had no such men, and its people persisted in their ways. Or did these loyalties have their roots in ancient tribal conflicts? At any rate, bitterness against Rutoši had accumulated since the withdrawal from Serbia. During my two-month stay in Nova Varoš, the Partisans were to take Rutoši at various times, though they never surprised the people or engaged them in a fight. But they always brought back plunder—cattle, dairy products, and grain. In time they got used to it and were glad to go "into action" against Rutoši whenever they ran out of food.

Tito came to Nova Varoš on December 15. We held a Central Committee meeting in the schoolhouse next to the church. We were all in a good mood, though with a touch of sorrow: we were parting. The greater part of the Serbian units, now reorganized, were leaving for Bosnia. Tito and the rest of the Central Committee were going with them—all except me. I was to stay behind with the party leadership of Serbia, to which the Central Committee was naming new members—among them Mitra—because some of the old members had been arrested in Belgrade and some had joined the Partisans elsewhere. Our assignment was to organize the units still arriving from Serbia and to return there with them.

Yet alongside this major and flattering assignment there were some smaller and not so flattering tasks. One was the execution of an Italian officer in reprisal for killing of the young mother at Drenovo, about which the Italian command had to be informed. The other was the proposed execution of Mika Djordjević, who had been accused of subversion within his unit and ousted from the party as a fractionalist.

Tito formed the First Proletarian Brigade out of the Serbian units and two Montenegrin battalions on December 21, Stalin's birthday, which was not a coincidence. However, this did not please the Comintern and Stalin; in a dispatch they protested against the formation of a "proletarian" unit. We justified it on the grounds that such a model unit would boost morale. The brigade was proletarian not in a literal but in an ideological sense: the workers in the brigade were a minority; the majority was made up of party members and Communist Youth. But the designation "proletarian" was a recognition of the ultimate goal. This was the first permanent unit whose territory included all Yugoslavia, and the first for which the party's aims were the only aims. It consisted of 1,200 men armed with automatic weapons—not a large number, unless one keeps in mind that they operated as an adaptable military organization. Its creation demonstrated Tito's talent for military organization.

The date of the founding of the First Proletarian Brigade was celebrated as an army holiday until 1948, when we quarreled with Stalin and therefore moved it up one day to December 22. Even when we were dealing with questions involving our own destiny—when we were valiantly struggling for life—we did so with Leninist motivations and under the shadow of Stalin. Along with many other ill turns, Stalin chose to be born on the very day which, sixty-two years later, would coincide with the need of Yugoslav Communists for a revolutionary army. I too was enthusiastic over the First Proletarian Brigade and the date of its founding. Are not the exploits and sufferings of our own flesh and blood also real ideas—the most real of all?

3

The conclusions of the Central Committee on the evolution of the struggle against the invader into a "class" civil war were formulated—not accidentally—on the day of the founding of the First Proletarian Brigade. This was done in the form of a letter from the Central Committee to the party leadership in Montenegro.*

The letter was written by Kardelj and signed by Tito, which was evident from the style and from the handwritten note at the end of the letter. It was a practice of the Central Committee—though admittedly not a frequent one—to have a letter written by the Central Committee member most involved with the relevant question, and then have Tito sign it. (Likewise, in 1943 in Jajce, I composed a letter that was sent to Svetozar Vukmanović-Tempo in Macedonia.) This detail is significant, for it reveals that in Tito's mind such an appraisal, though he agreed with it, had not yet matured into simple, practical conclusions. The letter criticized the party leadership in Montenegro, and Milutinović, for committing "new ultra-left and worse errors" (it follows that I must have committed "right" errors—M. Dj.). It then went on to reproach them with clever sophistry in treating the Partisan units as "an armed force of the party" rather than as a force under the party leadership, and the people's committees (the civil authority) as the creation of the Communists rather than as a "revolutionary democratic authority . . . under the party's leadership." They were also criticized for the defeat at Plevlja. The responsibility for disregarding Tito's warnings against frontal en-

* *Zbornik dokumenata,* Vol. III, Book 1 (1950), pp. 370–71.

counters was laid largely on Milutinović. The letter referred to Moša Pijade's letter to Tito of December 7, 1941, which called for an investigation into the defeat at Plevlja. I believe that Moša's letter had no essential bearing on the criticism of the Plevlja defeat, but it was like pouring salt on wounds, and cast doubt on the ability of the leadership in Montenegro. Tito did not think much of Pijade's judgments, but he did not ignore his reports either.

Finally, and most important of all, there were several general remarks in the letter regarding a number of changes and developments:

> The victories of the Red Army [reference is to the repulse of the Germans from Moscow—M. Dj.] have inaugurated a new stage in the development of the world political situation in the sense of a sharper class differentiation. This is evident among us as well as in the ever greater consolidation of reactionary forces against us. . . . Our party faces new tasks: the more intense unification of the working class with the petty peasant masses; the elevation of the role of the party; a greater popularization of the U.S.S.R. and the building of socialism; the internal strengthening and expansion of the party; the formation of our proletarian military units, which will have a decisive influence on the outcome of military encounters. In this regard the Central Committee has taken the first step: on this very day the First Proletarian Brigade was formed as the nucleus of similar formations in the future. The banner of that brigade is red with a hammer and sickle on it. This is in fact an armed force of the party [and yet the Montenegrins are reproached in the same letter for claiming the same for their own units—M. Dj.].
>
> In other countries the situation is taking a similar turn. Our strong units have remained in Serbia and are waging a struggle against the Nedić forces and the Chetniks. As for the Germans, according to our unverified reports, there are not many. It appears that we shall soon be in a position to undertake a decisive clean sweep of Serbia. Our units are now largely reorganized and battle-ready. The morale of the units is good.*

I received this letter in Nova Varoš and digested it with the leadership of Serbia and Sandžak, and they in turn with lower party organizations. It set our course of conduct in Nova Varoš—what other course could there be, for thoroughly dogmatized and trained functionaries to follow? And if our conduct was not as violent and bloody as in "dogmatic" and simple-minded Montenegro, this should not be ascribed to anyone's "humanitarianism" and "understanding"—not even mine—but to the absence of social and political differentiation. Our territory was small—a little town surrounded by mountain villages, most of them either recep-

*_Zbornik dokumenata,_ Vol. III, Book 1 (1950), pp. 366–71.

tive to us or inactive—so that we had no one against whom we could "correctly" apply "the spirit of a sharper class differentiation."

Even the most remote and backward little town offered some luxuries to guerrillas from the hills and mountains. That was certainly true when we took over Nova Varoš. That mountain town, whose Moslem and Orthodox inhabitants lived apart but in peace, welcomed the Partisans cordially, probably because—unlike the Ustashi and Chetniks—we did not threaten any creed or nationality.

However, the influx of Partisans did alter the town's routine, sluggish way of life. Our people were too advanced for such primitive conditions, and they threw themselves into making the town over. Our women could not accept the centuries-old "backwardness" of the Moslem housewives, whom they rounded up and dragged to conferences, while our activists were not happy unless they saw a Moslem imam and an Orthodox priest sitting side by side at public meetings. The most popular and best attended were the cultural events, with recitations, skits, and even ballet. There were also dances which lasted late into the night. In these activities a great regard was shown for religious differences and also for the sensibilities of party puritans; serious discussions were conducted in committees regarding the excessively short skirt of one ballerina, and my expert opinion was sought. The ballerina came out of it well, though the party's matrons continued secretly to carp at her. I actually had very little to do with this, other than to encourage a more intense and varied activity.

There were also executions, mostly of adversaries in the vicinity or spies from Serbia. The executed included people of both faiths, and the executioners were their co-religionists. There was no need for me to get mixed up in this, and so I occupied myself with propaganda and the organization of units. Someone told me that the legs and arms of the executed were sticking out of their graves, and I cautioned the Sandžak officials to dig more carefully, to avoid offending the local people. Though there may have been some improvement subsequently, I don't believe that anything much was done about past cases. Partisans didn't bury even the bodies of their own comrades with much care, unless some very important person was involved. Where their enemies were concerned, there was an antitraditional attitude toward death and dead people.

We sent a letter to the Italian command in Prijepolje informing them of the execution of the captured major in retaliation for the murdered young mother in Drenovo. I wrote most of the letter, and I cited the "unmilitary" reprisals of the Italian army against our population. The

Italians replied in detail and with such wit that we laughed with all our hearts. The Italians didn't justify their own misdeeds but listed ours for us. Thus they cited that during the July uprising, in Ceklin, the Partisans doused the dead Italians with gasoline and burned them, and then drove swine there to tear the corpses to bits. I later asked Peko Dapčević about this episode, since he had led our people in that battle, and he confirmed it: "Yes, we soaked them in gasoline and set fire to them to keep any infection from spreading in the July heat, and the swine came down from abandoned villages after we withdrew." Everyone sees and justifies cruelties and violence from his own point of view. To our demand that the recognized rules of war be observed, the Italians replied that we were not a regular army, and that we ourselves were responsible for the trials and tribulations of the people. They expressed wonder at our senselessness. "You do not have to listen to Axis radio stations," they wrote us, "but to the BBC to learn that the Axis armies are outside Moscow, that the fall of the Soviet Union is inevitable, that the misfortunes of your people are in vain." I had once seen the Italian officer in passing, while he was taking a walk in front of the prison; a good-looking man, in an Italian way, and dressed in a neat uniform. They told me that he behaved calmly and bravely at his execution—probably reconciled to his fate, that he was the one to pay with his life for the senseless murder of a young mother in some Balkan mountain wilderness.

When the unit arrived to which the accused Mika Djordjević was attached, I sent a courier to bring him to my office in the schoolhouse. Before me stood a rather tall blond youth. His metal-rimmed spectacles made his perturbed expression seem intelligent. I ordered him to put down his weapons and wait in an adjoining office because Cana Babović and my wife, both members of the party leadership in Serbia, had sent word that they wished to see me. This message had an official ring to it which struck me as odd because I shared a room with them, Mitra and I sleeping on the floor, and Cana on the bed.

As soon as the two of them came in, they went at me to release Djordjević. "After all, he's our comrade, even if he did make a mistake," Cana expostulated. And Mitra: "How can we go back to Kragujevac tomorrow and explain his execution to his neighbors, after the Germans killed off his own people?" And I said, when finally I got a word in: "But this is the Central Committee's decision. You must know what it means to violate the Central Committee decisions!" But Cana was adamant: "You can take that much responsibility on yourself!" And Mitra said, "If we execute him, it can never be remedied."

I finally promised that I would speak with Djordjević and see what could be done. They understood this to signal the success of their intervention. "All right now, what happened? What is wrong in your unit?" I asked Djordjević. "Maybe I did criticize," he replied. "There was some panic and doubt during the retreat. I made my criticisms to keep the unit together."

"All right," I said, "take your weapons and go back to your unit. But be careful in the future. You know that you were expelled as a fractionalist, so even your well-intentioned criticism can be taken amiss."

Djordjević proved himself a very capable man during the war, especially in administrative affairs. After the war he went into the judiciary. He was the presiding judge of the Supreme Military Court, and in that capacity conducted the trial of Draža Mihailović. He was also given other high court functions. We never spoke of his predicament in Nova Varoš, which almost cost him his life. But he was one of the few who greeted me in the street after my fall from power.

One day our people arrested a few Moslem merchants in Nova Varoš. The town was divided into those who were pro-Partisan, mostly the young people, and those who were anti-Partisan, mostly the small shopkeepers who were unhappy because we had requisitioned their goods. The arrests created consternation and horror. Some Communists came running to see me, followed by a group of townspeople. The arrested Moslems were among the town's most prominent citizens, and no one could accept the charge that they were spies for the Italians. That afternoon I went to the prison of the town command to look into the matter.

The commander of the regiment that guarded the approaches to Prijepolje had arrested a Moslem beggar woman who kept making trips from Prijepolje, and she confessed that the Italians were using her to carry messages into Nova Varoš. She also implicated one of the arrested merchants. The investigators in Nova Varoš further "elaborated" on this and arrested the others. I spoke first with the most prominent of the prisoners and gained the impression that he was innocent. Then I spoke with the beggar woman—a half-witted, frightened wretch who was nevertheless clever enough to guess at her investigators' hopes of exposing a "fifth column" and the "class enemy." Immediately I was struck by the weaknesses and contradictions in her testimony. I asked her, "How many of our mortars did you report to the Italians?"

"Ten!" she replied. We had only one.

Soon everything became clear. The woman had been frightened;

sensing what her investigators wanted to hear, she had tried to please them with her answers. The level of our investigation was not very high. They would have seen through this deception, had they not been obsessed by their dogmatic vision of the future and—already even then—by a professional deformation.

On another occasion Rifat Burdžović, secretary of the Sandžak Committee, told me in amazement that the fighting men of a Serbian battalion had exposed as a homosexual a certain Moslem, a good soldier and zealous Communist. Puritan that he was, Burdžović wondered whether he should execute this "freak." I did not know what the party's practice was, nor had Marx and Lenin ever written about such matters. Nevertheless, while my common sense led me to conclude that not only bourgeois decadents but proletarians too were subject to such vices, I decided that no perverts could hold positions or be party members. So Burdžović ordered the poor fellow to resign from the party. The fellow confided to Burdžović, in tears, that a dissolute Moslem bey had seduced him as a boy. I learned later that this homosexual, who was the very picture of masculinity, was exceptionally courageous and that he fell bravely. War confronts a person with questions which he would otherwise easily avoid. And the higher one is placed, the greater is the importance of such "trivia," the greater the glory, and the greater the room for mistakes.

Apart from all this, I spent most of my time working with the party leadership of Serbia in consolidating and preparing the units for their return to Serbia. The Party's hold over the army, the detailing of its functions and duties, and its constant watch over everything and everybody, had survived defeat, and was to give the Communists the advantage over their opponents. The organization and consolidation of the party within the fighting units was my most important task. I attended the meetings of regimental cells and regimental commands. I got a horse and made the rounds of the units in the field, only to return late and in the freezing cold, worn out by the dysentery which plagued me for two full months.

In the first few days we had to prepare the Kosmaj Detachment, which numbered around 130 men. They were mostly students and apprentices from the towns, even though the Kosmaj Mountain, which gave the detachment its name, was full of villages nestling in its groves and hillsides. These men were hardened not only by battle but also by having been deprived of their youth, their families, and normal existence. Their commandant was young and spirited, and their commissar was an old prison mate of mine and friend from Belgrade, the writer Djordje Jovanović-Jarac. He had been expelled from the party for his poor behavior while under arrest, and was not yet reinstated. Nevertheless, he was entrusted with the position of commissar, for he was capable, devoted, and well educated. He was a good speaker, witty and considerate toward his men.

We supplied the Kosmaj Detachment with whatever was available. Jarac reported directly to me. He came to my office on the morning of his departure. As usual, he reinforced his air of seeming indifference with jokes and swearing, though he was aware that I knew him to be an extremely sensitive person. In war he apparently found a confirmation of his imagined personality—fearless and daring. For a while we chatted about anything and everything, even our common frustration as writers. I lamented this, while Jarac pointed out ardently that this war against the most inhuman of forces was in fact the most complete fulfillment of every authentic intellectual. He spoke with caustic bitterness of his former Surrealist colleagues—Ristić, Vučo, Matić, and Davičo—who were vegetating in the security of cities instead of identifying themselves with the dangerous but total freedom of the insurgents.

I spotted a louse on his sleeve and said, "Why don't you get deloused? We have the facilities here."

He flicked the louse off and replied, "What for? I'll just get some more tomorrow. Lice are annoying only if you notice them."

"So you're still a Surrealist," I observed.

"Not at all! It's just that my men and I don't want to lend importance to things which aren't important today."

"Look here, Djoko," I said, as we embraced in parting, "if you run into lots of trouble, come back. The war isn't being fought only on Kosmaj!"

But for Djordje and his detachment the return to their revolutionary homeland—the heart of Serbia and Šumadija—was their pride and destiny. The detachment reached Kosmaj with great difficulty and was wiped out. Djordje sought refuge in Belgrade, only to return to his reconstituted detachment and perish with it in 1943. His mother was held a hostage in a concentration camp near Belgrade. A police agent told her one day, "I have good news for you, madam. You're going home. That criminal son of yours is dead."

Before departing, the Supreme Staff proposed that Sjenica be taken in order to expand our base of operations in Serbia. The campaign started before I had a chance to take part. It was conducted by an improvised staff consisting of men from Serbia and Sandžak. Preparations were made in haste and without adequate intelligence, and the enemy—the Moslem militia—was underestimated. The Moslems, from traditionally warlike Pešter, were defending their own territory. Besides, still exhausted and disorganized after the retreat from Serbia, our army was not eager for battle. Indeed, the battle didn't amount to much: the Partisans withdrew from the plain around Sjenica without serious losses. The prospect of victory is not the only motive for an army, but the victory must be worth the trouble and sacrifice. An insane form of human relations, war is nevertheless a motivated and extremely rational act.

Though the crackling of gunshots along the borders of our little oasis continued, there was a lull of sorts. It dragged on until the end of January, and we took advantage of it to get organized and to engage in political activity in the villages. Here Čeda Plećević came in very handy with his experience and oratorical skill, gained in many years of agitation in the villages of Šumadija. A long-time Communist, he left the party during King Alexander's dictatorship, but organized a left-wing democratic group which accepted co-operation with the Communists. This group was joined by Dr. Ivan Ribar, Lola Ribar's father and a distinguished member of Ljuba Davidović's Democratic party. In 1932, when I was a student, I met Plećević in jail. He remembered that encounter, and a mutual understanding grew between us, which was perhaps all the deeper because he showed no desire to rejoin the party, nor did I think it advisable as long as there was an effort to bring prominent non-Communists into the struggle.

The Partisans—particularly those from Serbia, who brought with them a self-sacrificing, conscious sense of discipline—were considerate, even filial, in their attitude toward the peasants and their property. Though Radojnja and Bistrica were exceptional in their support of the Partisans, we encountered hardly any opposition anywhere except in Rutoši, where the Chetniks had a foothold. Then, unexpectedly, a group of peasants from Haljinovići surrounded and disarmed one of our squads which was reconnoitering near Sjenica. Our surprise was all the greater since the squad was protecting Haljinovići from its traditional Moslem adversaries in Sjenica. With the squad also was Andra Petronijević, a royal officer. If his lack of courage contributed to the disarming of the squad, his family's reputation deterred the attackers from killing them on the spot. Petronijević was more like a priest than an officer—loquacious, cordial, but also honorable and unpretentious. The disarmed Partisans later reported that there had been a chance to resist but that Petronijević put them into an impossible position by his negotiations. The leaders of

Serbia and Sandžak were bitter, though there were a few peacemakers as well. We did not subject Petronijević to criticism. As an officer and non-Party member, he was important to us. Yet he himself felt that, because he was an officer, he had lost something in our eyes.

I believed that our reaction had to be implacable toward everyone who without reason attacked the army—our army, for we recognized no other. This stand was adopted. I personally led the campaign against the group at Haljinovići, with support from two battalions, and leaders from Serbia and Sandžak. Everything went according to plan, despite the Siberian cold for which the Sjenica plateau is famous. We crossed the mountain undetected and surrounded the village at night. But the dogs gave us away, barking at us on all sides—and the enemy got away. We caught three of the fifteen, but one of these then got away. We confiscated some cattle, and the next day held a meeting with the peasants. We executed the two guilty men on a hillside above the village. They were typical mountain people—rawboned, with their heads swathed in scarves, so that one could see only mustaches covered with frost, and little eyes which stared in disbelief into the rifle barrels. I felt sorry for them as I turned in my saddle and saw them fall. But I felt no pangs, rather the terrible compulsion which had mindlessly and irrevocably joined our destinies.

A merciless courage, a merciless unity permeated our forces. Two or three days later we began our march into Serbia—but cautiously, in order to find out whether conditions warranted the movement of all units and the party leadership. Our units were expertly led and they moved forward resolutely. But they were met with fire by village guards in every community. The guerrilla mode of warfare was no longer adequate for gaining control of this territory and winning over the people. Finally we were met by Nedić's organized forces, which were massing for an attack on our territory.

We leaders waited at Uvac with foreboding. After two days, having penetrated to Ljubiš, our units retreated in orderly fashion and without serious losses. In the winter dusk, with much anxiety and even greater sorrow, we watched our army descending the snow-covered hills. Somebody remarked that we would have to wait until spring to go back into Serbia. No one dared say that our chances might be no better then. The Partisans brought back ten prisoners—peasants from the guards and supply depots. We spent a long time questioning them about the conditions in Serbia, then fell asleep amid the soldiers in the schoolhouse on Kokino Brdo; the teacher had been shot as a Chetnik, and his family was crowded into a shed in the school yard. I was awakened by a sudden outcry: the chief of staff was shouting at a courier. When I asked what was going on, he said, "They haven't buried them yet!" Embittered by their failure, the Partisans had tried the prisoners in the night and shot them. "A bad Chetnik element. It'll teach them a lesson," the chief of staff said to me, when I remarked that they might have at least freed some. It would have been more advantageous for the Partisan cause had they released them all. But it is not only dogmatism and zeal that sow death, but also pragmatism—whether slight, moderate or excessive.

Three or four days later an offensive was unleashed at us from Serbia. The Italians had turned over Nova Varoš to the Germans, that is, to Nedić's Serbia. At the same time the Chetniks were reactivated in the neighboring villages. Our territory was so narrow that we could not maneuver, let alone pull back in depth. Without any major clashes, methodically our units drew back in the direction of Nova Varoš.

We worked out a plan: to group all our forces on nearby Zlatar Mountain, skirt around Prijepolje in a forced march by night, then cross the Lim River to liberated territory. The Supreme Staff was in Foča, which the Chetniks had abandoned, and considerable areas of Sandžak, Montenegro, eastern Bosnia, and Hercegovina were under our control. Such a plan meant the abandonment of our return to Serbia, but also the preservation of some 1,500 Serbian Partisans and 500 from Sandžak. I shared in the sorrow over Serbia, but did not waver in assuming this responsibility.

First, we had to care for the wounded. I summoned Dr. Julka Meš-tović, who was in charge of the hospital. At the time, however, she was busy with an operation, so I went to the hospital. There I was amazed at her patient courage, which I hardly suspected though I had known her for years. She was preparing to amputate a hand, using a saw in the absence of proper instruments, and brandy in the absence of ether. Taking her aside, I ordered her to proceed immediately with the hospital to Zlatar Mountain, and told her that she would get a unit for protection and help. "All right," she said, "as soon as I finish the operation."

The evacuation of the other services and civilians was carried out during the day in orderly fashion, while the leaders kept vigil in my office. The snow was deep, the frost prickly; there was fighting along the trails, and couriers kept coming and going. The ring around us was getting tighter. But there were no surprises. Ljubinka Milosavljević, a member of the party leadership of Serbia, was suffering from infectious jaundice. It was proposed that we leave her with a reliable person in some village. Even her mother wavered, determined to remain with her. But I wouldn't hear of it: "Let her die with us!"

It was eight o'clock at night, and machine guns were still clattering above the little town, when I set out for the mountain with a group of leaders. The night was cloudy but bright, and our column climbed in haste, single file. We were hardly twenty minutes under way when, from a wooded knoll about a hundred yards to the left, a submachine gun opened fire on us—a group of Chetniks cutting off our only way out. The entire column—about fifty leaders and escorts—ran off the trail down the slope. "Stop!" I roared, but nobody stopped, not even Mitra. Instinctively they herded together, and together they fled. I looked around: I was alone except for a courier who was holding on to a horse loaded with silver and documents. Just then a rear guard regiment caught up with us. The commander recognized me. Rushing through the knee-deep snow, I shouted "Comrades! Charge, charge!" In no time a whole formation was running with me, shouting, "Hurrah, hurrah!" The submachine gun was now silent and the commander sent a squad to reconnoiter the knoll.

With the regiment I continued up the mountain to the summer houses designated as our rallying point. In and around the houses were fires, kitchens, groups of soldiers, and refugees, but without commotion or noise. There was no trace of the comrades who had fled from the trail. Someone proposed that we send patrols to search for them. But I rejected this. "Where would they search? In any case, they know where we agreed to meet." I was convinced that they would find us. Danger steers men—like animals—to safety. And so it was. They arrived in the morning—exhausted, covered with frost.

We did not get much rest either. Some dozed in the houses or next to a fire. We waited until noon, until the units assembled and had a good meal. Though everyone was after us—Nedić's forces, the Chetniks, the Italians, and the Moslems—up there on the mountain, with its thick snow and blizzard, we felt safe at least for the while, as if in a fortress. A common and even cheerful courage filled us all—fighters and civilians, the wounded and the leaders. Everybody knew how difficult—almost hopeless—our position was, but everyone felt that, being united and determined, we were indestructible.

The march up the mountain proceeded without trouble: we had local Partisans to act as guides, the cold subsided after noon, and we felt confident. At dusk we emerged from the pine forest according to plan, and made our way above a valley with hamlets. We knew there were Chetniks in the village of Kosatica, which was on our way, but we hoped they would not dare attack us. Indeed, our long column passed by undisturbed—all except our rear guard: the staff and the Second Užice Battalion. The Chetniks fired on the staff, which, bunched together, was moving along at a rather leisurely pace. The firing of about thirty rifles from a crag to the left surprised us and we threw ourselves flat on the trail. A bullet hit next to my left ear, so deafening me that for a moment I thought I had been hit in the head. Somebody shouted "Fire!" We all fired together, then tore across the snow-covered slope as if it had been a sheet. Somebody shouted, "Second Užice Detachment, forward!" The battalion behind us rushed over the snow-covered waste, crying "Hurrah, hurrah!" and swung around the crag. The Chetnik rifles suddenly fell silent. We captured no one, we suffered no losses.

The march continued with no more incidents. Only two pack horses, left with us by the Posavina Regiment when they returned to Serbia, slipped on an icy crag and plunged down a precipice. It seemed as if, with them, Serbia had disappeared beyond not only our reach but our dreams. That night no one slept. It was the second sleepless night for the army, and for me the third—except for a brief nap here and there. Lack of sleep was a big problem for many of us during the war, but not for me. An hour or two of sleep or an occasional catnap was enough to get me through a night's march. But we were not all built the same.

With the cold misty dawn we descended a jagged ravine into the village of Lučice. The more the light of day revealed of our destination, the less I liked it. A few hamlets were tucked into the slope above the Lim, on the other side of which the highway ran between Bijelo Polje and Prijepolje. Prijepolje was only five or six kilometers away, and an Italian garrison was stationed there. We were exposed, though the Italians would hardly be in a better position if they came up the highway. Above us was a rocky ridge. We had to occupy it to protect ourselves from the *hodža* of Pačarizi. It was hard to ask our men, tottering from fatigue and lack of sleep, to move on. Even so, a Serbian battalion made its way up to the ridge, while the Sandžak units moved along the Lim to a nearby village, so we would not all be crowded around the same ford that evening. We were afraid it would slow up our crossing and cause us to be caught at dawn. The Lim had risen and there was only one crossing to Lučice. We settled in a log cabin and everyone fell asleep promptly, except for me. I lay down but couldn't sleep. I may have dozed off briefly, only to awake with a start, worried. I woke up the rest of the staff.

Among them was Milinko Kušić, with whom I had the greatest friendship and understanding. He had been in prison until the beginning of the war, and during the uprising was commissar of the Užice Detachment. But I didn't get to know him until we met in Nova Varoš. We took to each other at first sight. I was drawn by his fine intelligence and an irony which did not stop at taboos. His pale, thin face with its dark eyes, elongated nose, and lips set in mockery, reminded me of etchings of Robespierre or Saint-Just. He was enterprising, thoughtfully brave, and strict—equally with himself and others. He rode his officers, though not arrogantly or by pulling rank, but rather by his persistence and knowledge. Perhaps it was this secret, inborn sense of authority that drove him to immerse himself in operational details. He did not succumb to routine, or to fancy and chance, which maybe brought us together all the more.

That morning Kušić got up slowly and reluctantly. But I was excited: "What if the Italian tanks seize the crossing over the Lim? We have no alternate plan! Our men aren't worried, but that's because they trust us! Who can take responsibility for the destruction of so many good men? We must consider all the dangers and possibilities!" Kušić calmed me down, but he agreed with me. Meanwhile the other leaders were getting up. We held a meeting in the little room, its windows looking out over the Lim, and over the highway from which we expected an attack by the Italians and the Moslem militia.

Sure enough, around ten o'clock the Italians opened machine-gun fire from a hill above Prijepolje. They were too far away to fire accurately, and though an occasional bullet came whizzing by, we didn't pay much attention. A little later the Moslem militia also opened fire on the Sandžak troops on the other side of the Lim. But that was not too serious either, though like the Italian attack it lasted some three or four hours. We thought of the Germans: had they been in Prijepolje, they would have blocked our crossing of the Lim.

We conferred: as a last resort, we would break through along the right bank of the Lim, across the mountains which descended like ribs into the Lim Valley, through hostile Moslem settlements. We were strong enough to defeat the Moslems in any initial encounter, but since we had no influence among them and little chance of winning them over, we would be shedding our blood in vain. And we still had to cross the Lim, which was rising steadily, in order to reach liberated territory. Nor was time on our side. The Italians would mobilize their garrisons on the Lim, and the Chetniks and the Moslems too, and effectively block our way. Careful analysis showed that this other course of action was unacceptable, except in desperation; we decided to stick to the plan adopted in Nova

Varoš, that is, to force the Lim on the spot where we had camped that morning.

The discussion of the latter course of action made us realize the seriousness of our situation and work out in detail our forcing of the Lim. Kušić had the best ideas and I saw to it that they were implemented: in fighting the Italians, it was crucial to engage in close combat; they preferred to fight only from fortresses and under cover. While we were making decisions, our men slept. We ate late. That afternoon two armored cars stopped a short distance from the village where we were. Several Italians got out and surveyed the scene carefully. We had issued orders that no one was to leave the houses. The Italians spent a whole hour in their reconnaissance. Just before dark, while our units were gathering along the Lim, the Italian artillery opened fire on our battalion on the ridge overlooking the village; several of our men and the commander were seriously wounded.

During the day our patrol had tried to ford the river. We were afraid that the Moslems, if not the Italians, would block the crossing at night. We posted some fifteen submachine guns at a bend above the ford to distract the enemy and silence him with heavy fire. But the Sandžak troops informed us that they would join us in the crossing, since the Moslems had been firing on them throughout the day and would probably block their way. That forced us to reorganize the crossing, but we decided to stick by the existing plan and let the Sandžak troops bring up the rear; however, we sped up the operation. There was a good deal of comradely criticism of the Sandžak contingent on all levels.

As soon as night fell, the crossing took place silently and in haste. One battalion of Čačak troops forded the river in their clothing and occupied the highway, while the remaining two battalions undressed. The women went to one side and stripped to the waist, the men pretending not to see: holding hands, one after another whole regiments waded through the icy river. Fifteen minutes later they dried off in peasant houses nearby. Ljubinka Milosavljević was taken across on my horse by a courier, who then returned so that I too could cross on horseback. On the other side I found Ljubinka lying in the snow with her mother; again I gave them my horse and courier. While they were resting in a peasant house, exhausted and frozen, a villager stole the horse and took him off in the night.

And so we forded the Lim, in tense expectation of a battle. No one was drowned. While our units were moving up into the hills, we leaders returned several times to arouse soldiers who had fallen asleep along the trail, lest they freeze or be captured by the enemy in the morning. That night of February 7–8 we spent marching in freezing weather and in anxiety. Morning found us climbing up Kamena Gora. To our left, from a hill overlooking the gorge, the Moslems of Komaran fired on us sporadically. Bullets whistled by our heads, although the odds of being hit were small because of the distance. I was riding a horse which somebody had lent me. The officers urged me to get off, but I didn't: Montenegrin vanity overcame fear. Firing at us were militia men from the same villages that had attacked the Sandžak troops the day before. We vowed that, as soon as we reached the forest below the peak and had a rest, we would go after those riflemen from Komaran who, out of traditional religious intolerance, joined every invader of the Serbian lands.

Around noon, as soon as the units assembled, we discovered the disappearance of Miša Pantić, a prominent physician and long-time Communist from Valjevo. Having a headache, he had lain down to rest in a peasant house, and so did not cross the Lim. That evening we sent a patrol to look for him in Lučice, but he wasn't there, so we lost track of him. After the war we heard from Slobodan Penezić that in late 1943, in Prijepolje, one of our officers had come across a peasant wearing a Serbian peasant vest. During the war Serbian leaders liked to wear peasant dress, and so did Dr. Pantić. When questioned where he had acquired the vest, which had bullet holes in it, the peasant was evasive. One word led to another, doubt piled up on doubt, and the peasant was arrested. He disclosed that there had been an illegal Chetnik group in the village; after we had crossed the Lim they killed Pantić and threw him in the river. As it turned out, there was also an illegal party cell in the same village. Some of its members had decided to cross with the Partisans, and took with them the captured leader of the village Chetniks. He was executed as soon as we arrived at Kamena Gora. It seems as if these feuding peasants could hardly wait for the invasion of their country so they could hunt down and kill one another.

We never did get a chance either to avenge ourselves on the Komarani or to win them over; we had to rest, eat, and then get moving. We sent the wounded, the sick, and the noncombatants on to Foča, while we stayed with the units to await word from the Supreme Staff. It was there that Mika "the Jew" gave me two gold coins for the army, in exchange for the horse we had given him for his sick wife. And I released Hasan Srna of Prijepolje. The Partisans had captured him along with some cattle he had bought for the Italians. They kept him in jail in Nova Varoš and thought of executing him. When we evacuated Nova Varoš, I ordered that he be brought along; he came, and day and night helped with the field kitchen. When I told him he could go home, he burst into tears of joy. In 1946, when I was in Prijepolje with Tito's motorcade, Hasan Srna rushed up to me with open arms. He had been interned by the Germans because he refused to go along with propaganda lies about the Partisans. We met once again in 1954, after my fall from power, when I spent two days fishing at Prijepolje. He invited me to have coffee in his home, while the "socialist" hotel manager refused me lodging.

I asked Hasan, "Does UDBA [the secret police] know that you invited me?"

He replied, "I can tell you: yes! I wouldn't have dared invite you without their permission."

Though our mountain people received us well, they were unwilling to give us their cattle voluntarily. Cattle was their livelihood. In the village of Obardi several of us leaders sat late into the night, trying to convince the leading peasants to donate their cattle to the people's army. Nothing helped, not even the class approach of pitting the poor against the rich. One peasant said, "If you take cattle from the rich, from whom are we poor to earn any money?" Finally another peasant responded to my relentless arguments, "He's outtalked us, this bastard. I'll give a calf!" Others followed suit. But all this could sustain us for no more than two days. A hungry, tired army looked dejectedly at their officers, and the officers looked at me. Until that time the Partisans had supplied themselves by voluntary contributions and plunder from enemy forces. Our money was used only for special purchases in the towns. Orders were so strict that in Serbia a soldier was executed just for picking plums from a tree. To the joy of the troops, I decided to resort to requisitioning—only from the more prosperous peasants, of course, and in consultation with them. Several days later we received an order from the Supreme Staff which called for requisitioning: the same troubles pressed the army and the people everywhere.

One day we were surprised in a village by an Italian ski patrol. The rifle fire penetrated the wooden huts and mortally wounded an elderly man. One of our patrols came on the scene unexpectedly and opened fire, forcing the Italians to retreat.

We could not maintain ourselves for long by requisitioning in those poor mountain villages. We expanded our sources by actions against the more prosperous Chetnik villages in the plains around Plevlja. There were few fights and plenty of booty. However, this easy life soon came to an end: an order came from the Supreme Staff for the units from Serbia and the leadership to proceed to Foča. In Čajniče on March 1—several days after our arrival—Tito proclaimed these units to be the Second Proletarian Brigade. On my recommendation, Kušić was appointed political commissar of the brigade. Though I knew it to be only an oversight, I was sorry that Tito didn't invite me to the ceremony: the Second Proletarian was the result of my efforts. But I took pride in the words of praise which Tito, as supreme commander, had expressed in the *Bulletin* of the Supreme Staff for our units and leadership in the withdrawal from Nova Varoš.

4

Covered with orchards and rising from the confluence of two mountain streams, the still undamaged town of Foča seemed to offer charming and peaceful prospects. But the human devastation inside it was immeasurable and inconceivable. In the spring of 1941 the Ustashi—among them a good number of Moslem toughs—had killed many Serbs. Then the Chetniks occupied the little town and proceeded to slaughter the Moslems. The Ustashi had selected twelve only sons from prominent Serbian families and killed them, while in the village of Miljevina they had slit the throats of Serbs over a vat—apparently, so as to fill it with blood instead of fruit pulp. The Chetniks had slaughtered groups of Moslems whom they tied together on the bridge over the Drina and threw into the river. Many of our people saw groups of corpses floating, caught on some rock or log. Some even recognized their own families. Four hundred Serbs and three thousand Moslems were reported killed in the region of Foča. Yet, judging by the devastation of a large number of villages, it seems that many more Serbs were killed. In the hotel in which I was put up—not, of course, until after our sanitation corps had steamed my clothing—I was shown a room with a bullet-riddled ceiling; the room in which Major Sergei Mikailovich, the Russian émigré who conducted the massacre of the Moslems, had boozed and caroused.

How did it all come about? The killing of Serbs was carried out methodically—first the leaders and the more prosperous citizens, then others right down the line—by small groups of déclassé failures, largely led by the Ustashi: returned exiles from Italy or Hungary. Men and movements had come to the fore whose vision of the future called for the extermination of other faiths and peoples.

In the beginning, at any rate, the Chetnik massacres of Moslems demonstrated the revenge and bitterness of those whose relatives the Ustashi had killed. Still weak, horrified by the massacre of the Serbs, the Communists at first collaborated with the Chetniks, but tried to dissuade them from punishing all Moslems and Croats. But the Chetnik leadership was taken over largely by officers who believed in the higher nationalist aim of exterminating the Moslems. Characteristically, the Chetniks were not united or consistent in this either. For us Communists the Ustashi were a totally alien enemy force, and the Chetniks a conglomeration of Serbian liberal nationalists, terrified peasant masses, Serb chauvinists, and fascists. At the start, the various Chetnik movements differed from province to province: in Bosnia and Croatia they stood for self-defense against extermination; in Serbia, for the restoration of the monarchy and hegemony over other peoples; in Montenegro, for counterrevolution. But all had their roots in ancestral traditions, in village life, and in national and religious myths. In such soil and such a consciousness—particularly after the Ustashi massacres of Serbs—the urge to exterminate other faiths and peoples was ever present. Yet tradition preserved some tolerance and humanity, so that enlightened and liberal ideas also survived. Not many Chetnik officers, let alone the Chetnik peasant masses, were obsessed with the ideology of extermination. Therefore we Communists tried to differentiate between traditional ideas and Greater Serbian chauvinism, and to attract or at least neutralize certain segments, particularly among the peasantry.

What about the people of Foča? What did they feel? There were few of them around—occasionally an old man, but mostly women and children. They rarely ventured out, except to go to meetings. There was no ray of warmth or curiosity in their expressions, which remained apathetic, dull, inhuman. They were emaciated and yellowed, and dressed in rags.

The Partisan authorities were moderate but determined. There was hardly anyone around to prosecute, so they put their energies into making use of local supplies. For example, they invented a bread made of dried pears and barley which was indescribably vile, though it contained vital nutrients. Recuperated Communist forces bubbled with life and activity in this half-dead little town, amid a people whom senseless extermination had made indifferent to death. Partisan institutions in Foča did not have the formal dignity of those in Užice, but were more supple and realistic. Not that there was any lack of statutes and decrees in Foča. There were jokes about Pijade's decrees concerning people's committees—the very institutions which learned conformists in Yugoslav universities today hold up as the model beginnings of the new "constitutional" order. Arso Jovanović worked up a statute for the proletarian brigades, but that too proved unrealistic. However, even these activities—in addition to the practical ones of establishing local authorities and winning over supporters—demonstrated our inexhaustible efforts in building a new order.

Kardelj and Ribar went to Zagreb to co-ordinate our organizations in occupied territory. Having accomplished his mission with some success, Ribar returned to Foča on April 2, 1942, and told us of the terror of living in occupied territories. Kardelj went on from Zagreb to Slovenia, where he stayed until the fall of 1943. Thus Ranković and Žujović took charge of almost all the organizational work—Ranković in the party, and Žujović in local government and supplies. Žujović had just recovered from his wound, and threw himself into a variety of activities as never before, though he was dynamic by nature.

Tito also had his illusions, but they were, so to speak, realistic ones. To the designation "Partisan Detachments" he now added the designation "Volunteer Army," in order to win over and encourage the peasant masses in eastern Bosnia, who were ready to fight against the Ustashi and defend their own villages. The Partisans and Chetniks in eastern Bosnia continued to co-ordinate their activities even after the rift between Mihailović and Tito in Serbia. Yet suspicion and discord began to arise, especially after the German offensive in January 1942. In the beginning of February 1942 the Chetnik commander in eastern Bosnia, Jezdimir Dangić, negotiated in Belgrade for the annexation of eastern Bosnia—between the Bosna and Drina rivers, including Sarajevo—by Nedić's Serbia. The Germans agreed to it but Pavelić and the Hungarians opposed it, and Dangić returned to Bosnia. We did not know this, but we suspected something of the sort from the hostile behavior of the Chetnik leadership toward us and a certain restlessness among the peasants. Eastern Bosnia under the Ustashi meant further resistance by the Serbian peasant "volunteers." To my comment that the Partisans were not only a volunteer army but the most voluntary of all armies, Tito added, "Yes, but these peasants actually call themselves volunteers. They are not Chetniks, and they don't want to be Partisans wearing five-pointed stars."

No such volunteers existed anywhere else in Yugoslavia, and those in eastern Bosnia turned out to be unmilitant, and often unreliable and pro-Chetnik. In the spring of 1942 they fell apart during the Italo-

Chetnik offensive; soon thereafter Tito dropped their designation and came up with the name "National Liberation Army and Partisan Detachments." Yet there was some benefit after all from our co-operation with the Volunteer Army: the bogey of Communism and hatred for Communists were dissipated. I had occasion to observe this personally while passing through eastern Bosnia in the spring of 1943.

The leaders stayed in a hotel, except for Tito who lived in a small villa beyond the railroad station. Tito isolated himself there because a nearby gulley offered protection from air raids. Moreover, the isolation suited the meditative mood which Tito fell into on rare occasion.

I arrived in Foča, got cleaned up, and went to see Tito. Our meeting was, one might say, tender. I felt for the first time that Tito needed me. Although it was rather cool and damp, we took a walk around the villa. Tito's keen observations brought out a torrent of my own thoughts, which for a long time I had had no one to share with. Tito was largely dissatisfied with the Communist "sectarianism," particularly in Montenegro, where the Chetnik movement was growing rapidly and openly collaborating with the Italians. On the other hand, I showered Tito with my views on our domestic situation, namely, that the collaboration of the government-in-exile and its army with the Italian occupation forces could go above and beyond relations between the U.S.S.R. and the Western Allies. That meant that we had to limit ourselves to war against the fascist powers and not get bogged down in an exhausting civil war. Today it is evident that the government-in-exile had approved Chetnik collaboration with the Italians, against the advice of the British government. My ideas on this subject were also recorded in Vladimir Dedijer's *Diary,* under the entry dated March 10, 1942: "This evening a second council of the party cell was held. Mitar [Bakić] spoke of the international situation. . . . Djido [Djilas] set things straight. The alliance between the U.S.S.R. and England is strong and will grow even stronger. The development of relations in a single country can outstrip the international situation. Therefore we must adjust. We must not allow anyone to impose a class war on us—which is what the invaders want. They wish to strip us bare. We are waging, and must wage, a war of national liberation! . . ."*

Tito asked to see me once again. We spoke of many other things as well, and I cannot say in what measure my views may have influenced his conclusions, though he had a way of quietly absorbing the ideas of others only to put them forward later as his own, when they had passed through

* Vladimir Dedijer, *Dnevnik* [Diary], 2nd ed. (Belgrade: Jugoslovenska knjiga, 1951), p. 81.

the sieve of his pragmatism. However *Proleter* (The Proletarian), the Central Committee publication, soon published his article, "The Communist Party and the Allies of the Invaders," which set as the most important objective the struggle against the invaders. He was to define this even more clearly in "The Central Committee's Open Letter to the Organizations and Members of the CP of Montenegro and the Bay of Kotor": "In spite of the provocations by the invaders and their allies to incite a civil war, the Communist party of Yugoslavia today makes every effort to rally all patriotic forces into a single front of national liberation. . . . The essential guarantee of success in this struggle is the energetic elimination of careless attitudes on the question of strengthening the leading role of the CP in the National Liberation Struggle. . . .*

This was a departure from the previously cited letter of the Central Committee from Rudi, dated December 21, 1942. This departure did not alter our stand with regard to the Chetniks or the form of government which the Communists sought, for this did not depend on us alone. But it did make everything hinge on the struggle against the occupation—including even the strengthening of the party, a concern which has been present in all Tito's declarations and acts right down to the present day. This is what gave direction and substance to the militancy of the Communists, many of whom had joined the party during the war and by way of the army, and did not favor a fratricidal struggle, though quite prepared to give their lives in the fight against the invader. Thus a new revolutionary ideology merged with a heritage. Our struggle was firmly rooted in the well-trodden path of our forebears. Death and victory became natural and just. In short, Lenin's "main link" was revealed, and now had to be grasped in order to set the chain in motion. From that time until the end of the war, nothing essential was changed in the party line and tactics: the revolution had been born in the struggle against the invader. Even though I was convinced at the time that this was the road to power for the Communists, today it is clear that our victory would not have been so inevitable, had the ruling forces of the Kingdom of Yugoslavia been active also against the invader, instead of against just the Communists. Even the idea of creating a "national government"—premature and unrealistic in Tito's thinking—was subordinated to the fundamental task: the struggle against the invader.

Thus everything at home was clear, or at any rate seemed clear to us. But with respect to the Comintern—that is, the Soviet government—Tito was at times driven from outrage to despair. True, in the Soviet Union

* *Zbornik dokumenata,* Vol. III, Book 3 (1950), pp. 45–49.

Moscow had begun to operate "our" radio station, "Free Yugoslavia," but it continued to pass over in silence the Chetniks' collaboration with the invaders. Soviet dispatches even expressed doubts about the "correctness" of our stand toward Mihailović and the royal government. In vain did Tito invite a Soviet government mission to assure itself on the spot of the truth of our assertions—just as he did in 1948, regarding the untruth of Stalin's charges concerning the "degeneration" of the Yugoslav party. He also sought in vain for Soviet intervention with the British government, to make it stop glorifying Mihailović as the sole leader of resistance in Yugoslavia. So, too, our expectation of help from Soviet planes on Durmitor fell through. In 1944, in Moscow, I learned that this help was to take the form of Yugoslav émigrés who were in fact Soviet informers, rather than the munitions which were needed for the survival and morale of our troops.

Although baffled and upset, Tito—like the Central Committee—knew that the Soviets viewed our struggle, and our relations with the royal government, in the light of their relations with Great Britain. Though in May 1942 we scoffed at Mihailović's claim that the Yugoslav Communists were Trotskyites because they didn't listen to Moscow, the suspicion was born in us that not only was this diabolical brew concocted by the British Secret Service, but that Moscow knew of it and chose to ignore it. All this was in time relegated to oblivion, only to be resurrected when Yugoslavia and the Soviet Union parted ways, and particularly when they crossed each other.

My own thoughts concerning the Soviet Union—along with my romantic and dogmatic enthusiasm for the system and its creator, Stalin—are best reflected in the following entry in Dedijer's *Diary:* "Saturday, March 7, Foča. The sixth party meeting on organizational questions was held. Djido [Djilas] was present. He said that the celebration of Red Army Day was poorly organized. Politically incorrect. Freedom does not come free. If we ourselves do not do our best in the struggle against the invader, if we sit with folded arms, we cannot expect the Red Army to liberate us. . . ."* That was a summons to sacrifice and hardships, but also to the realization that victory came through one's own efforts. I do not claim that I alone thought and spoke in this vein, but this view thereafter predominated in the party and the army.

Three events are linked in my memory with Foča: the death of Draža's Major Boško Todorović; the arrival of the British mission under Major

* Dedijer, *op. cit.*, p. 80.

Terence Atherton; and the withdrawal of the remnants of the Valjevo Detachment from Serbia.

Draža had sent Major Todorović to Hercegovina to organize the Chetniks. The Partisans found out about it and surrounded him in a house at Kifino Selo. He fought bravely—threw back a grenade that had been thrown into the house—and they could do nothing short of setting the house on fire. He rushed out and they killed him on the spot. His diary came into my hands. I note all this because of that interesting diary. Todorović was intelligent and educated—indeed, overeducated for a royal officer. The diary reflected his doubts, his despair over collaboration with the invader. There was even a certain longing for death on that hopeless, unwanted, and uncertain journey. Todorović was not alone among the Chetnik officers. Draža mourned him as his best officer. Would Draža have mourned Todorović, had he read his diary? And who was really guilty for the despair and sense of futility of countless and nameless men like Todorović—Draža or the émigré leaders? Or the bloody whirlwind that caught up the leaders, who were incapable of resisting it? Or was it everybody and everything?

We received Major Atherton well. He even went with Tito to the front, though there was more than ample reason for suspicion. In Foča he made contact with the royal General Novaković, to whom Tito showed deference more for his title than for his supposed inclination toward the Partisans. Then one night Atherton stole away with Novaković—though he could have gone to the Chetniks openly—and disappeared without a trace. The Chetniks spread rumors and Draža informed the British that the Partisans had killed Atherton. To this day Walter R. Roberts maintains uncertainty concerning this.* According to what I later heard from the Yugoslav secret service people who investigated the case, Atherton was killed by a Chetnik bandit who was after his gold coins. The Chetniks were so disorganized and undisciplined that even bandits could join them.

The surviving fighters of the Valjevo Detachment—about seventy of them—were included in the escort battalion of the Supreme Staff. I detected in them a certain quiet resignation, the result of the heavy losses and catastrophe experienced in the winter of 1941–1942. We in the Central Committee did not like to speak of their suffering and of the destruction of their detachment, as if evading pangs of conscience for not having gotten them out.

* Roberts, *Tito, Mihailović and the Allies,* p. 55.

I stayed in Foča for two weeks. Though a lot of things happened during that period, my conversations with Tito stand out as the most significant and the warmest recollections. At the end of two weeks Tito ordered me to Montenegro—truly a surprise, after the criticism of my "errors" in Užice—because the situation there was worsening.

"All right," I said.

He noted my lack of enthusiasm and added with a laugh, "You want Mitra to go with you, is that it? Well, why not?"

I laughed too and said, "It isn't so much that I want her to come with me, but I think she can be useful there, while here she's bored."

Ranković also proposed Mitar Bakić and Svetislav Stefanović. Thus a team was formed with myself as head. Before our departure we had a meeting with Tito. I believe that Ranković, Žujović, and Arso Jovanović were also present. The offensive against the liberated territories in eastern Bosnia, Hercegovina, Sandžak, and Montenegro had already begun. The Italians played the principal role: the liberated territories—except for eastern Bosnia, where the Germans operated as well—were in the Italian zone. But for us the most vital danger was the Chetniks, because they established their authority in the villages as they advanced with the Italians. The danger loomed of losing Montenegro, our most important base, and of its final separation from Serbia. An analysis was made of the situation and, naturally, of "errors." All along we were troubled by the peasants' excuse that they were going over to the Chetniks for fear of having their houses burned down and other reprisals. This issue came up at the meeting with Tito, and the following argument developed: If the peasants realize that if they go over to the invader we will also burn their houses, they will change their minds. This argument seemed logical to me, too, though I did not support it resolutely. Finally Tito made up his mind, though hesitantly: "Well, all right, we can burn a house or a village here and there." Later Tito issued orders to that effect—orders that were fairly bold, by virtue of being explicit.

Armed with this bold new decree, we set out on March 15, 1942, for Montenegro. On the evening of the sixth day, having crossed several canyons and mountain ranges, we arrived in Piperi—at the headquarters of the "errant" Milutinović, still located in the same house which I had left after my own "errors."

5

It was a fanatical conviction that brought me, exhausted and ailing, to Montenegro. I gave careful thought to my duties in a complicated new situation. I was frequently beset by a feeling of alienation from the region, the people, and their speech. It was a world which was no longer mine, and never could be again. In this alienated world, death was nothing unusual, whereas life had lost all meaning apart from survival. Yet life must have some meaning. That meaning had to be found in a heightened belief in ideas, in justice, in the rationality of dying. Later I accepted that alienation. I even made fun of it: yours or not, I would say to myself, this is the only world you have in which to survive.

On our way to Montenegro I noticed that the peasants treated us with a marked air of confidentiality—as if they had something important to tell us but didn't dare, or else didn't know how to express it. At the same time there was a noticeable reserve and awkwardness over our arrival among the Communists, particularly in the lower ranks. A kinsman of mine, an elderly peasant, joined us along the way: he walked in front of me the entire time, holding on to my stirrup, obviously hoping to make an impression on the authorities. This made me uncomfortable, for I knew nothing about his conduct. As it turned out, he was not guilty of anything. He even ventured to tell me that in the command they were executing people "for nothing." During bivouacs at night or while marching, we chatted about everything and everybody, and as the impressions and perceptions piled up, I was able to get the feel of things.

Until the beginning of 1942, although they bickered and even fought with the Chetniks, the Communists did not regard them as enemies and collaborators of the invaders. In fact, despite their verbal acrimony and moral revulsion, until the end of the war the Communists regarded the Chetniks in a special way, that is, not as fascists but as "domestic traitors" —adversaries who, for reactionary and chauvinistic reasons, were led into agreements and collaboration with the invader. On March 25, 1942, the third day after my arrival at the main headquarters in Montenegro, I wrote the following to Peko Dapčević, Mitar Bakić, and Milinko Djurović.

The traitorous gentlemen are forcing a class war on us at a time when our task is the war of national liberation. We must not fall for this tactic, we must carry on the war of national liberation ever more resolutely. . . . The struggle which is now being waged in Montenegro boils down, in fact, to the struggle for the middling peasantry. . . . This struggle must in no way lead us to an erroneous conclusion about poor Anglo-Soviet relations—a conclusion which many of our comrades draw from the fact that the émigré government in London, through its supporters in our country, has begun to collaborate openly with the invader. The Anglo-Soviet alliance is directed against Hitler's imperialism. . . . To protect their privileges, the Greater Serbian gentlemen in London have begun a class war, their tactic being to destroy the most dangerous opponent—that is, the Communist party and the Partisan movement—while temporarily collaborating with the remaining opponents. . . . The adherents of the London [émigré—M. Dj.] government have had to take the path of open collaboration with the invader because of the force and scope of the national uprising against that invader. Had we sat with folded arms, there would have been no conflict, and we too would be stuffing ourselves with roast lamb and drinking plum brandy. . . . Naturally, we cannot win this kind of war with military skill or textbook military tactics alone. Such a war can be won only by combined military and political action, and on fronts combined with partisan fighting."*

The weaknesses of the Chetniks were fairly clear to me. They sprang largely from their inconsistent, contradictory stand: they made themselves out to be nationalists, adherents of the Allies, who were opposed to the invader "in principle," while at the same time they accepted weapons and supplies from the Italians and co-ordinated their operations with them against the Partisans, most of whom were not Communists but a patriotic peasantry led by Communists. The Chetniks, or rather the regime for which they nostalgically yearned, had lost the war without a fight. Such a stand inspired no authoritative leadership or spiritual cohesion. All these weaknesses provided advantages for the Communists,

* *Zbornik dokumenata i podataka* (1953), pp. 223–25.

who were confident and resolute, and had political leadership at the head of their military forces.

It became increasingly clear to me that our imprudent, hasty executions, along with hunger and war weariness, were helping to strengthen the Chetniks. Even more horrible and inconceivable was the killing of kinsmen and hurling of their bodies into ravines—less for convenience than to avoid the funeral processions and the inconsolable and fearless mourners. In Hercegovina it was still more horrible and ugly: Communist sons confirmed their devotion by killing their own fathers, and there was dancing and singing around the bodies. How many were executed in Montenegro and Sandžak at that time? I don't know, but several hundred doesn't seem exaggerated. All too lightly the Communists destroyed the inherited, primeval customs—as if they had new and immutable ones to replace them with. By retrieving the bodies from the ravines and giving them solemn burial, the Chetniks made impressive gains, while pinning on the Communists the horrible nickname of "pitmen." But on the whole, though they were quite bloody, clashes between Chetniks and Communists in Montenegro—unlike those in Serbia, where the Chetniks earned the name of "cutthroats"—occurred without the killing of the young and old, or the raping of women: ancestral norms restrained ideologies. And though I didn't have sufficient maturity to express it, I sensed that the Chetniks also profited from the Communists' excessive stress on Montenegrin, as distinct from Serbian, nationality. I thought then, and I believe to this day, that this explains in part why the Chetniks gained their most vital and broadest support from the Vasojevići, a region and clan which from ancient times has looked to Serbia for leadership. Serbianism was the most vociferous and emphatic sentiment of the Montenegrin Chetniks—all the more so in that the Montenegrins are, despite provincial and historical differences, quintessential Serbs, and Montenegro the cradle of Serbian myths and of aspirations for the unification of Serbs.

Yet the fundamental causes and roots of the Chetnik movement in Montenegro were not quite clear to me at the time—nor are they clear to official politicians and historians even today. The Communists had in fact gained power in the uprising of July 1941—a power which they then consolidated and institutionalized. The opponents of the Communists—mostly from the royal government—could do nothing but rebel or submit. These rebels did not have sufficient strength or means to resist without Italian support. In essence this was a counterrevolution pitting itself against a revolutionary regime. But the Communists and the Chetniks were deceived by their own propaganda—the former claiming that their main aim was the struggle against the invader, and the latter, that they were saving Serbia from annihilation. For people with such single-minded, heady views, all traditional values take on a one-dimensional, distorted aspect. The Communists in Montenegro celebrated Christmas Eve in early 1942 (January 6 N.S.) with recitations against aggressors and traitors from Njegoš's *Mountain Wreath,* while the Montenegrin Chetniks celebrated that same holiday the following year by the "inspirational reading" of the episode of the Massacre of the Renegades—the Moslems—from that same *Mountain Wreath.*

We were studying the situation in the main headquarters when the devastating news of still another unexpected Chetnik success arrived. Bajo Sekulić and Budo Tomović had perished in a surprise Chetnik attack at Kolašin. I had grown close to both in party work. I knew Bajo from Belgrade, where we had lived for a while in the same garret. Loyal and intelligent, he had been sent at my suggestion into Old Montenegro during the July uprising. Later he was moved to the main headquarters, and Milutinović succeeded him. Steadfast, attractive, and refined, he must have died disappointed in "the people," yet enthusiastic about Communism. As for Budo, when I came to Montenegro he was the secretary of the Communist Youth and a member of the Montenegrin party leadership. His bulging eyes and stocky frame were the natural wellsprings of his intelligence and wit. Although I worked most closely with Blažo Jovanović, it was with Budo Tomović that I had the best political rapport. I believe that Sekulić and Tomović were the only two men who measured up to their jobs in those unexpected, unschematic conditions. Though I mourned Bajo, his heroic personality gave his death a natural quality, whereas Budo's did not have any "reasonable" justification.

The deaths of Bajo Sekulić and Budo Tomović were the consequence of a series of oversights and failures. The leadership in Montenegro did not comprehend the personality of the Chetnik commander Pavle Djurišić or the peculiar nature of his troops. In the minds of the leadership, Djurišić was just another royal officer incapable of going beyond the stereotypes of the military academy, demoralized by the collapse of their state and the tactical collaboration with the invader. Djurišić had distinguished himself during the July uprising in the battle at Berane, where the worst fighting took place. Deserting the Partisans, he had gathered a largely volunteer force of experienced men and had established military disci-

Seventh SS Prinz Eugen Division advancing toward Čemerno, May 1943

Djilas with his brother, Aleksa, in Montenegro, July 1941

Ivo-Lola Ribar and Djilas, 1942

The Partisan leadership in Montenegro, May 1942. In the back row are: Blažo Jovanović (second from left), Mitar Bakić (third from left), Peko Dapčević (fourth from left), and Ivan Milutinović (fifth from left)

Draža Mihailović and Dragiša Vasić, his political adviser, on Mount Zlatar, July 1942

Germans near Prozor, March 1943

Tito (fourth from left) and Djilas (seventh from left) with the Supreme Staff, Bihać, December 1942

Germans escorting Montenegrin peasants to their execution, May 1943

German, Ustashi, and Italian officers, Pljevlja, May 1943

Sick and wounded Partisans in Bosnia, June 1943

Partisan units crossing the Drina, May 1943

Funeral feast for a dead Partisan, Sandžak, May 1943. At the head of the table are Dr. Ivan Ribar (left) and Dr. Sima Milošević (right)

The Prinz Eugen Division crossing the Sutjeska, June 1943

Partisans crossing the Piva, May 1943

pline. He made use of guerrilla tactics—surprise attacks carried out mostly at night. He felt no dilemma in the methods he chose or in collaborating with the invaders, first the Italians and then the Germans. Along with Keserović in Serbia, he was among the few of Draža's commanders who earned the Partisans' respect as well as hatred.

Along with territorial military units, we decided to create mobile units composed largely of young people and party members. This decision was enthusiastically supported by the unit commanders. On March 24 I informed Tito that I had ordered the formation of two battalions of 200 Partisans each, including 120 party members. I planned to take 2,000 Partisans out of the territorial troops and to send guerrilla groups to the enemy's rear. Tito approved the creation of the mobile shock battalions and the dispatch of the groups as well. Those 2,000 men were eventually to be separated into two brigades, the Fourth and the Fifth. "It is not necessary," I wrote Tito in my report, "for you to send two battalions of the Proletarian Brigade, because even if both Proletarian Brigades were to come, matters would not be resolved in our favor without radical organizational and political changes in Montenegro. We have here the manpower for two brigades no less militant than yours, and a good portion of the party cadre is still intact. But we do not have the brigades themselves."*

I understood Milutinović's difficult, discouraging position. He was a man of patriarchal virtue and revolutionary asceticism, but uninventive and administrative in his approach. Differences arose between us. He regarded certain clans—the Vasojevići, for one—as reactionary. I belittled the recently held assembly of patriots at Ostrog, his most important and favorite contribution. Regarding the clan question, I was right: the decision by the majority of a clan for or against the Partisans was a temporary and deceptive phenomenon. Regarding the assembly, he was right: although the emphasis was on military reorganization, and the Ostrog assembly was attended by a number of people who subsequently went over to the Chetniks, Milutinović had a knack for attracting non-Communists, especially older people. His manner of speech was simple and folksy, he was sincere and open. In fact, Tito took his side in the question of the Ostrog assembly, and I soon realized that I had acted in haste.

Otherwise, Milutinović and I were generally able to work together harmoniously. He too recognized the unfitness of the Montenegrin party leadership. The comrades did not lack devotion, but were inadequate

* *Zbornik dokumenata i podataka* (1953), p. 219.

and ineffective. We agreed to propose to the Central Committee a change in the leadership, and in staff of the main headquarters as well. We divided the responsibility between us: I took on the army, and he the party. The party had broken down to such an extent that a few Chetnik agents were among its members. We also gave out assignments to the members of my team. Bakić and Mitra were sent to the Katun district, around Cetinje, and Stefanović stayed with Milutinović.

The Piperi clan, on whose territory the main headquarters was located, and to which Milutinović and the entire leadership as well belonged, sided with the Partisans. Even so, I was adamant about having the staff and the leadership move into the more secure rear. But Milutinović, in his offended virtue and honesty, did not wish to face up to the danger of a surprise night attack. Finally he agreed. I set out for Gornje Polje, the headquarters of the Nikšić Detachment of Sava Kovačević, while Milutinović followed three or four days later, after securing the transfer of the printing press and other auxiliary services. That my evil forebodings had a basis in fact was shown soon after Milutinović's withdrawal: the Chetniks broke in so suddenly that Savo Brković, a member of the leadership, was unable to rescue the archives. He was reprimanded by the party because we Communists, unlike the Chetniks, guarded our archives like holy objects.

On my way to Gornje Polje, at a rocky mountain bend we came face to face with a group of armed men whom we recognized only after three or four moments as Partisans. They were led by a stocky fellow with a big mustache, a fur cap, and artillery boots. "Are you Djilas?" he asked. When I said yes, he smiled joyfully and said, "I am Sava Kovačević!" We embraced and kissed on both cheeks. This was our first meeting. On hearing that I was on my way to his headquarters, he had come out to meet me. And from the very beginning, during that journey, we became cordial friends.

Communism and the war provided an incentive for a close friendship with Sava, but this was not the chief reason. Perhaps we would have been even closer in other circumstances and with other incentives. I was known as a party intellectual, and he as a glorious hero. We discovered in each other what we lacked in ourselves or secretly yearned for—Sava my intellectuality, and I his epic heroism. He was a veteran prewar Communist, full of fighting spirit and self-sacrifice, though the Montenegrin leadership had a hard time convincing him to recognize the Montenegrins as a separate nationality. In his region of Montenegrin Hercegovina, no separateness among Serbs is recognized. The war gave final shape to Sava's personality, and he fulfilled himself in it, not only by his guerrilla prowess and his charges against Italian tanks, but by the inner fury and drive which he inspired in others. I imagined the leaders of our national uprisings to be just like him. His very appearance suggested rebellion—strong, rough, restless, but open and friendly. And with a powerful voice: he often shouted his commands from mountain to mountain.

Some false notions about Sava have persisted—that he was merely a resourceful and courageous peasant. He was that, but he was a construction worker as well, and even something of an intellectual: he had a high school education, he read a good deal, and his writing was quite literate. As an artillery man, in the army, he made it to reserve noncommissioned officer, a rank that only the brightest and most diligent attained. He had the mental capacity and the education to be a commander, and everyone accepted it as perfectly natural. Yet his rashness and violence have been passed over in silence, probably because that part of him has been misunderstood. Yes, he could be rash and immoderate. But without malice—like a child—in the terrible game with death. He looked after his men and did what he could for them. And when he chewed them out, his rage lasted only as long as the shouting. He felt that the party lost too much time in discussion and persuasion. He co-operated with political commissars, but he would have approved their abolition as well. Inside him—beyond consciousness and ideology—warfare and struggle were inevitable, particularly on the soil which bore him and at the time he lived in. Sava Kovačević was the most striking of those rebels who cropped up and made themselves leaders and symbols apart from the party. Quite apart from official apologias, his legend lives on in the memory of the people, who never forget rebellions and dream of them.

I was not the only one who even then had some perception of this personality and was attracted to it. I believe that he was conscious of himself as well. Sava treated me as someone who had to be protected and cared for. I was one of those rare people who could persuade him easily—in all matters except that I too should suffer deprivation and rigors like everybody else. During our journey to his headquarters he blurted out through a broad grin, "You look skinny, starved—I'll fatten you up in Gornje Polje!"

6

Indeed, one ate well at Sava Kovačević's headquarters. But there were also a fair number of "boarders": in addition to the committee members and staff personnel, all kinds of patriots, couriers, and important guests. There was something ceremonial about the meals: everyone waited around a long table until the commander, Sava Kovačević, came in, and he in turn waited at headquarters for Milutinović and me, so we could all arrive together.

At first it was a big mystery what food would be served, and particularly how that food was obtained. Actually it was taken from the supplies which the Italian command in Nikšić provided for the Italian prisoners held by the Partisans. There were seventy or eighty prisoners, and Sava requested the Italian command to supply food for them, because he didn't have enough to feed his own men. The command agreed and at a fixed hour once a week Sava's couriers, properly dressed and armed, took their horses and mules to Nikšić to get rations. The rations were then distributed, roughly half to the prisoners and half to Sava's staff and hospital. With some additional meat and potatoes, this was good nourishment. I expressed a certain embarrassment over this, but Sava retorted with laughter, "It's better for the prisoners to go hungry than for us to! Who asked them to come here anyway?" The logic of war was on Sava's side, as well as the feelings of those who ate the macaroni with tomato sauce.

About a hundred yards from Sava's headquarters in Gornje Polje were the Vidrovan springs, which supplied the aqueduct that took water to Nikšić. Whenever Sava got angry at the Italians, he closed off the pipes and Nikšić was without water. He couldn't resort to this technique for long because it hurt the townspeople as well, but it came in handy as a warning to the Italians not to violate established military procedures.

I relayed to Sava the new position of the Central Committee regarding the Chetniks, and Milutinović's and my decisions. It did not take him long to begin implementing them.

Sava's detachment had constant trouble with the Chetniks from the village of Ozrinići, near Nikšić. Its stone houses had become fortresses from which the Chetniks carried out attacks, sometimes alone and sometimes with the Italians. When he heard of the Supreme Staff decision that a house or small village could be burned down if it served as a stronghold, Sava immediately thought of Ozrinići. I approved the attack, as well as the burning of the village. A surprise attack was scheduled for March 27, and Ozrinići was taken without a heavy struggle. The booty was considerable, but they couldn't get it all out. Ozrinići was burned to the ground and the "stronghold" was destroyed, yet as often happens when commanders give in to bitterness, the reaction of the people was negative. Though quite a few of them took joy in the misfortunes of Ozrinići, and understood the military reasons for our action, the peasants simply couldn't get it into their heads that the Communists could act like the invaders and the Chetniks.

The reaction was still worse in Old Montenegro. Bakić had ordered the burning of Donji Zagarač because it had "gone over to the Chetniks." If the destruction of Ozrinići could in some way be justified by its having been turned into a stronghold, that of Zagarač could not: it was far removed from any Italian garrisons, though Chetnik authority had replaced the Partisans.

Milutinović and I immediately recognized the evil in this act: it turned undecided, vacillating peasants into bitter adversaries—into Chetniks. On April 20 Milutinović sent the following message to the Supreme Staff: "Thanks to the quick intervention of the CC [referring to himself and me as Central Committee members—M. Dj.], the massive burning down of houses—whole villages were being put to the torch—was stopped, while through the prompt intervention of Comrade Djido [Djilas] frontal struggles with the Chetniks have been halted, and the partisan method of warfare adopted."*

Upon arriving in Montenegro, I had quickly informed the Supreme Staff on March 27 that unjustified executions were being carried out.

* *Zbornik dokumenata,* Vol. III, Book 1 (1950), p. 92.

"Don't such death sentences cause fear and dread not only among the population, but within the party itself? How is one to explain that on territory liberated by the Partisans, the Chetniks have been successful in inciting such panic that the Partisans don't encounter a single soul?"* On the basis of my reports, the Central Committee reiterated this point in its letter of April 12, 1942, to the Montenegrin Communists. The executions were stopped—in large part on my initiative—except for those of prominent ringleaders. It was on my orders that all twelve Chetniks captured in the taking of Župa on April 15 were freed.

Yet even if I had been able to attend to everything, there were other leaders involved besides myself. On April 1 over thirty conspirators were executed in Šavnik, of whom a large number were of the Drobnjak clan of Karadžići—the clan of the reformer of the Serbian language, Vuk Karadžić. If I remember correctly, twenty-seven Karadžići were executed in two groups. There were other executions there as well. The repercussions of this one were particularly painful, not because there had been no plot—a plot did indeed exist—but because even those who had merely known of the plot were executed, though they had learned of it only through clan affiliation. Pavle Pekić told me about all this. He conducted the investigation and reported in detail to Moša Pijade, who had moved to Žabljak largely in expectation of Soviet planes, and who completely dominated the local committee and command by virtue of his legendary reputation. Pijade, of course, kept the Supreme Staff informed, but I don't know what he reported. Pekić stressed that it was "Uncle Janko" (Moša Pijade's *nom de guerre*—M. Dj.) who insisted on so thorough a purge. Pijade was prejudiced against the Montenegrins, particularly after the factional strife with Petko Miletić, but in this case the prejudice was intensified by his rash temperament, his lack of feeling for the region or the situation, and the bitterness of so many years in prison. Yet Pijade was no more bloodthirsty than the others. What was bloodthirsty were the ideas, the attitudes, and the compulsion to make life serve dogmatic class or national aims.

It was shortly thereafter, on April 8, that Tito approved the replacement of the Provincial Committee for Montenegro, the Bay of Kotor, and Sandžak proposed by Milutinović and myself—mostly myself, since I was better acquainted with the cadres. Of the old leadership only Blažo Jovanović remained, and also—if I remember correctly—one worker, because we had to have a worker. Radoje Dakić, prewar secretary of the Belgrade Local Committee, became secretary. This leadership was also

* *Zbornik dokumenata,* Vol. II, Book 3 (1950), p. 231.

joined by Veljko Mićunović, a prominent Communist from Belgrade. Another period of Montenegrin Communism had come to an end; insofar as it revived after the war, that only serves to show how uncreative and unoriginal is the stereotyped Russophile outlook of the inhabitants of this "land without justice." Soon afterward the following persons were appointed to the main headquarters of Montenegro: as commander, Peko Dapčević; as commissar, Mitar Bakić; as members, Sava Kovačević for his merit and fame, and Savo Orović as a patriot and former colonel of the royal army.

Immediately we got to work on forming youth battalions—a variant of the shock battalions—as the youth were the most resilient and most dedicated. Later a significant part of this manpower filled out our brigades. I presented one such battalion with its standard—the Serbian-Montenegrin tricolor with a five-pointed star.

A battalion came through Gornje Polje from the front at Kolašin. It was composed of men from the district of Katuni. They were demoralized by hunger, exhaustion, and confusion, but loyal to the uprising and traditionally the most militant of Montenegrins. Sava Kosanović lined up the battalion, strode in front of them all bristling, his glance shooting through each of them, then stopped and thundered, "Lay down your arms!" Reluctantly, the soldiers laid down their rifles. Standing next to Sava, I explained as calmly as I could, "Comrades, you know that we are now forming permanent battalions and that we need the weapons. Those of you who don't want to stay in these battalions don't need their guns at home."

One of the men summoned enough courage to say, "We want a furlough."

Calmer now, Sava retorted, "Your rifle doesn't need a furlough!"

Among the fighting men was a kinsman of Krsto Popivoda, a high functionary. He recognized me, felt ashamed, and kept his arms. So we disarmed about half, while the other half kept their arms and stayed in the shock battalion. The Partisan army in Montenegro was crumbling, yet at the same time its core was being cemented by Communist ideology and by an inherited pride and defiance of aggressors.

I was in Gornje Polje when the couriers brought Sava Kovačević a report that Borač had been taken from the Moslems, and a gift as well—a captured amber cigarette holder. We were all joyful, and Sava most of all: the Serbian rebel against Turkey awoke in him. He presented the cigarette holder to me and I have kept it to this day, despite the vicissitudes of war and my reluctance to hang on to things.

The operation at Borač began on April 17, 1942, and was planned by Arso Jovanović. There had been some talk of it while I was in Foča, and the Central Committee regarded it as significant. Borač, a Moslem settlement at the source of the Neretva River, cut so deep into liberated territory that it practically cut off communication between our units. The Ustashi from the liberated regions, among them a number of their notorious leaders, had jammed into Borač and superimposed their own bloodthirsty fanaticism on the militant and traditionally anti-Orthodox population. In fact, Borač was transformed into a haven from which vicious and reckless bands made forays. The men of Borač believed that they were invincible: those who could beat them didn't take the trouble, while those who wanted to beat them couldn't. The story was told—and it was true—that Prince Nicholas had come to Borač with his Montenegrins during the war of 1876–1878. So did the Hercegovinian rebels of that time. Borač had its own real and mythical history.

The Chetniks yearned to attack Borač, but they had no motive other than national and religious hatred. We Partisans were drawn into this venture partly because of its popularity with the Serbian Orthodox population, and even more because of growing Chetnik influence among that population. Plans called for the participation of some 2,000 local Partisans and one battalion of "proletarians," against 1,200 Ustashi and townspeople. However our actual number was higher, for when the Orthodox villages got wind of our intentions they rose up, young and old, for vengeance and booty. The battle was sharp and bitter. The general order was to kill all adult males, drive out the population, and burn down the houses. Only the last two tasks were carried out. The main body of the enemy got out by going through our volunteers. It was not until June, when forced to retreat from Borač, that we grasped the farsightedness of that undertaking—at least, to the degree that we were saved from depradations and atrocities. We were to reflect on it again in June 1943 when, though surrounded by the Germans, Italians, Bulgarians, and Ustashi, we observed that the ring around us wasn't so tight: our enemies lacked the knowledge of the terrain and the militancy which the men of Borač might have lent them. When I was in that part of the country in 1948, I was told that Borač was practically restored. I was glad to hear it, and hoped that the people of Borač would forget their history, or that they might even have learned a lesson from it.

The Supreme Staff insisted that we take Kolašin, because it was an active and exposed Chetnik center. It was the Chetniks of Kolašin who climbed up Sinjajevina Mountain and in the early hours of April 16 killed or captured an entire regiment of our young people. The taking of Kolašin required preparatory operations, including strong and sudden attacks in the surrounding area. One such area was the Upper Morača.

The attack on the Upper Morača, which took place on April 23, 1942, was planned and executed by Peko Dapčević. I was there not so much to raise morale as to gain knowledge of Chetnik territory. Peko's attack was sudden and well thought out. Realizing that they would be surrounded in the canyon of the Upper Morača, the Chetniks fled in all directions. There were no large clashes or great losses: it was like some war game in which our units were slated to "win," by virtue of being made up largely of younger men who had been honed by suffering and their ideology. Although no blood was shed in this battle, the effect was inconceivably horrifying and majestic. For hours both armies clambered up rocky ravines to escape annihilation or to destroy a little group of their countrymen, often neighbors, on some jutting peak six thousand feet high, in a starving, bleeding, captive land. It came to mind that this is what had become of all our theories and visions of the workers' and peasants' struggle against the bourgeoisie.

The mountaintops were still buried under a layer of snow, but upon descending just two or three hours into the gorge of the Morača River, we were greeted by the budding spring, and banks of primroses and violets wherever the sun touched the earth. But this transformation was of no importance; youth and life itself had been absorbed into the war, and we had to round up the fleeing Chetniks and secure ourselves against a surprise attack. In the hamlets we were met by women and children, silent and sullen despite the loud propaganda of the local Communists, who were themselves taken aback by the coldness of their fellow villagers. The men had fled to the wooded ravines, so impenetrable that only hunger or a guarantee of their lives could draw them out. We realized that we could not venture there, and that it was dangerous to hang around much longer. And as we were descending into the Polje Dragovića, we thought of the pasha in the folk epic who understood, only after having been surrounded, that "it is easy to go down to the Morača, but it is hard to get out of the Rzača."

It seemed like a good opportunity for my proposal to split off the peasants from the royal officers and the Greater Serbian ideologists. I immediately began talking with a group of captured peasants, alternating my reproaches with an understanding of their fears and wants, and in the end I let them go. They were all peasants who until yesterday, so to speak, had been on our side. Yet they warmed up to me very slowly, not just because they were afraid of one another or were for the Chetniks, but because they had doubts about our staying there. These doubts were the real reason for the peasants' attitude. I knew that we wouldn't stay long, but was ashamed to lie to the peasants. I told them that we might lose this or that territory, but that the Chetniks would perish shamefully with the invaders and that the final victory would be ours. It was obvious that, Communist propaganda notwithstanding, what flickered in them still was patriotism—the struggle against the invader. But this was not the nature of struggle at that time. The rise of the Chetniks and the harsh Communist countermeasures had brought the war into the villages and made the peasants reticent and double-faced: they sided with whoever came along, and tried to wriggle out of any risky commitment.

Yet all this demonstrated that the Chetniks had not evoked enthusiasm or sunk deep roots: the peasants didn't understand or accept the tactic of collaboration with the invader. The Chetniks took pains to demonstrate that their rule was a continuation of the old regime, which had alienated itself from the people in an unlawful and arbitrary form. The core and real strength of the Chetniks were the so-called regular Chetniks, composed largely of former police agents, minor bureaucrats, anti-Communists, and other trash. But there were few of these regular Chetniks; only Djurišić succeeded in forming units among the Vasojević clan whose strength approached that of the Partisan brigades. The Chetniks didn't have the enthusiastic manpower or the weapons for a sizable regular army. The invaders didn't trust the Chetniks, therefore gave them no weapons, only ammunition for small arms, of which the Chetniks had so much that they contemptuously called the less amply supplied Partisans "five-bulleteers." Because of their tactic of collaboration, the Chetniks never succeeded in being anything more than an auxiliary army of the invaders, even though they maintained a relative "independence" organizationally and politically. Unlike the Germans, the Italians didn't even look upon them as potential enemies. The Chetniks acted independently only in isolated villages, and occasionally in massacres of Moslems. In war, particularly in an occupied land, everything but warfare is barren and senseless.

After that the peasants gave us food gladly, though they were in need themselves. The next day the peasants who had fled to the ravines returned. Moreover, one of them revealed the cave in which Ljubo Bašić, the former inspector in the Royal Ministry of Finance and a Chetnik ringleader, was hiding. It is impossible to say who sentenced Bašić to death—Dapčević or I, the local Communists, or relatives of those who died in battle. Bašić was doomed by his role, his high office, and even his beard, which he had let grow to confirm his loyalty to the Chetnik cause. I didn't forego the opportunity to have a talk with him. It was a disappointment. Instead of a bitter and voluble opponent, I was confronted by a confused and uncertain civil servant. He was so nondescript that I can't remember when he was executed or how he behaved at the time.

We were forced to retreat. There were two ways out, but we were afraid that the unprotected exit along the Rzača might be cut off by the men of Rovci. Their commander was Captain Vuk Bećković, my classmate in the Kolašin high school and the father of the poet Matija Bećković. Vuk's sole concern was for his men, so we retreated comfortably along both paths from the Morača gorge.

Dapčević then went to the Kolašin front and I started back to Gornje Polje. On my way through Boan, local Communists told me that Professor Stojan Cerović wanted me to meet him in the village of Tušina. Cerović was a man of repute and capable. He published a liberal opposition weekly in Nikšić called *Slobodna Misao* (Free Thought), to which I had contributed in my rebellious youth—before the doctrinaire puritans set me straight. The weekly was sloppily edited and even more sloppily printed, but it was the only journal at all in Nikšić, and the only one in Montenegro worth anything. Cerović had also initiated new economic ventures, and generally raised the educational and cultural level of the town. He visited me in prison in 1933, when I submitted to him a collection of ten stories (all now lost). The local Communists had had a fuzzy attitude toward Cerović, their intentions ranging all the way from collaborating with him to executing him. There were also timid hints that I ought not to enhance Cerović's reputation by visiting him, since he was a known reactionary—even though he had belonged to the democratic opposition, and the Italians had arrested him after the uprising.

I didn't hesitate, but set out to see Cerović. He was living in a stone house with city furnishings. I stayed with him for four hours, in conversation and over a good meal. Cerović showed signs of mild depression, if not outright fear, but drew no attention to his past tolerance of the Communists or to his present inclination toward them. He was just what he appeared to be: a liberal and a patriot with a point of view that was independent, though not flagrantly or pointedly so. The important, in fact crucial, thing for him was the improvement of both material and intellectual conditions. Such men are rare in Montenegro, which is known for its extremes. He saw that a revolution was taking place, notwithstanding my explanations that this was primarily a fight against the invaders. He was not enthusiastic about the revolution, but couldn't accept the occupation either. Finally he asked me if the Italians might break through, and what I would advise him to do in that event. I said that it was entirely possible for the Italians to break through, and urged him to withdraw with the Partisans. And so he did: he retreated with us into Bosnia and became a popular, beloved figure—the professor in the hat with the five-pointed star, carrying an umbrella. He was also made a member of the first Partisan assembly (the Antifascist Council of National Liberation of Yugoslavia, or AVNOJ), held in Bihać on November 26, 1942. But he had once had tuberculosis and in 1943, in the course of the Fourth Offensive, his lungs collapsed and he died. Today Nikšić takes great pride in him.

Upon my return to Gornje Polje I continued reorganizing the units. It was there that I celebrated May Day, which assumed great significance for us because Stalin mentioned in his order of the day that the occupied parts of the Soviet Union and Yugoslavia were covered with the flames of partisan warfare. This was the first indication of a more definite position by the Soviet government regarding our struggle. But if we took pride in it and it intensified our conviction, nevertheless it was our own situation and our own capabilities that determined the course of events.

7

The changing of the party leadership at that time and the attendant purge also involved Radovan Zogović, with just a little help from me. Zogović had got into a fight with Milutinović, largely as the result of differences in temperament. Zogović was annoyed by Milutinović's unintellectual hairsplitting, and Milutinović by Zogović's sarcastic intellectuality. Once the clash had erupted, Milutinović began to discover "a leftist tendency" in the pages of *Narodna Borba* (The People's Struggle), which Zogović edited with Masleša. There were indeed such "leftist tendencies" here and there—unfortunately, very explicit. I was perhaps the first to call it "Zogovićism." Zogović came to me to hurl insults at Milutinović. However, not only did he fail to win me over, but when a party commission investigated the case, I submitted testimony concerning this criticism of Milutinović. It was a matter of disloyalty to a friend and of loyalty to the Politburo. True, I did not bring out all the details, but—bitter as I was over the dilemma in which Zogović had placed me as a friend—I was so moved by my own eloquence against him that I even "uncovered" his Trotskyite methods! Zogović did not know that, after my "errors" in Montenegro, I regarded my own position in the Politburo as insecure. And since he was not in the upper echelons, he didn't have an adequate feel for the full degree of frankness which the Politburo had achieved in its midst. Nevertheless, when the party sentence was being handed down, I defended Zogović from expulsion. He was given only a serious reprimand. How far matters went is evident from the attitude of the goodhearted and comradely Radoje Dakić, who had just been named party secretary of Montenegro: he actually blurted out that Zogović ought to be executed! My relations with party comrades were always open, and with Zogović even cordial. Now a shadow fell over the relationship with Zogović which could never be dispelled, though we worked together and were friends. Following this incident I always felt guilty about him. Yet we never spoke of it—how could we? Weren't these "errors" and "sins" something deeper, inaccessible? The revolution is hardest on revolutionaries.

The truth of this was demonstrated most cruelly in those very same days with regard to the Tadić family. Tadija Tadić, a student, was a party member and commander of a Partisan battalion. His uncle Spasoje, a police agent, had gone over to the Chetniks. The Partisans didn't yet have an organized intelligence service, but the people were so divided by bloodshed that almost nothing could be concealed. Thus it was learned that Spasoje was hiding in a hut in the woods, and a squad was sent to kill him. His nephew Tadija was able to warn him, however, and instead of running away, Spasoje got hold of a submachine gun and like a mad dog killed two Partisans and wounded several others. Someone denounced Tadija for warning his uncle of the raid, and Tadija was arrested. Then the Tadić family rose up, a whole regiment of them. The Partisan command quelled the mutiny and arrested the ringleaders, all of them party members.

The arrested Tadićes were taken under guard to Gornje Polje. The investigation was conducted by Radoje Dakić and Sava Kovačević, but Milutinović and I were kept informed and we approved the sentence: three Tadićes were sentenced to death, I believe. The sentence was too severe for one of them—a raw and agitated youth—even though he had said, "A bomb should be thrown at the Committee!" Ljubo Tadić, today a well-known leftist philosopher, barely got out of the mess through "self-criticism." I believe he was the Tadić with whom I had a talk in my final review of the case.

It was decided that the Tadićes would be executed by a firing squad on a secluded spot by a cliff. Two royal officers suspected of a plot were to be executed with them. This was not done merely for economy but to associate the fate of party enemies with that of outside enemies. At their execution the Tadićes, particularly Tadija, shouted party slogans; one of the officers cursed and the other scowled in silence.

On their way back from the execution, the Partisans were somber but not depressed. They didn't resent the sentence passed on the Tadićes, and didn't give a damn about the officers, yet everyone was struck dumb by the reality of what they had helped to create. It was as if something terrible had happened to them. Even though our army was crumbling, everyone knew that the Tadićes were not our enemies, that they had been moved by powerful and primeval clan ties. Tadija had been put through school by that uncle, and was himself a model family head. In all likelihood Tadija would not have warned his uncle had he realized how he would react. The Tadić familv continued to fight as Partisans and many died for the cause, but they did not forget the terrible fate of their kinsmen. Blažo Jovanović told me later that the Tadićes wanted to claim the bodies of their executed kinsmen. Though our authorities don't permit it even to this day, I didn't oppose this request. I felt this encounter with the Tadićes to be a part of my own fate.

In Gornje Polje I got to know Savo Orović and Marko Vujačić, and tested Milovan Šaranović. Orović had been through the wars of 1912–1918, became an outlaw during the Austrian occupation, and rose to the rank of colonel in the Royal Yugoslav Army. Sava Kovačević once wanted to test Orović's courage, and dared him to go through machine-gun fire to help some woman and her children. Orović was exceptionally brave. On one occasion I was making the point that we Communists worked together with non-Communists such as Orović, to which he shot back, "By God, I'm as much of a Communist as I can be!" He came to Communism by way of inherited ideas—Panslavism, Serbianism, Yugoslavism, the fatherland. He accepted Tito, the party, and the new regime much as he had once accepted the King and the unification of Yugoslavia. As for Vujačić, he was striking and eloquent, kindly and all too accommodating. A blend of village tribune and demagogue. For him nothing was of a permanent value. He liked his comforts, and the fact that he joined the Partisans and endured risks was less the result of his enterprise than of his heritage, his knowledge that the struggle against the invader is never in vain.

Šaranović was a young royal officer distinguished for his bravery and daring. But officers were deserting and causing mutinies in our units, so it was deemed necessary to pin him down. I didn't even get a chance to reveal my intentions, however, before he declared, "I vowed to myself that I'd stay in this fight to the end, and I will—have no doubts about it!" That ended the conversation—more awkward for me than for him—and he returned to his unit. Late in 1942 he was sent on a mission to Slovenia, where he perished—faithful to his vow.

One evening Sava Kovačević took me on an inspection tour of outposts near Nikšić, using a motorcycle captured from the Italians which Sava had immediately learned to ride. The very next day Italian artillery turned the rocky peak of Uzdomir, our key position, into smoke and fire; an attack followed, even though our unit had already withdrawn. With the taking of Uzdomir the Italians had gained a vantage point; a day or two later they were pounding us with their mortars at Gornje Polje, while the Chetniks of Nikšić stepped up their attacks in the villages. An offensive was in the offing; the enemy was securing starting positions while destroying our bases. We knew that this was part of a larger offensive already in full swing at Foča, in eastern Bosnia, and in Sandžak.

This forced us to step up our activities. The most pressing task was the arming of newly created regular units, made up largely of young people. A large number of rifles were in the hands of peasants who occasionally answered our appeals to protect their own area, but avoided more distant campaigns or regular duty. But how were we to take arms away from our own peasants, who had fought against the invader from the beginning, and only yesterday helped drive the Chetniks from their soil? Sava Kovačević proposed an extreme measure: to call them together, line them up, and simply take their arms away. But what if they resisted—not just those few who secretly favored the Chetniks, but the majority who were for us and would feel hurt and endangered without weapons? I didn't agree with Sava. It was decided that I should speak to the peasants and persuade them to lay down their arms.

An assembly was called. The peasants—some four hundred of them, largely from the district of Nikšić, my ancestral home—didn't know why they had been summoned. If they had, many wouldn't have shown up. We set up two men with submachine guns on a mound in the middle of the meeting ground, just in case. Sava was also there, greeting the peasants and joking with them. He knew many by name.

Standing on a boulder, I spoke for nearly an hour. Had anyone recorded that speech, probably I too would laugh at it and wonder at its demagoguery and frankness. First I made them laugh, which dispelled their distrust. Then I praised their heroism and self-sacrifice, even above the heroism and self-sacrifice of their ancestors. We were no better equipped than the insurgents of olden times against the Turks, yet we had to face an enemy who fought with planes and tanks. I told them that we were their children, reminded them of their murdered relatives, brothers, sons, and neighbors: our blood and lives could not be separated, even if they renounced this glorious struggle—the most glorious and most difficult in the memory of our mountains. I threw in verses from the folk epics and Njegoš, and I lashed out at the Chetniks: we could understand, I said, why they were against us Communists, but we couldn't understand why those lackeys murdered with the invader's own guns their brothers who rose up against that invader. And though victory was certain for us and the great Allies, I said finally, the fickle fortunes of war might indeed force us to abandon this or that village. Even so, we would never abandon the war throughout the homeland, and that was why we were forming regular units. These units needed weapons and we had none to give them except those borne by heroes: heroes who for reasons of their own—always just and honorable ones—were unable to join the units which would go wherever the peoples of Yugoslavia might summon them. I added, half in jest, that it would be good for them to give up their rifles, just in case the Chetniks came into their villages; they could tell them that the Communists would not have disarmed them if they had any confidence in them.

The peasants calmly gave up their rifles, though here and there individuals scowled. But I don't think my speech would have had such success, had Uzdomir not fallen.

While the peasants were piling up their weapons, a stout, elderly peasant with a brand-new rifle over his shoulder came up to me and took a pot of cheese out of his bag: "You don't know me, but we're relatives. I'm a Djilas, too, Stevan Djilas. I heard you would be here, so I brought some cheese for you and your friends." He realized I was silent, that my glance was connecting the rifle with the cheese, then added timidly, "Let me keep my rifle. No one in our clan has ever been disarmed by our own people."

I cut him short: "Put down the gun *and* the cheese."

Sava Kovačević was standing there, watching us with curiosity. I then had a pleasant chat with my kinsman and my other relatives. Sava said to me, "I thought you'd kick that pot of cheese and take his rifle away!"

A couple of days later I set out for the battlefield at Kolašin. The Supreme Staff had ordered the capture of Kolašin, a key stronghold of the Chetniks. Two or three days earlier we had sent two battalions there. Without even knowing it, they broke through the circle that the Chetniks had formed around Dapčević, who was in the gorge of the Tara, in the village of Bistrica. My escort was exhausted by a two-day march across the mountains, so beyond Bistrica I pushed ahead on my own. A few Chetniks were still in the woods, but I knew that they wouldn't venture to attack. At dusk I arrived in Polje, at the schoolhouse which served as

Peko's headquarters. Peko had to go to Gornje Polje, to take over the main headquarters for Montenegro, so the operation at Kolašin was taken over by his chiefs of staff Velimir Terzić and Savo Drljević.

That same evening the peasants and party members came to me with complaints against the excessive, insane executions. Some may have been encouraged to do so on the theory that it is easier to talk more freely before leaders, or because they were acquaintances of mine or of my father. The next day such complaints increased, as though by tacit agreement.

The night before my arrival Lazar Tomović, former president of the commune of Polje, had been shot through a window while talking to a Partisan quartermaster about requisitions. The quartermaster assured me, "I know nothing about it. I felt terrible. A single round of bullets, and he fell right there in front of his own people." My father had baptized Lazar's children, and I had known that serious, capable, and honorable peasant since my childhood. It was not this spiritual kinship that upset me; rather I was convinced that Lazar was not guilty of anything but a lack of enthusiasm for the Partisans. A couple of days before Lazar's murder, Sekula Bošković, who was over eighty years old and an invalid, had been killed, also in Polje. They had dragged him from his bed, killed him, and thrown him in the Tara River. Sekula was the brother of Lazar Bošković, the leader of a local uprising in 1876, and the uncle of Boško Lazarev, a renowned military leader in the wars of 1912–1918. What could this old man from a legendary past be guilty of in this unexpected, unpredictable present-day reality? All that came to my mind was the old man sitting by the hearth, saying something like this: "By Saint Basil of Ostrog, the Italians are better than these hotheads of ours. At least the Italians recognize God! These other fellows don't even recognize the King, yet they want to lead the people!"

Something even more painful happened the day before my arrival; Jovančević had been executed in Mojkovac, despite the fact that he had escaped from a Chetnik prison in Kolašin, where he had been sentenced to death. Jovančević was of the Vasojević clan, among whom the Chetniks first prevailed. He was a party member, but had surrendered to the Chetniks. The party committee had passed a decision that every party member who surrendered should be executed, and carried out that decision regardless of the circumstances–including even the fact that Jovančević had corrected his error. A peasant said to me, "It's unheard of to kill a comrade who's escaped the enemy's gallows." And still another said, "If you treat your own people like that, what can we expect?"

Though embittered by the Jovančević case, I also could feel for the committee that sentenced him: defeated and driven out of their own clan and region, the Vasojević Communists nurtured a conscious heroism toward the enemy, as well as a mindless hostility within their own clan. By suffering and dying and not wavering, they had come to regard their own judgment as a higher law.

The Vasojević Partisans captured a priest and eight peasants, all Chetniks, while they were bringing in hay from the mountains. A merciless fate awaited them as well, but the execution was postponed when the committee found out about my displeasure. The Chetniks were placed in the schoolhouse in Polje and were well guarded. The day after I arrived, I went to see them, though I had already made up my mind, on the basis of the reports, that the priest and one of the peasants, a militia commander, should be executed.

I spent some three hours in conversation with them. I immediately realized that the priest didn't fit the "death category." Pale, shriveled, and submissive, he pleaded that he was only carrying out his priestly duties and that, if permitted, he would conduct burial services for the Communists as well. Moved by the priest's appearance and demeanor, and wishing to make a good impression, I set him at liberty. The militia commander, on the other hand, filled the bill perfectly: pugnacious, cocky, and well-to-do. Nevertheless, in the end I decided to free him as well, but not until he had promised in front of everyone that he would no longer serve as a commander. The Vasojević Communists told me that the release of these two prisoners left a generally favorable impression in their region, which had been ravaged by Chetnik raids.

I don't relate this incident to vaunt my own humanity; every revolutionary leader can cite such cases, just as his opponents can charge him with crimes. Rather, I wish to point out to what extent the destiny of men in war, particularly a civil war, can depend on circumstances and the temperament of decision makers—and how complicated it is to be a leader. The decisions of every leader have to take into account the challenge of potential opponents from within the revolutionary movement. Thus Pijade—in a letter of May 8, 1942, addressed to Milutinović, Bakić, myself, and Tito as well—protested against my releasing the Chetniks in Polje: "I disagree with the decision to release rather than shoot the five Vasojević Chetniks [the number is wrong—M. Dj.]. The people in the Šavnik region will not understand this. So many in our district have been shot and mutilated by the Chetniks, that there is the utmost bitterness against them, above all against those of the Vasojević clan. . . .

I wouldn't give a straw for their story that they were 'led astray.' . . ."*

The next day, before setting out toward Kolašin, I rode on horseback to visit my mother's sister, Aunt Mika, who lived not far from the school. This simple peasant woman was shocked and bitter at the fratricidal struggle, but she was glad to see me. In answer to my question as to who was worse, the Partisans or the Chetniks, she replied, "Well, you are fighting for a just cause, but you are harsh and bloody."

The attack on Kolašin was prepared carefully and in complete secrecy. It was worked out by Velimir Terzić and Savo Drljević; I signed the order as well, to lend additional authority. Our forces were three to four times more numerous than the Chetniks. The basic idea was to avoid a clash along the way, then launch a surprise attack on Kolašin at night. Accordingly, the main body of our troops was sent along the left bank of the Tara, through the forest, to break into the town before dawn.

I set out along the right bank of the Tara, with the Lovćen Battalion of the First Proletarian Brigade, sent us by the Supreme Staff as a model unit for Montenegro. The commander was Pero Ćetković, a former officer, and most energetic. Though he accepted a position in the lower ranks, he was soon promoted and would have gained the highest positions had he not been killed in 1943. His commissar was Jovo Kapičić, a Belgrade student who was also brave and resourceful, but more cheerful and cordial than Ćetković. The qualities of these two very different and complex officers seemed to carry over to the entire unit. I never saw men who were so contemptuous of death, yet who so mourned their dead and cared for their wounded. They were students and peasants, most of them party members. Yet the party ideology was but the token and expression of traditional Montenegrin sentiment. Those few days I spent with them reassured me that the legendary Montenegrin heroism still lived, and that it was a conscious and cheerful sacrifice to the struggle. It is no accident that it was precisely their own region, with Mount Lovćen as its center and symbol, that gave rise to the restoration of the Serbian lands. It had an abundance of individuals consciously prepared to defend their soil and nationality, even when fainthearted and confused leaders were a hindrance to them.

The sixteenth of May was a clear day, the leaves were young in the cool beech forests. The Lovćen Battalion climbed up above the Chetnik rear guards and by early afternoon reached Markovo Brdo, a hill overlooking Kolašin. Up till now we had found the stubbornness of the Chetniks suspicious. From Markovo Brdo we could see the Chetniks in

* *Zbornik dokumenata,* Vol. III, Book 3 (1950), p. 168.

the villages below, grouping for a counterattack. Our forebodings were confirmed: on the mountain peaks on the other side of the Tara, through our binoculars we observed a column in retreat. We waited until dusk. A Chetnik mortar began to lob grenades into our midst. Kapičić ran up and reported the death of a soldier, a renowned hero. Ćetković ordered a retreat. Then Ćetković and Kapičić sprang up, grabbed me by the arms, and dragged me back to the forest. No sooner had we entered the forest when the Chetniks attacked Markovo Brdo amid exploding bombs and heavy fire. But there was no longer anyone there: our battalion had pulled back in haste.

The units designated for the attack on Kolašin reached the town by daybreak, but retreated at the first firing, without a fight and without losses. Many convincing reasons were given for that failure: exhaustion, tardiness, lack of co-ordination, an unexpected flank attack, and so on. Tito characterized these operations as "stupid." Dapčević later remarked to me, "The plan was too complicated. We should have gone directly on the highway, with flank guards." My own feeling then was that our failure resulted largely from the doubts of our men about the wisdom of that battle. Everyone knew that the Italians and Chetniks were advancing on our flanks and to our rear—from Nikšić, Plevlja, and Foča. Even had we taken Kolašin, we would have had to abandon it soon. Supreme Staff later confirmed that the fall of Kolašin would not have settled the Chetnik problem in Montenegro. In fact, the letup in the uprising in Serbia was slowly making itself felt in Montenegro, Hercegovina, and eastern Bosnia. The leadership was not yet reconciled to this fact or able to see it. The rank and file could feel it because they had no strategic aims or ideological concepts.

During the night we arrived in my village of Podbišće. We set up battalion headquarters in the house of my uncle, who had fled to Kolašin with his family and cattle. He had left some grain, which was distributed. After lunch we started up the mountain, and a Chetnik gun began to pound us; like the Italians, the Chetniks tried to incite fear with their firing, and taunted us with their abundance of munitions. My fellow villager and school friend Grujić was wounded. He pleaded in terror, afraid that we might leave him behind, but we didn't. He is still alive today, though his brothers perished.

On Sinjajevina Mountain I dismissed Milutin Lakićević, a political commissar, on the spot, because he had slapped a young soldier who had collapsed from exhaustion. Several days later Lakićević died in the battle, and the story spread that he had exposed himself intentionally. He was in fact exceptionally brave, and my dismissal of him was only a temporary measure to bolster the role of the commissars in fostering comradeship, political education, and the suppression of injustice.

I did not stay long on Sinjajevina, where the front became momentarily stabilized. The defeat at Kolašin, for which there were no clear military reasons, filled me with foreboding. I decided to go to see Tito and the Supreme Staff—to warn them and to receive new orders.

8

Early one morning I was riding on horseback over Mount Durmitor, all alone. As I was climbing up by way of the bend at Štulac, a long column of mounted men approached. It was Tito and his men, retreating to Montenegro after the fall of Foča. Tito immediately noticed my anxiety. Though I impatiently explained to him that things were going badly, he didn't show much concern. He may even have said, "Well, we'll see."

We descended into Žabljak and set out for Crno Jezero. Before we reached the lake, a plane came upon us and we took cover in a stand of fir trees. I took out of a saddlebag a ham I had brought as a treat for Tito, and Tito, Arso Jovanović, and I had a feast, while the plane showered machine-gun fire all around us. The delicious food, the bright day, and Tito's self-confidence restored my good humor, though the onslaught of dangers and failures never left my mind.

For weeks on end all of us had eaten nothing but meat, usually unsalted, and for months we were lacking in ammunition. These were our most acute troubles, as well as the most convincing causes of our failures. Yet one could, and did, see these causes in different ways: did they not arise from the shrinkage of our territory and the conflagration of the civil war? Still, we leaders were sufficiently alert, and our men sufficiently wary, not to be drawn into heavy fighting. But we sensed that the biggest, most far-reaching setback was still in the offing—the separation from Serbia, without which it was impossible to win, or even to preserve the movement from disaster. Tito saw this, too, perhaps all the more acutely in view of his greater responsibility. In maintaining his own confidence, he at times reacted hastily and angrily. Yet even while the setback in Montenegro filled everyone with uncertainty and despair, Tito was already planning new brigades which no one would be able to vanquish.

We stayed at Crno Jezero for two or three days. While Ranković remained behind to rally the Sandžak units on the Tara, Tito called a meeting at which the decision was reached to move the party leadership and a large number of party members to the rear of the Chetniks in Montenegro and Sandžak. Prior to this Milutinović and I had been moving smaller groups, but it was only at that meeting with Tito that the final, crucial decision was made. I offered to stay in the Chetnik rear—I regarded the failure of the operation as my own as well, and was prepared for sacrifice. But Tito would not permit it; Politburo members shouldn't expose themselves to unnecessary dangers, he said. The taunts of Mitra, Koča Popović, and other comrades from Serbia that my Montenegrin heroes, so greatly extolled, self-glorified, and indomitable, were now rushing to submit to the invader, even if made in jest, might also have goaded me into "showing them" what we Montenegrins were capable of.

All the work in connection with the move to the rear was done by Milutinović and myself. We were told that the fear of Chetnik terror was so widespread that brother turned brother from his door; we felt a deep concern and sorrow in sending our people into an uncertain and ugly situation. Yet no one asked to be excused and no one held back, not so much because prospects were no better for those who stayed with the units, but because the fighting and reprisals were so fierce and destructive that we had no choice but death or victory. Yet that decision proved to be wrong. Just as the failure in Montenegro was but an extension of the failure in Serbia, so our leaving many Communists and sympathizers in the rear was but the extension of similar measures in Serbia. Although our small groups kept the Chetniks worried about the resurgence of Partisan activity, the Chetniks almost annihilated the Communists by their constant raids and terror, and delivered many sympathizers to Italian camps. The suffering of these Partisans can only be compared with the suffering of the outlawed Chetniks after our victory—in that same region and in the same way. A year later, the conviction took root within me that it would have been more sensible and effective if we had abandoned the idea of the little groups, and instead got out as many people as possible. That would have made possible four or five Montenegrin brigades instead of two, and fewer awful deaths and much less horrible suffering in prisons. But we Communists neither wanted to nor knew how to accept the reality of the changes and defeats brought on by the civil war.

Alongside the heroic and tragic undertakings, there were also the grotesque and the preposterous. Pijade concocted the idea of establishing on Mount Durmitor animal farms that would be modeled on the *sovkhozes* or Soviet state farms described by our Soviet-trained "Muscovites," and stocked with cattle seized from Chetnik peasants. With his lively imagination, Pijade threw himself into a detailed inventory and disposition of sheep, cows, bulls, sheds, pens, shepherds, milkmaids, herders, monthly and yearly yields of wool, milk, and meat. There was a fair supply of those animals—some twelve thousand sheep alone, with hopes for their increase along with improvement in planning and organization. For a beginning Pijade engaged Mitra, happy that he would at last have an intelligent and resourceful helper who, on top of everything, could take shorthand. True, neither of them knew much about farm animals. Mitra was from a small town where the well-to-do kept animals, so she might conceivably have known how many teats a cow had, but in the case of Moša Pijade even this was unlikely. However, that seemed unimportant for the job at hand. Who can expect herders to manage farms? Moša and Mitra zealously organized shepherd brigades, administrators, and inspectors. Provision was also made, to be sure, for competition between camps. Moša established strict economy and discipline, yet the animals kept disappearing, and the yields fell short. Mitra made fun of the whole venture with a merriment no more restrained than her diligence and devotion to the job. Even so, this project might have survived until the end of the war, had not the Chetniks and Italians swept down upon us, and the Partisan units and the peasants appropriated the animals. Overnight, everything disappeared except Moša's and Mitra's saddlebags crammed with regulations, inventories, decrees, and orders.

Tito withdrew to the village of Plužine, where I joined him on May 24. He stayed in the house of Governor Sočica, who had been killed by Montenegrin separatists. His house and large aviary were maintained by the Adžić family, also well-to-do, and favorably disposed to the Partisans. The bees buzzed in the fruit trees as bad news came in from all sides. Tito's Zdenka was in such a vile temper that she snapped even at Tito. During enemy offensives she always behaved as if the main objective of the Axis Powers was to destroy her personally. We did not lack for food, but since the Piva was full of trout I wheedled several sticks of dynamite from Vlado "Rus"* to kill fish. The war and our want mitigated the cruelty of this sport. One day I climbed a tree growing out of the side of a cliff, in order to toss the dynamite at just the right spot. The branch collapsed, and had I not been able to hang on to the tree trunk, I would have found myself within the radius of the explosion.

One day Zdenka really blew her top. Ashamed, Tito asked me in embarrassment, "What the devil is the matter with her?"

"I think she's in love with you," I said. "She's just carrying on."

Tito laughed mischievously. When I mentioned this to Ranković, he said with a guffaw, "You must be the only one in the whole army who doesn't know about their relationship!"

Tito had an even more grotesque experience in connection with Zdenka. Unnerved and humiliated by Zdenka's abuse of the escort battalion comrades, Tito turned to Djuro Vujović, an elderly Montenegrin, a Spanish War veteran and Tito's personal bodyguard.

"Tell me, Comrade Djuro, what am I to do with her?"

Djuro replied without hesitation, "I would have her shot, Comrade Tito!"

This reply of Djuro's was recounted with relish on all sides.

The region was incredibly beautiful: a canyon rising to the sky above a swift stream coursing between conical cliffs and pines nestling on the bluffs, and springs welling up in radiant green and blue eddies. Yet this stunning beauty failed to refresh me, but rather stirred my resentment. This was for me a new feeling: I beheld that manifold and inconceivable beauty, yet it was as if it wasn't I that was seeing it, but someone outside myself. I almost said aloud, "I can't experience this, I'm incapable of grasping the wonder of nature!"

Tito was also disturbed, but only over the prospect of abandoning Montenegro and deciding which lines of retreat to take. There were three possibilities: Serbia, eastern Bosnia, or western Bosnia. The cadres from Serbia yearned for their homeland, and the leaders believed Serbia to be crucial. Eastern Bosnia was convenient as the gateway to Serbia, but was a devastated land with a drained population. Through radio contact with Croatia we learned that there were considerable free territories in western Bosnia, as well as a massive Partisan movement. Such a distant march offered prospects of replenishing our forces, but also meant aban-

* Vladimir Smirnov, a Russian émigré, was an engineer who joined the Partisans and became an expert in destroying bridges. Later he became the chief of a unit of engineers.

doning, if only temporarily, a territory that was crucial to the outcome of the war.

Though Tito shared his concerns with me all along, this was particularly true now, when the other Central Committee members were scattered far and wide on other assignments. Tito also liked fishing, and joined me one afternoon. We were descending through the woods toward the mill built by the governor's father, and I was thinking out loud about the incredible position of the royal government, sitting in London and handing out commissions and the highest decorations to the allies of our country's invaders. Stopping in the middle of the slope, Tito said with conviction, "They shall never return to our country. The people will never forgive them."

These tensions sparked not only his farsightedness, but his rashness as well. When Tito got a report of a fairly easy breakthrough by the Chetniks on Pivska Planina and of the Second Proletarian Brigade's losses, he grew so angry that he ordered his escort battalion out, and placing himself at its head, set out for Mount Durmitor to show them in person how to fight. I knew Tito well enough by then to know that he was reluctant, for reasons of prestige, to go back on a decision. I set out with him. "At least let's die together!" I said, half in jest and half in earnest.

We descended to the Piva River, and as we moved along I kept trying to convince Tito to turn back. In his haste he had started out without a horse, but the three- or four-hour march over a rocky incline in the heat of the sun was no deterrent: he was sturdily built and an excellent walker. He had left Zdenka in the village, so I couldn't exploit her hysteria. I invoked the Comintern and the Russians: what would they say to this rashness? But this didn't sway Tito either. In the end, bitter and reckless, I told him that as supreme commander he had no right to push into battle at the head of a mere hundred Partisans, and leave the entire army without a head. That seemed to hit the mark. He slowed down, sat on a rock as if to rest, and said, "We also have wounded back there who haven't been evacuated." Without further explanation, Tito turned back.

In those days a very special relationship developed between Tito and myself—small irritations and an easy familiarity, as between father and son. I was the only Central Committee member who protested and came out openly against him. But I believe that I was also the most sincere toward him.

The very next day, if I remember correctly, I set out with Vlado Dedijer and Arso Jovanović for Pivska Planina. The Chetniks had broken through in considerable force. We had only two small battalions left with which to confront them. But we did secure the withdrawal of our wounded, around six hundred of them. That evening the clouds grew red over Durmitor's peaks, as Pijade and Pekić set Žabljak on fire to keep it from serving as a Chetnik stronghold. That operation did not seem justified to me even then; the Chetniks were also a guerrilla army, so for them every village could be a stronghold. On the other hand, it was justified not to accept the Chetnik takeover of a town which had been ours for so long and with such devotion. Though there were regrets over the burning of Žabljak, there was no criticism.

On June 9, 1942, at sunrise, Tito set out across Volujaka Mountain for the Sutjeska River. I pulled out to attend to some minor matter. On the way, I stumbled on a young peasant who was arguing with a group of soldiers. He caught sight of me, with my crisscrossed bandoliers and better clothing, and called out, "The staffers are on the run, aren't they?" The Chetniks used the term "staffer" to make fun of Communist officers. In my anger I slapped him. But this didn't fluster the peasant, who with a curse lunged at me; only my pointed pistol stopped him. The incident would have ended there had not the peasant been the brother of an officer, and had he not been talked into lodging a complaint with Tito. I had just barely got back when the peasant came running to Tito.

Tito was angry and upset: "When the Central Committee members act like this toward the people, no wonder the Chetniks grow stronger!"

I too was enraged, perhaps all the more because I felt ashamed and guilty. "He got off easy, after such a provocation!"

All that day and the next Tito and I didn't speak, and we avoided each other. Our closeness and mutual cares suddenly turned into anger and resentment. But we had to live together and fight together, so after we made camp in Grab, near Sutjeska, on June 11 or 12, Tito began to speak to me again, all the more gently and generously, as my silent reserve scarcely concealed my pangs of conscience and remorse.

At Čemerno near Gacko, Petar Lalaković lost his life. He was a peasant from Serbia, the bearer of the highest Serbian decoration in World War I, the Star of Karadjordje, and the first bearer of the highest Partisan decoration, the newly inaugurated National Hero of Yugoslavia. Stocky, prematurely gray, fairly well read, he was one of those capable peasants who distinguished themselves in great wars and political party struggles in Serbia. I had grown very close to him during our retreat from Nova Varoš. He remarked to me then that had our officers, during the retreat across Albania in 1915–1916, had as much understanding of their men as we did now, fewer would have perished. We respected him for his past and for his qualities. Yet who could protect a warrior who had brought all three of his sons into the war, and who looked upon war as a human condition and the fate of his people? His departure from the scene was all the more painful in that we never found out when he died or where.

News came also of the death of Ranković's wife, Andja. I was on cordial terms with her, and respected her as a persevering and courageous worker. I had met her in Foča. I believe she wanted to complain to me about her husband's liaison with an enterprising young secretary, but I kept out of this. She was assigned to a regiment as party secretary, so as to erase the impression that higher officials protected their kin. But she may also have taken this as a maneuver to get her out of the way of her husband and his secretary. Though Ranković and his secretary didn't have intimate relations, their association led Andja to something more than a suspicion. I believed that Andja died harboring doubts which her simple and honorable mind couldn't resolve.

Tito was also touched by Andja's death—especially because of Ranković, who collaborated with him most devotedly and had for him the most patient understanding. "We have to let him know! But how will we let him know?" asked Tito. I offered to do it, and Tito said, "You're the closest to him, so you'll be able to write it best."

I wrote Ranković a letter dated June 12, 1942—if I remember correctly—on a page out of a memo pad:

My dear Leko,

I have taken it upon myself to give you some painful news, the most painful. This is hard for me because I consider myself your closest friend. . . . Your faithful comrade Andja, who loved you so much, and who deserved that we all love her, died a hero's death on June 11 at the village of Dulić (between Antovac and Golijski Krstac), fighting a Chetnik band. Many other comrades died with her.

I will not try to console you. You are not the man for this, nor are these the times. But I do wish to say that I know you will bear this greatest of sorrows staunchly and calmly. True, Andja deserved your eternal mourning. You may perhaps find another comrade in your life and be happy with her—one cannot rule this out. But you will not find another Andja, who was so devotedly yours. Nevertheless, be strong and steadfast. And do not forget—your loss is also the party's loss, and we all share in your sorrow.

Comrade Leko, I have no more to say in closing, except if any friend's love

can ease this sorrow, then this would be mine and from us all. I feel this love for you today, all the more in that your sorrow—and mine—for Andja is so great.
12 June 1942

Your Djido

Italian planes were observing our positions, so early the next day we moved down the Sutjeska. The staff first made a stopover in a little forest above Tjentište, and two or three days later moved into a forest near the village of Vrbnica. All the villages in the Sutjeska Valley had been destroyed. First the Ustashi burned down the Orthodox villages, and then the Chetniks burned down the Moslem villages. The only houses and people left were in the neighboring hills. The devastation was all the more horrifying in that here and there a shaky door frame, a blackened wall, or a charred plum tree stuck out of the tall weeds and undergrowth. Though lush greenery swayed in the cool breeze on either side of the swift river, my memories of those days are weighed down with bitterness, hurt, and horror.

On the afternoon of June 14 I set out to attend a conference concerning the wounded. Just as I reached Tjentište a couple of Italian bombers (Savoia-Marchetti) flew overhead. Men and horses jumped off the road; after tying my horse to a tree, I also ran to the Sutjeska River. I sat by the edge of a gulley with a soldier, while several old Montenegrins clung to the bushes along the steep riverbanks. The planes were flying around the mountain peaks; I waited until they had receded beyond Vučevo. One dropped a string of bombs that formed an arch whose top was falling straight at us. I had seen bombs falling before, but never a whole row of bombs coming straight at me so swiftly and unrelentingly. I cried out, "Get down! The bombs are coming at us!" I threw myself into the gulley as one after the other the bombs exploded from the direction of the river and along the bank. One fell five or six yards away. I thought, If there's another, it'll hit us directly. But there were no more bombs. Rather, a cloud of dust settled over us. I got up. My neck was stiff. The soldier got up, too, with a wound on his cheek. "My neck is stiff!" he moaned. I patted him on the back: "That's nothing. It'll go away."

Down by the bank, the Montenegrins crawled out. One of them stretched himself out on the grass with a bloodied thigh. I knew him—the uncle of Budo Tomović. He showered me with a traditional pathos: "I don't matter—I've lived and fought long enough! Let the youth live! May my wounds bring good to the army!" Apparently the target wasn't us, but nearby campfires and the road with the horses. Unlike the

German pilots, the Italians aimed so badly from their antiquated planes that it was dangerous to find oneself outside their target area.

I ran to my horse and raced to get the doctor. Her name was Saša Božović, the wife of Dr. Boro Božović, today a professor and academician. Saša's own leg was wounded; she could hardly walk, so I brought my horse for her. I helped her mount and led the horse across the meadow, though bombs were still falling some distance away. Saša was not at all impressed by the Montenegrin's rhetoric. She examined the wound, stuck on a bandage, and said, "It's nothing—just a scratch! You can get up and go." And the Montenegrin got up and went, but with an ostentatious limp.

I took care of my business and returned to headquarters at dusk. Everybody had watched the bombardment and the people rushing away from the cloud of dust. I told them that I, too, had been there. This didn't particularly excite anyone, as if they wouldn't have missed me much if I'd perished. In war only great and unexpected disasters are exciting, and then only momentarily. I lived to experience many bombardments, but none that horrified and paralyzed me quite like this one.

I also had several unpleasant experiences in the party, but none affected and seared me as much as my punishment some two or three days after that bombardment. Tito called a Central Committee meeting at which he reported that our failures in Montenegro and Hercegovina had placed us in a serious predicament. "Comrades Milutinović and Djilas are much to blame," he concluded, "and for this I believe they deserve *na vid.*" He used the Russian expression *na vid,* which is equivalent to a reprimand—the next to the worst of the party's five punishments. Pijade and Žujović were quick to agree, though they were not members of the Politburo. Ranković was silent while Lola Ribar, who was keeping the minutes, recorded the punishment.

No analysis was ever made of what constituted Milutinović's "errors" and mine. They were not the same, yet it somehow came out that they were. In truth I didn't make any grave errors in my second campaign in Montenegro, nor has anyone ever showed me one. And if I had made grave errors earlier, why did they send me to Montenegro again? I took the punishment as something hasty and unjust, not so much because I thought I had made no mistakes, but because they were no greater than anybody else's, and were never spelled out in detail. Milutinović, however, was cut to the quick. He knew nothing outside the party, and now that party was punishing him unjustly. His death in 1944 and his elevation into a party saint helped my status and position: nobody ever mentioned these reprimands. Apparently Tito was the first to forget them. He never brought them up again, and Milutinović and I continued to enjoy his trust. In 1952 Ranković told me that Tito had found a record of that reprimand among his papers, that he had forgot all about it, and that the whole matter should be annulled. It was not annulled, however, or brought up again either—until my fall from power. At the Seventh Congress of the League of Communists in 1957, Ranković did not fail to mention that I had received a "high" party reprimand during the war, though he said nothing about Milutinović.

Milutinović and I were working on rallying the Montenegrin non-party patriots. The bulk of the work was done by Milutinović. I helped him by organizing discussions and writing a resolution. The resolution denounced Draža Mihailović's collaboration with the invader, and was signed by well-known public figures. Tito sent the resolution to the Comintern on June 17, 1942, but it was not broadcast by Radio Free Yugoslavia until July 6—presumably, after being "screened" by the Soviet government. The resolution had a special significance: for the first time the Western press wrote openly about Mihailović's collaboration with the Italians, and about the Partisan movement's activities.

Our headquarters were moved to a forest near the village of Vrbnica. Two Montenegrin brigades, and one Sandžak and one Hercegovinian brigade, were formed. Peko Dapčević was made commander of the Fourth Montenegrin Brigade, and Sava Kovačević the commander of the Fifth Montenegrin Brigade. In my talks with Sava I detected more pride on becoming commander of a brigade than dejection over the retreat from Montenegro. I asked him about the Italian prisoners, half of whose rations we had kept for our staff in Gornje Polje. He said they had been executed. At my remark that this should not have happened, he replied, "The Italian command would have found out anyway that we were confiscating half the food. Besides, they were so weak they could hardly move." At Supreme Staff, too, they were not very happy over such treatment of prisoners, even though the Italians and Chetniks hardly spared ours. Yet no one reproached Sava; our people were so embittered that excess seemed more understandable and justified than laxity.

We assembled enough manpower in the Sutjeska Valley for four brigades, not counting the First and the Second from Serbia. Such a force would have been adequate to retake Montenegro from the Chetniks, had the Italians not helped them with munitions, food, and troops, and had in exhausted and frightened people not come to believe that the Communists were losing. A decision had to be made as to the direction of our retreat from this devastated region where we lacked everything, especially food. We could no longer survive on only unsalted meat, a diet which depressed and wearied us.

On June 19, 1942, in the woods near Vrbnica, a Central Committee meeting was held at which it was decided to move four brigades to western Bosnia, but to keep the Hercegovinians and Sava Kovačević's Fifth Brigade where they were in order to operate in the area after the Italo-Chetnik offensive had subsided. That decision—which was later shown to be very significant indeed—was proclaimed rather than proposed by Tito; he was apprehensive that it might be opposed. But nobody opposed it, for his stand was apparent from the discussion. I went along because my experience in Nova Varoš had alerted me to conditions in Serbia. Those members who were sentimental over Serbia and its importance were persuaded by Tito's authority and rationality. Tito mentioned after the war, half in jest, that on that occasion he had pulled rank. Undoubtedly he was determined to impose this decision, all the more so in that the Comintern and reason stood behind him.

Following this meeting, Tito and Arso Jovanović held a consultation with the officers. But no meeting of the Supreme Staff was held, either then or in the course of the war. Tito personally appointed the Supreme Staff members, in the name of the Central Committee, from among the distinguished leaders of the uprising—largely party leaders, and among them myself. The original Supreme Staff established in Foča was enlarged. However, this was a representative body which never met. The Staff was, in fact, Tito. Not even those members of the Central Committee who might have been on hand during military councils played a greater role than the commanders, and their role was probably smaller. Everyone could put forward his opinion, and if it met with Tito's approval it would be adopted most often tacitly. Tito had frequent differences with Arso Jovanović, largely over incidental matters. Arso had no recourse but to give in. With Terzić, who replaced Arso for a while, there was very little friction, because Terzić accepted Tito's views without much discussion. Jovanović and Terzić were both good and brave officers, but they had little standing with the commanders. The commanders themselves preferred to deal with Tito. And Tito in turn was not happy to have anyone interfere with his command. He cultivated personal, though not cordial, relations with the commanders. After the fall of Italy, when the army got bigger and operations more complicated, the technical services of the Staff proliferated and acquired a greater role. None of the Central Committee members except Žujović had exhibited any particular military bent. But all learned something about warfare. Tito did not support the military ambitions of party leaders, nor encourage a greater political role for the commanders. Although there was no conflict between the party and the army, Tito combined both powers and thus maintained a balance. The party was the more important. The commanders and most of the officers were party members; ideology commanded the rifle.

The New Army

1

The "Long March" for the Supreme Staff began early in the morning of June 23, 1942. The day before, King Peter II had visited Roosevelt, and the next day Churchill joined them. The King's entourage concealed the facts from Roosevelt, but Churchill was kept informed by his missions and certainly apprised Roosevelt in turn. Although the Comintern's agencies, or rather those of the Soviet Union, had finally mentioned us publicly, we were still an army of "outlaws." Our sufferings and feats were ascribed to our opponents, whose enterprise had largely consisted of concluding alliances with the invaders against us. But injustice abroad and defeat at home—a defeat all the more unjust in that it was inflicted by an opponent who made up for his own weakness with aid from the invader—gave rise to inexhaustible spiritual energies within us.

Before his departure for the new territories, Tito issued an order which prescribed the severest punishment for the least offense against the population or against property. The Communists implemented the order, and the army, hungry and in rags, accepted the order with selfless pride. This ideological sharpening had led to a clarification of feelings toward the people. On the second day of the march, a Montenegrin peasant woman whose cow had been seized by the Partisans came to the Supreme Staff. Tito was horrified: "Don't our comrades know that in the village a cow is like a member of the household?" So I went from unit to unit until I found a cow good enough to satisfy the woman.

During the first days of the march, Lola Ribar and I wrote a pamphlet about Soviet-British relations. Obsessed by the recent party chastisement, I certainly did my best to demonstrate my party loyalty to myself and others. The pamphlet was printed and reprinted and evoked a fair response, for it showed the advantages to us in Soviet-British relations. When I came across this brochure after the war, I was stunned by its stylistic and conceptual clichés: for fighting men, simple "relevant" formulas are more important than complicated but "irrelevant" truths.

Tito too was struggling within himself. His were the greater troubles, since his was the greater responsibility. His problems were not only those of his own land, which could be discussed, but those unmentionable, irrational ones that came from Moscow. Whenever Tito woke up from a nap, he would emerge through the door of his little tent, look around, then get angry and rail at silly, petty, and nonexistent violations. This would last a couple of minutes, and then he would recover and withdraw in shame into the tent, only to re-emerge calm and collected after a rest. By tacit agreement Ranković and I saw to it that no one disturbed him during such scenes. After our arrival in Bosnia these outbursts of Tito's ceased and never recurred, not even during the heaviest fighting.

The four brigades, along with the refugees and wounded, moved separately in two groups. The northern group was made up of the Second and Fourth Brigades, and Arso Jovanović and I were assigned to it as military-political co-ordinators. The southern group was composed of the First Proletarian and Third Sandžak Brigades, and was under Tito's direct command. But the two groups were not separated until we reached the mountains above the Sarajevo-Konjic railroad. The march lasted around three weeks, with short interruptions, and took us across mountains or, as Arso insisted, along the Central Bosnian Divide, the longest in the Balkans. We proceeded without any serious encounters until we reached the railroad and Krajina.

The first clash took place one evening at Kalinovik, with Chetniks from Serbian villages, and the next morning with an Ustashi Home Guard garrison. The Chetnik attack provided our people with a good excuse to help themselves to the food in the Chetnik houses without violating Tito's order. That night, while crossing the road, Tito, Zdenka, and I lost contact with our escort. We were concerned about the unfamiliar territory and the possibility of running into the enemy. Finally Tito decided to wait while I went ahead to make contact, which I soon succeeded in doing.

At dawn we found ourselves on the meadows below the peaks of Treskavica. We heard firing from the direction of Kalinovik, so Tito sent Arso and me back to see what was going on. We climbed up a cliff by the side of the road. Kalinovik lay below us; our units were withdrawing in orderly fashion in single file, zigzagging amid the grass and rocks. Bullets whistled all around, but I was seized by a wicked urge to test Arso, a former royal officer, and he went along with it, probably because he didn't want to appear overcautious before a Central Committee member. The soldiers shouted at us to take cover, and Arso finally agreed. "After all, how can I justify my allowing a Central Committee member to expose himself to danger needlessly?" he asked. In all that host of people, so they informed me, it was my cousin Stana who lost her life. I felt sorry for her, though I regarded this as the best way out for her, since she was half-blind and well along in years.

We spent all our time in the woods. Our surgeon, a captured Italian who had conscientiously attended to our wounded, ran away. A couple of days later we set out before sunrise. It was a radiant morning beneath the peaks of Treskavica. Two Italian bombers caught us in a clearing. The command "Take cover!" was heard. But the planes darted over the peaks without noticing our columns, and dropped their loads on the woods where we had spent our days.

Suddenly darkness fell over the mountain and it began to shower. By afternoon the sky had cleared, and the Fourth Brigade surrounded the village of Rakitnica. The brigade staff had moved ahead. We stopped just short of the village, from which bullets began to fly, until the brigade rushed forward across the wet pastures and grain fields. The hair of our girls streamed behind them. With the other staff members, I hastened to the village in bold excitement. The Moslem peasants met us apprehensively. We captured a gendarme and shot him, as his wife wailed with an infant in her arms. Peasants can instantly gauge what kind of an army they are dealing with; these villagers understood that we prohibited looting, and soon showered us with complaints about it. A patrol brought in a Montenegrin, a mustachioed fellow from Morača. He admitted nothing.

"We can still see the cream on your mustache!" I said.

"I was hungry," he said.

"We'll have you shot as an example!"

"Comrade Djido, don't let me die shamefully over a jug of milk, but in battle as befits a man."

"Take care, next time."

Ashamed, I let him go. A few days later, two Montenegrins from the First Proletarian Brigade were shot for a similar offense.

This side of Igman we met up with a unit which included a friend of mine from the university who was now a functionary, Boriša Kovačević, and also a former officer, Rade Hamović, who became a chief on the General Staff after the war. The unit had been shattered and was down to some twenty men. It was divided within: Boriša insisted that it should first be brought up to strength and then go into action, while Hamović claimed the opposite. I was sorry for Boriša, who was disciplined and felt dejected, but in council I supported Hamović: the unit could be strengthened only in action.

It was here that our groups split up. The southern group, Tito's, saw more action. But ours did not lag far behind. We were closer to Sarajevo, a strong center of resistance. One night we descended on the railway at Hadžići. The attack was led by Ljubo Vučković, the commander of the Fourth Brigade, and later a chief on the General Staff. Peko and Arso poked fun at Ljubo: "His girl friend is there. He'll capture the place, all right!" The success was hardly significant: the station and a nearby sawmill were burned down. But before we could tear up the railroad tracks an armored train came at us, illuminated the area, and sprayed it with machine-gun fire. We took cover behind a cluster of haystacks. I remember World War I veterans telling me that bullets didn't penetrate hay or wool. We also captured a real live Ustashi. I asked him if he would join the Partisans. He replied, "Why not? You eat lamb too, don't you?"

On July 6 we took the little town of Kreševo without a struggle, and captured so much food that we had to burn the better part of it. This was necessary because later that day the Ustashi from Sarajevo counterattacked, and we drew back into the hills.

On the plateau overlooking Kreševo, Svetozar Vukmanović-Tempo appeared with a squad of ten. The last time I had seen him was before my departure from Belgrade. I tried to have a friendly chat with him and learn about conditions in eastern Bosnia, but he was on edge and wanted to talk only with Tito. We gave him directions and he went off. However, that afternoon we ran into him on an even higher plateau: he had been slowed down by encounters with the Ustashi along the way. In the meantime he had relaxed a little, but that night he set out again to find Tito. In his memoirs Vukmanović says that Tito told him, "With only one squad you passed almost openly through villages that you'd abandoned. Just try that in Montenegro or Hercegovina!" Tempo could indeed move about "almost openly" here, because the Ustashi regime was without much military force.

On Vranić Mountain we seized a herd of oxen and cows and some cheese from a man who traded with the enemy. Boško Djuričković, a Belgrade student and battalion commissar, took the share of the cheese he thought should go to his battalion. The Serb commented with innocent mockery that the Montenegrins were always first when it came to booty.

I threatened Djuričković: "I'll dismiss you and disarm you!"

He retorted, "You may dismiss me—you have the power. As for disarming me, you didn't give me my rifle, so you can't take it away!"

It flashed through my mind that this is how his grandfather might have answered a Montenegrin chieftain, and I withdrew in silence. I always had great respect for Djuričković, who attained the rank of general.

The Ustashi at the Šebešić station, in a wooded area near Bugojno, fled before us. We burned down the station and distributed among the inhabitants all the food supplies we couldn't take with us. The fire was brightly reflected against the hillsides. I helped a slender little woman lift a sack of flour on her shoulder; she bore it away with an incredible ease.

Sharper and more stubborn resistance met us as we entered the valley of Bugojno. Hostilities were keener there: the Croats and a considerable number of Moslems sided with the Ustashi, while the Serbs were for the Partisans, and a few here and there for the Chetniks. Some also were passive—Moslems when alone with the Croats, and Serbs too close to the Ustashi. The war had simplified relations, but our impressions were only superficial because of the shortness of our stay. Fearful of the highways and railroads, we skirted around the valley to the north—an unnecessary effort, but we had plenty of food.

The Ustashi conducted raids out of Bugojno. While we were sitting in a log cabin, a grenade exploded nearby and a piece of shrapnel mortally wounded the battalion's political commissar, a worker from Serbia. At the station at Šebešić the battalion commander Duško Dugalić, a Serb, also died. I had quickly become friends with this easygoing student with Tatar features who had once supported himself by selling books in cafés, and who waged war in the mountains of Sandžak and Bosnia as if making merry at a wedding feast. It was rumored in the brigade that Cana Babović was partly to blame for his death. Cana was acting political commissar of the brigade; hard-boiled and hard-driving, she couldn't stand the sharp-tongued, lively Dugalić. It had been suggested at headquarters that he be replaced as commander, but he was defended by the political commissar Kušić, because Cana may have gotten into his hair, too. Two or three men lost their lives at the station in Šebešić, and Dugalić, they said, wanted to counteract Cana's "criticism" with some exceptional act of courage. However, the Ustashi had strung wires across the windows to repel bombs. Dugalić died there without accomplishing his act of heroism. Cana later mourned him and stressed his good qualities.

On the night of July 15 we forded the Vrbas River, and the next morning came to several Serbian Partisan villages; the peasants were delighted that comrades in arms had come to them all the way from Serbia and Montenegro. They regarded our arrival as help, not a retreat, and that was how it all turned out.

We took the little town of Prusac. The Moslems offered no resistance, but the young men fled to Bugojno. We established contact with Tito's southern group. They had taken and then abandoned Prozor, and had burned down Šćit Monastery—along with the monastery library, which contained some rare books. The monastery was burned because the prior had called in the Ustashi, who posted themselves in the belfry, and killed and wounded several Partisans. As a rule the Partisans never touched churches. But they were more than happy that the necessity of destroying the Ustashi meant burning down a monastery, rather than just a secular building.

The Ustashi carried out frequent scouting raids which always ended in their fleeing to Bogojno. On July 16 the Second Proletarian Brigade occupied the village of Hurije, and on the night of July 16–17 the Fourth Montenegrin Brigade attacked Bugojno, but was repulsed with losses. On the next day, July 17, the Ustashi conducted attacks in small groups which would suddenly dart out from the stream and then just as suddenly dart back again.

On the morning of July 18 we received a report that the Ustashi of Francetić's Black Legion had slaughtered the population in the village of Hurije, and I went there to investigate. The village was Serbian and "guilty" in the sense that our units had spent the night there, despite the villagers' protests that they might have to suffer for it. Ten days later, when I met Tito, I told him about it. He was horrified. Vladimir Dedijer begged me to include my account in his *Diary,* so the description of the event was preserved. All those the Ustashi found—men, women, children—were killed. There is no mention in my notes of Communists, much less of Partisans there. My notes end with the following words: "I felt nothing. I did not feel sorrow even for my dear friends who perished in that village. It seemed to me that I, too, could die just as easily, without emotion, not dodging the bullet even if I knew its course. No feeling of oneself or of one's own life. Here the peasant women didn't flee, no one wept. Yet a person could just as coldly take revenge. Truly there is only one thought in my brain and in my heart, in my whole being, and surely in others, too: it isn't worth living on this earth as long as there are men who commit such atrocities. There is no other way: it is a question of either us—or them."

This note, slightly edited, was soon thereafter published in the *Military and Political Review* of the Supreme Staff, and after the war made its way into school readers. However, after my name had been blotted out, the note also disappeared and fell into oblivion.

The Supreme Staff ordered another attack on Bugojno for July 20–21. The Second Proletarian Brigade took part in that attack and I went along. The attack was badly co-ordinated and poorly led, so it failed. During the retreat that night, I lost my way and the staff informed Tito. Tito was quite disturbed, I was told, and said he wouldn't let me leave his side if I came back alive.

The Second Proletarian Brigade proceeded to Kupres on July 31, 1942, and with the help of the local Partisans occupied Zlosela—a cluster of villages inhabited by Catholics, whose men had fled to Kupres and whose women met us with silent hatred. It was there that I met Simo Šolaja, a legendary hero and the local leader of the rebel Serbian peasants. He was blond, sturdily built, sincere, and rather simple. A bit abashed over the slowness and lack of organization of his own fighters, he was dazzled by the presence of the Proletarians and the party chiefs.

Supreme Staff had now brought all units under its own direct command, and by the next day I again joined Tito—somewhere on Malovan Mountain. Arso Jovanović had arrived before me and attacks were being planned—first on Livno, and then on Kupres.

2

The Supreme Staff itself was moved to Cincar Mountain.

On the night of August 4–5, 1942, we attacked Livno; by morning the town was ours, except for a single building where the Ustashi were resisting. Ranković left for the town the next day and I followed—he to mop up the fifth column, and I to gather propaganda material.

In actual fact we had taken what was easiest to take—points of resistance manned by the Home Guards, that is, the regular army and local militia. This exposed the rear of the Ustashi unit that had drawn back into Dr. Mitrović's villa, which was fortified and well located for a defense. The Ustashi had killed Dr. Mitrović and his family and turned the villa over to German mining experts. I made my way to the other side of the town to the Catholic monastery of Gorica, where the staff of the First Proletarian Brigade was located. The monastery was built of stone with beautiful and comfortable chambers, and big shady trees in the courtyard. Its belfry had been used as a machine-gun nest at the request of the prior or his assistant. The Catholic Church—and particularly the Franciscans, its most active element in Bosnia—accepted Pavelić's state as legitimate and their own. Nevertheless, we did not molest the monks, who were stunned; it was enough of a blow for them that atheists had taken over the sanctuary. Only Koča Popović poked fun at one of the monks, who was so senile he didn't know what was going on.

Livno is predominantly Moslem. Not many Livno Moslems had helped the Ustashi, so they met us with fear but not hatred. The captured Home Guards—a whole battalion of some 700—were put in a camp but not strictly guarded. They had shot at us—there had been a good deal of firing—but hadn't inflicted many casualties. The minute the Partisans had approached their positions, they had thrown down their arms. Some even boasted that this was the third rifle they had handed over to the Partisans. They were mostly peasants from northern Croatia, peace-loving and industrious: the prewar core of followers of the Croatian Peasant party. They were the best-natured and weakest force we were to fight. The Ustashi called them—and with good reason—the "Partisan supply unit." Soon they were all released. They didn't join the Partisans—they said they would do that when they got home!—except for one. His name was Ivo and since he was from Zagorje, Tito took him as his groom. Ivo was pleased with the cordiality and equality among the Partisans, but never attained this himself. He served Tito faithfully and diligently until the Fifth Offensive was completed in the spring of 1943, when he asked to go on leave to his village, and was shot there by the Ustashi.

On the morning of July 7, the day of my arrival, the Ustashi and the German units in the Mitrović villa surrendered. They would have held out longer, had the Partisans not knocked down the wall of the adjoining house and begun to demolish the villa with direct artillery fire. Faced with annihilation, and having our promise that they would be judged by a court-martial, they surrendered. The Germans, who were not military personnel but mining experts, had apparently exerted pressure on the Ustashi to surrender, but not before the Ustashi had hit a group of Montenegrins with a mortar shell and killed ten.

I had not seen Mitra since the battle of Sutjeska. She had been assigned to the newly formed political section of the First Proletarian Brigade. These political sections were modeled on similar units in the Red Army. Since we had too many officials, Ranković created these political sections, which would concern themselves with building up the party in the army. Mitra was more than pleased with life in the unit, all the more so because, as a result of her efforts, the party organization had revived and had come to dominate the "military." She had become estranged from me, as if I weren't irreplaceable for her.

Mitra and I stepped out of the monastery to have a talk. We were strolling along the road which led to the monastery, when a group of Ustashi prisoners guarded by two Partisans came by. They were being taken to the monastery, now a prison and interrogation center. As the group entered the monastery grounds, one of the Ustashi jumped through the wire fence along the road and rushed across the meadow toward a pile of boulders. The guards turned to fire at him at random, and the rest of the Ustashi began to undo the ropes around their wrists with their teeth. I pulled out my revolver, with the barrel struck the nearest Ustashi on the chin, and shouted that I would shoot if they didn't all stand still. Then I grabbed a rifle from one of the guards—he relinquished it in confusion—and jumped out on the road. The Ustashi fugitive had almost reached the boulders. I took aim and fired. He continued to run. I knew that the soldier had fired two bullets and I one, so there had to be two more; but I had to aim carefully—he was more than two hundred yards away. I leaned the barrel on the fence and fired. The Ustashi staggered. I fired again. The Ustashi fell into the ferns. Meanwhile, led by Mijalko Todorović, couriers came running from the monastery, shooting wildly. "One got away!" Todorović was shouting. I told him that the Ustashi was lying in the ferns. Todorović went over to take a look and told me he was breathing his last. I felt no regret, nor pride either. I had done my duty, like any soldier.

After lunch I went with Mitra and Dedijer to requisition a printing press and some paper. The press was antiquated and poor, but good enough for Partisan needs. Along the way we stopped to look at the Mitrović villa. Its owners had obviously enjoyed a comfortable life as well as the beauties of the region: below flowed a swift stream which sprang from a nearby cliff. Inside, the spacious living room was so crammed with half-naked corpses that it was difficult to pass among them. In the summer heat the corpses had begun to decompose. To this day it isn't clear just how they all died. It was said that they were killed by a shell from our artillery, but there was no trace of such a blast in the room. My theory is that the wounded were killed off by the Partisans or else by the Ustashi, who wanted to protect them from torture.

Throngs of peasants from the Serbian villages began to arrive at the monastery, bringing specific charges against the Ustashi. Ranković patiently collected all the facts and sorted out the prisoners, including ten civilians and ninety Ustashi. I was staying at the monastery, so I was able to get a look at them—all young men in fine physical condition, but with unpromising former occupations: store clerks, barbers, waiters, high school dropouts, village toughs. That night I was assailed by insomnia: for the first time I thought of the Ustashi as my own people. In the morning I expressed my doubts—I no longer remember to whom. Were all these young men criminals—lost to the people, the struggle, and the future? Couldn't at least some be singled out, rehabilitated, and saved?

Did all of them have to be jammed into Room A for execution, rather than Room B for further investigation?

I entered Ranković's office and laid all my doubts before him: my concern lest the Ustashi be treated as mere stereotypes. Ranković didn't lose his calm. He may even have said, "Now you'll see for yourself." A guard brought in the next Ustashi—a well-built young man, blond, with fine features. Ranković glanced at his record and told the Ustashi that witnesses had charged him with slitting the throats of six people. The young man replied readily, "It isn't true—I slit only two!"

Ranković asked the Ustashi two or three more questions regarding his participation in the fighting, then ordered him taken to Room A. "You see," he said, turning to me, "there's nothing I can do. They're all like that."

I fell silent. Yet I was not ashamed or disheartened because of my illusions. What Ranković was doing I too would have done, perhaps more expeditiously.

I was also present at the interrogation of a camp counselor of the Ustashi youth, a tall, dark-eyed girl. To my query as to why she wouldn't join the Partisans, she replied, "Because it would be immoral to change one's views!" She stood up for herself bravely. However, Ranković later reported that she weakened at her execution, and was weeping and trembling.

All the Ustashi were sentenced to death, but none was tortured. The civilians were all released. The executions were carried out by Montenegrins, who volunteered so as to avenge their comrades killed in the mortar fire. The condemned were led away at night, in groups of twenty. The Ustashi and the Montenegrins were agitated; some were smoking nervously. At first glance they couldn't be told apart, except that some had rifles and stars, while others had wire around their wrists. The executions took place in a nearby gulley; the shots were heard in the monastery. As usual, no effort was made to bury them properly: their legs and arms stuck out of the mound. Civil war has little regard for graves, funerals, requiems.

The case at Livno was an exception. The Ustashi did not generally surrender in groups, but only as individuals here and there. When we took Jajce on September 25, some fifty Ustashi committed suicide, many by blowing themselves up with hand grenades. In Šćit they came out onto the burning belfry to be cut down by Partisan machine guns.

Why all this hatred for the Serbs, Jews, and Gypsies, and such personal desperation, I asked myself in wonderment. Obviously this was not just because of "Greater Serbian oppression." Serbian children were hardly able to oppress anyone, and the Jews and Gypsies—who were also being destroyed—had never ruled over old Yugoslavia. Why did Pavelić agree to stop the massacre of Serbs converted to Catholicism, if he hated Serbs and considered them an inferior race? The Ustashi were not a party, but a military organization of the select. There were supposed to be 32,000 Ustashi, but there were more like 15,000. Can lawlessness and murder make an ideology? The Ustashi didn't strike me as having a coherent ideology. They adopted Nazism and fascism out of weakness, but behind this stood something more basic and of longer memory. What? Catholics and Orthodox were not at such odds in these parts in ages past—perhaps because Islam, the common enemy, had oppressed both. Yet they had developed separately—within the same country, but as if in separate lands. The Ustashi concocted a vocabulary of their own which no one ridiculed more than the Croats themselves, but from which there sometimes emerged a strikingly beautiful word. Hitler's invasion unearthed the long pent-up shadows of ages past and gave them a new dress, a new motivation: neighbors who might have lived out their lives side by side were now all of a sudden plundering and annihilating one another! I wondered if this were not some kind of socially conditioned human malady which would disappear in a society free of oppression. That thought was adequate for fighting a war, but not for one's peace of mind.

One hears to this day that without the Ustashi massacres there would have been no uprising of Serbs in Pavelić's state. This line of thinking never occurred to anyone at that time, and belongs to a subsequent stereotyped approach to political realities. Of course, had Pavelić's state not been what it was—had it had any legitimate, rational foundation—the uprising would not have flared so swiftly. But one kind of guerrilla movement or another would have developed as the Axis Powers weakened. The Serbs wouldn't have accepted any alien state, not even Croatian, unless they had to. And the Communists were the most self-sacrificing and resourceful in arousing and organizing the Serbian masses in Bosnia and Croatia.

As for the Germans captured at Livno, it was taken for granted that they would be shot. However, in the course of their interrogation the idea came up that, since they were civilians, we could offer them in exchange for our captured comrades. One of the prisoners, the engineer Hans Ott, was particularly energetic in this regard. A rather portly man in his fifties, Ott exhibited an unusual flexibility and even understanding for the Partisans, while he spoke of the Ustashi and the Chetniks with contempt and disgust. Vlatko Velebit, because of his adaptability and excellent knowledge of German, spent a good deal of time with Ott. The German command agreed to an exchange. Dedijer's wife Olga came back as one of those exchanged. She was especially welcomed by our medical corps as a good and tireless physician.

It was at that very time—on August 3, 1942—that Moscow reached an agreement with the Yugoslav émigré government to raise the legation in Moscow to embassy rank. Tito had experienced so much trouble and bitterness that on August 11 he sent the following message to the Comintern:

> Can nothing be done to better inform the Soviet government of the traitorous role of the Yugoslav government and of the superhuman sufferings and hardships of our people, who are fighting against the invaders, the Chetniks, the Ustashi, etc.? Don't you believe what we are telling you daily? . . . [The legation's elevation to an embassy—M. Dj.] can have terrible consequences for our struggle. We emphasize: the Yugoslav government is collaborating openly with the Italians, and covertly with the Germans. That government is a traitor to our people and to the Soviet Union.*

Negotiations over the exchange of the German prisoners were still in progress when our side made attacks on Kupres—the first on August 11 and the second on August 13. Kupres was a wedge inside liberated territory. It was a small town of some fifty scattered stone houses, converted by the Ustashi into a fortress. They possessed considerable forces: around 1,500 Black Legionnaires, the militia from the surrounding villages, an armored car, a battery of guns, and a brave and clever commander, Jure Francetić. The first attack revealed that without heavy weapons our units couldn't take Kupres. They reached the town easily enough, but once there they couldn't establish a foothold because of the heavy machine-gun fire. Arso Jovanović pointed this out, but Tito was determined. In the second attack the soldiers from Krajina also took part, eager to show that they were better than the Proletarians. But it was this attack that cost us the greater losses: about 100 dead and 150 wounded all told—far too many for a guerrilla engagement. Infuriated, the Partisans set fire to Zlosela. Still, this was not like the defeat at Plevlja; the army had been purged and strengthened, and could withstand any defeat and losses.

* *Hronologija*, p. 315.

The arrival of the Supreme Staff brigades meant a stepping up of the fighting. Before long we moved to Glamoč. Dušan Mugoša got through to us from Albania; the uprising was burgeoning there as well, though Tito pointed to an undeveloped party as a basic weakness and danger in Albania. Mugoša stayed till October, when he set out again for Albania with Blažo Jovanović.

At a Central Committee meeting it was decided that here, too, in Krajina, a greater effort should be made to create brigades and organize a government in the rear. At the end, under "other business," Tito proposed that Krsto Popivoda and Lidija Jovanović be ousted from the party, because in Montenegro they had demobilized a detachment of soldiers who were later rounded up in their homes by fifth columnists and some of them shot. Ranković was silent; Pijade agreed. Only I objected: I said Krsto had done it out of ineptitude, and Lidija went along because Krsto was her senior. Tito rejoined, "For such an offense even shooting wouldn't be too great a punishment!"

I said to Ranković later, "Why kick Popivoda out when we'll just have to take him back in again?"

"Nothing we can do about it now," he replied.

Popivoda was also demoted to the rank of private, and in that capacity displayed great heroism. Ranković later put him to work in the Central Committee; in the following year, little by little he came back to the party.

Lola Ribar returned from unliberated territory accompanied by his father Ivan. Old Ribar, who had fled from Serbia, brought us nothing but bad news. Tito decided to send me to work in Belgrade. I began growing a mustache as a disguise. I was filled with apprehension and a high sense of responsibility.

The arrival of the Supreme Staff brigades in Krajina, though a matter of necessity, had proved advantageous for everyone. Peripheral and sporadic wavering toward the Chetniks ceased in the area. Local detachments gained enthusiasm, and the local government and party grew more systematic. Dalmatia, where the uprising had flared up, acquired a firm center of resistance with the fall of Livno. A direct link was established with Croatia, and an indirect one with Slovenia. The Central Committee and the Supreme Staff found themselves in a more stable position and with better prospects ahead.

At a council meeting in Glamoč with military and party leaders from Croatia on August 25, 1942, a more specific and energetic course was adopted for the formation of brigades and divisions, and for a greater activity around the Una and the Sava rivers, and in Dalmatia. The result was that on November 1, 1942, the Supreme Staff issued an order designating the National Liberation Army as a regular army, thereby relegating the Partisan detachments to a secondary position both in name and in practice. It was here that Tito's most essential quality came fully to the fore, in the creation of new forms of organization. He did not "manufacture" ideas of his own; rather he accepted or rejected ready-made ones easily and deftly. It was as if for him ideas did not exist, except as road signs to reality—or perhaps, to put it more precisely, to some part of reality which could be shaped and mastered by him. So, too, the idea of transforming a multitude of diverse detachments into regular army units arose not only out of the ways and means of warfare, but from a perception of future possibilities.

The various sections of the Supreme Staff also proliferated, especially the education of cadres under Ranković, and sanitation under Dr. Gojko Nikoliš. On Pijade's initiative, directives were issued concerning the organization of rear-guard commands and the structure and tasks of the civil authority. An Agitation and Propaganda Section was also created under Supreme Staff, with myself as chief. Until then, no such agency had existed; whenever the need for it arose, one of the leading comrades—most often Kardelj or I—would take on a specific task. Formally, the Agitation and Propaganda Section was a military affair, but in actual fact it was an arm of the party, and it continued to function irregularly—which is to say, whenever conditions warranted—until we entered Belgrade in late 1944, when it evolved into the Agitprop of the Central Committee, which I managed until my fall from power in January 1954.

In the end, even the priests in the brigades came into their own. In Krajina and Lika all the local Orthodox priests had been killed or had fled, so our own priests performed marriages and baptisms in a big way. And wherever church bells existed, their ringing was accompanied by the people's wailing. The priests were moved, not only because they believed, but because they were witnessing the renewal of faith. Vlado Zečević was conspicuous in this. He fulfilled his priestly duties all the more zealously because he was a member of the Supreme Staff, though he was by then no longer a believer. We teased him about it in headquarters, without malice. However, no one denied the tradition and vital meaning of religion and of the divine services for the people, or their benefit for the Partisans.

One afternoon at Glamoč, I met Sava Kovačević and his staff. A month after our departure from Montenegro the Chetniks had surrounded Sava's units, so he and his men had had no choice but to follow us. Several days before I ran into Sava, Peko Dapčević and Mitar Bakić were at a meeting at the Supreme Staff. They made all kind of charges—Mitar discreetly but emphatically, and Peko casually and openly—against Sava and his staff: they were inept and disorganized, and had panicked; they had got out by stealth, as if privileged to do so because, until recently, they had been chiefs in the Main Headquarters for Montenegro. Tito listened without comment—and for us who knew him well, without assent. Ranković's and my own rejection of such "appraisals" by Mitar and Peko were more open. After Mitar and Peko left, Tito concluded that Sava had done well to get out: "Not even we ourselves with several brigades were able to hold our ground there!" Sava was asked to make his report. When I encountered him, he was riding with head bent low and frowning. As soon as he saw me, he reined his horse and blinked as if to ask, "How does it look for us over there?" Understanding him, I immediately replied cheerfully, "Don't worry, everything is fine!" Sava's face widened to a grin. Tito kept Sava for supper, and everything proceeded as if nothing had happened. Sava was ordered to give his brigade a rest.

Soon thereafter, the Supreme Staff moved into a forest at Mliništa, to be nearer to the fighting around Jajce and more protected from aircraft. We stayed there until the rains and the first frost. I didn't have much to do, so I wrote an article, "Noble Hatred," inspired by the writing of Ilya Ehrenburg, and the horrors I had witnessed and heard about daily. Nearly every Serbian village had its own ditch where the frenzied Ustashi had thrown their victims. The joint German and Ustashi offensive on Kozara Mountain had just ended; around forty thousand Serbs in the villages at the foot of Kozara had been captured and sent to their death at Jasenovac, probably the most horrible of the camps of World War II. In the article I introduced a new approach, treating hatred as a noble impulse against fascism. I gave it to Tito to read before publication and he liked it. The article was printed and reprinted. It was more expressive of existential than ideological conflicts.

At that time the Germans were frantically storming Stalingrad, which the Russians were defending as the mainstay of their hopes and survival. It was marvelous, inexplicable, that we regarded the defense of Stalingrad—like no previous event of the war—as something exceptional and crucial. Stalingrad was the first thing people asked about at meetings, and soldiers at gatherings of units. Slowly the conviction grew that not only would the Russians hold out, but that the Germans were getting bogged down in a vast and decisive defeat.

Our organizational and political work was accompanied by continuous fighting and the taking of towns. It was becoming increasingly evident that our enemies were unable to contend with our burgeoning forces or withstand them. It was also clear that, given the diverse territory now under the control of our forces throughout Yugoslavia, there was a growing need for political unification and direction. At the time of the fall of Livno there was talk of reviving the newspaper *Borba* (The Struggle), but it was put off because of technical shortages and unstable conditions. By the end of September, this too was put into effect. The original idea was that Moša Pijade should edit *Borba;* indeed, the first issue came out under his supervision. However, though he respected and liked Pijade, Tito was not sufficiently confident of the keenness and moderation of his interpretations. For this reason he decided against my going to Serbia, and asked me to edit it. I took over *Borba* in early October and remained on the job for over three months.

3

At first we settled in Petrovac—by Tito's orders "on the outskirts," because of the air raids. Soon we moved into the village of Drinice for greater convenience. *Borba* came out weekly, with a team of editors almost large enough to put out a daily. Our technical facilities were meager; we used manual power. But the crew was so professional and dedicated that we published all sorts of things—periodicals, Stalin's *Foundations of Leninism,* directives, brochures, and forms.

The task of editing was a collective one; I exercised a supervisory role through discussions with other editors. We depended heavily on the news which the Supreme Staff and the Central Committee received through their monitoring services. At the start we had correspondents in the elite units, but we never got a chance to develop this. We ate out of one pot, all of us together—the editors, the workers, and the guards. To do the hard chores in the kitchen, the cooks took in a peasant girl, illiterate and dirty, whose family had been killed by the Ustashi. She learned to read and write, and after a couple of years became neat and dependable, and acquired a rank. The workers' apprentice was a little girl who had hidden in the bushes while the Ustashi burned her family and fellow villagers alive in a stable. And I brought along a young boy by the name of Vuk, who was entrusted to me by a fellow villager while fleeing from Montenegro. The boy was bright and diligent, and soon mastered the typesetter's skill. He was killed in an air raid in March 1943. Frequently it was no safer with the Supreme Staff than in the lower commands.

Here in rebellious Krajina we didn't need the guard except to cut wood and bring food, and to keep watch in case some enemy group got through the forest. There were no informers in Drinići or the surrounding villages. Not once did the planes that constantly cruised overhead threaten us. There were no political groups in the villages save the Communists—an ideal situation; the Communists and the people, solidarity in sacrifice and hope. The local Communists told us with malicious joy how the Ustashi had first killed off the bourgeoisie in the towns—priests, merchants, political party leaders—so that we were left with the people pure and simple. The Chetniks who showed up there with the Italians were left stranded and weak when the Italians drew back into "their" Dalmatia. The local Chetnik leaders, Radić and Drenović, therefore collaborated with the Ustashi, who went on slaughtering Serbs wherever they could. Serbian peasant wisdom was incapable of interpreting this collaboration as cowardice or treason, but simply saw it as madness and degeneracy.

The special suffering and turmoil of the Serbs of Krajina and Lika gained historical significance when they took up arms as a people—the only group to do so. Thus even in the intellectual offices of *Borba,* Tito's article in *Proleter* of December 1942 was received with enthusiasm.

> The Serbian people have given the greatest contribution in blood to the struggle against the invader and his traitorous assistants Pavelić, Nedić, Pećanac, as well as Draža Mihailović and his Chetniks. . . . The Serbian people know well why our national tragedy occurred and who the main culprit was, and it is for this reason that they are fighting so heroically. It is therefore the sacred duty of all the other peoples of Yugoslavia to take part in the same measure, if not even more, with the Serbian people in this great war of liberation against the invader and all his minions.*

The editorial staff of *Borba* instantly drew close to the peasants, not only because of vital everyday connections, but because of something new in our perception of them: we finally recognized the enormous "new" role of the peasantry in the revolution and in revolutionary wars. In our case the workers, such as there were, were not in the ranks of the Partisans, except for small class-conscious groups or party functionaries. Our army was predominantly peasant. This was a frequent topic of discussion in the Central Committee, of course, with the aim of giving this a Marxist explanation and Marxist generalization.

The peasants of Driniće also understood that something important was happening in their village. Yet they didn't receive us so cordially

* Josip Broz Tito, *Borba za oslobodjenje Jugoslavije* [The Struggle for the Liberation of Yugoslavia] (Belgrade: Kultura, 1947), pp. 138–39.

merely because we appreciated their contribution, but because they knew that the Communists, as the military and intellectual backbone of the struggle, kept them from chaos and destruction. The Communists immediately instituted an authority and fellowship: an order, an economy, transportation, schools. And for the young and for women, they offered prospects of enlightenment.

Mitra and I moved in with an old couple and had a little room with one bed. This was no exceptional luxury, but we slept soundly because we were exhausted. Nevertheless, I managed to reread Tolstoy's *War and Peace,* by the light of a bedside lantern. In its planning and its battles, Hitler's invasion resembled Napoleon's. But that was a mere game, compared to the current combat unto death of nations conquered by fascism and the technology of mass extermination.

Borba was frequently visited by functionaries, particularly army political workers, who came to use the press or to confer about propaganda work. Most interesting of all was a visit by Branko Ćopić and Skender Kulenović, agitprop officials who became famous writers and public activists after the war. We spent the entire day in meetings with them. Ćopić and Kulenović were already well known—particularly Ćopić, whose humorous little poems the people recited as their very own. In 1943, when I was in Slavonia, I found his poem "O falcon, my falcon" being used as a folk song. Kulenović had survived the siege of Kozara, and it was by mere chance that the fascists hadn't discovered him. One could detect in him, in his gentle sadness, the horrors which he had survived in that protracted disaster and fortuitous salvation. There were conflicting views of the tragedy on Kozara Mountain. Djuro Pucar, then regional secretary, spoke of panic and confusion at crucial moments, while other surviving leaders spoke of the myopia and naiveté which had permitted the encirclement. In party circles, at least in the Central Committee, Kozara was never discussed—self-contained and unfathomable in its tragedy. Kulenović undoubtedly knew of many faults and weaknesses in connection with the tragedy, but he never mentioned them. Kozara was for him something beyond ken, beyond ideology and party—a tragedy of the country and the people, an immeasurable calamity and folly. Ćopić, on the other hand, was cautious and humorous, more so than in postwar years.

The Regional Committee was not very pleased with Ćopić and Kulenović. They were especially put out by Ćopić's practice of camouflaging his biting satire as jokes, but did not deny their worth. Secretary Pucar encouraged Ćopić and Kulenović to visit us in the hope that we might have a positive influence on them. But nobody had any influence on Ćopić and Kulenović. We talked about art, the extraordinary nature of our war, the future of the world. Unfortunately, we did not have a copy of Kulenović's poem "Stojanka, Mother of Knežpolje," the most moving and unsurpassed wail of an endangered people, written with vivid horror and passion. Nevertheless we discussed it. Radovan Zogović made a couple of abstract, hairsplitting remarks. I then relieved the tension, praising the poem moderately so as not to offend Zogović, even though I sensed that only in Kulenović's poem did the tragedy of Kozara receive its true expression and testimonial.

Individual comrades from *Borba* were sent briefly on special missions by the Supreme Staff, now established in some remodeled lumber railroad cars in Oštrelj. Tito decided to send Vukmanović-Tempo to Belgrade and Arso Jovanović to Slovenia. After three months of fighting the Italians had put an end to the uprising and eliminated free territory in Slovenia, and Tito felt that the Slovenes were lagging behind in military organization. Relations between Tito and Arso were good—as always, when they were not directly involved in operations. Tito even remembered to give Arso some woolen socks before he left.

However, I was not present when the final decision was reached to take Bihać. The move was supported at the center and among the provincial leadership, and was an important step in the preparation for the First Session of AVNOJ (the Antifascist Council of National Liberation of Yugoslavia). The idea was that the taking of such a large center, particularly in the middle of Pavelić's "state," would give the AVNOJ session a certain status at home and abroad.

As a part of the Bihać operation an attack on the environs of Grahovo was also planned, to prevent the Chetniks from endangering our weakened rear. The operation was not considered a difficult one; Tito sent the Serbian brigades, which we tried to spare as much as possible. We decided to deal severely with the Chetniks: to burn down their houses and take no prisoners. Attacked suddenly by experienced units, the Chetniks offered little resistance and fled into fortified Grahovo, where the Italians were stationed. The Plain of Grahovo was put to the torch and plundered. It was reported that the fire almost swept the birthplace of Gavrilo Princip, who had assassinated Franz Ferdinand almost thirty years earlier. Our losses were small, and the Chetnik losses high. It was also reported that some three hundred Chetniks were executed, and that Djuro Pucar's own brother barely escaped execution.

Bihać fell on November 3, after a fierce struggle. Both the Ustashi losses and our own were high. The booty was enormous. Bihać had some fifteen thousand inhabitants at the time, a developed commerce, trade, and rail connections. The brunt of the battle fell on the Krajina Brigade. Veselin Masleša was there for *Borba.* He recounted that hundreds of peasant carts from all over Krajina waited at the approaches for the town to fall. The peasants considered it their duty to get out what booty they could for their army; of course, in the process they kept some of it for themselves, even though our units suppressed plundering—albeit only here and there, and not too strictly.

As expected, the fall of Bihać caused great reverberations; it made up for the defeat on Kozara, though not for its tragedy. The liberated territories of western Bosnia, Lika, and Dalmatia—comprising fifty thousand square kilometers with around two million inhabitants—were now linked.

The Supreme Staff and the Central Committee moved into Bihać. Only I stayed put, with *Borba,* having replenished my supply of paper and ink through confiscation. But soon thereafter I too rode off to Bihać, arriving during the meeting of AVNOJ. I did not attend the session, nor the meeting of the Central Committee either—if there was one—which had decided to hold that session. However, I had been informed about this during a visit to the Staff at Oštrelj, so AVNOJ would have met, even had Bihać not fallen. This First Session of AVNOJ was completely of Tito's making—both the ideas and the decisions. This does not mean that there were differences in the Central Committee—though the name AVNOJ did seem awkward to me. On the other hand, my "leftist" inclinations were satisfied by the word "Council," a translation of the Russian word *soviet,* which many people were not aware of then or later.

The positions which Tito initiated, and in whose formulation Žujović played the most active part, were so broad that they could have been accepted by all opponents of the occupation—had the opponents not known that those positions originated with the Communists and so concealed ulterior motives. The editorial staff of *Borba*—even Zogović—concurred in this latitude as a necessary "stage." The vastness and stubbornness of the uprising in Krajina, Dalmatia, Lika, Kordun, and Banija—which enabled us Communists to create an army permeated with our ideas and led by our party—convinced us that through struggle against the invaders we were resolving the basic questions of the revolution. This came out loud and clear in Tito's article, "The Struggle against the Invader Is Our First and Most Important Task" (*Proleter,*

December 1942). It was as if, now that the revolution had found its vital concrete path, it could conceal itself.

Tito was of the opinion—and the rest of the leadership supported him—that some sort of government should be proclaimed at the AVNOJ session. But Moscow objected. The creation of a government seemed premature to me, but I didn't oppose it because I believed that some organ of civil authority had to be instituted for such a large territory. Tito found a formula which satisfied Moscow and us—an Executive Committee which would not be a government, but would decide all questions of state and concern itself with the rear and supplies.

There were fifty-four delegates at the session, largely from Serbia, Bosnia, and Croatia. Ranković and I were not among them, so as to avoid too heavy a Communist representation. There were no elections, nor could there be; rather, Ranković compiled a list which the Central Committee supplemented and confirmed. I did not attend the opening session. The discussion abounded in emotion but brought forward no proposal, let alone criticism. Only the older peasants spoke in a lively, unconventional way, in a language that reflected their age-old, nonideological patriotism. Everything was arranged in advance by men who were in complete agreement with the leadership. Yet this is not why the First Session of AVNOJ did not approach the far-reaching significance of the Second one a year later; rather, it was because the tide had not yet turned in favor of the Partisans, and the enemy offensive soon thereafter precluded any Executive Committee activity, so that all responsibility again fell on the party and the Supreme Staff.

Andrija Hebrang had also made his way to Bihać. He had been part of the recent exchange for Pavelić's officers. Tito and he were frequently together. With Hebrang was his young and beautiful wife, who had also been exchanged. I got to know her at a dinner at Tito's. Hebrang assumed the leadership of the Central Committee of Croatia: the Central Committee did not decide this formally, but everyone accepted it as natural, because Hebrang was an old Communist, Tito's long-time comrade in arms and friend. This brought an end to Vladimir Popović's function in the Central Committee in Croatia: the Central Committee had appointed Popović operational chief because of the trouble and confusion which a Soviet informer had provoked in the Central Committee of Croatia at the beginning of the armed struggle. Hebrang's meeting with me was restrained, but without a trace of malice: I was not put out by Hebrang's "enthronement" because it was my hope that, if I survived the war, I would be rid of officialdom and able to return to literary work.

Tito's and Hebrang's friendship accounts for the fact that Hebrang did not go through the party commission—a procedure established for those who had been arrested or served a prison sentence before the war. When Hebrang declared himself for Stalin in 1948, Ranković reported to Tito that his agents had evidence of Hebrang's unworthy behavior while under investigation.

Tito asked Ranković, "And why didn't you question him them?"

Ranković replied with a reproachful smile, "I thought that you had done that yourself in Bihać."

In early December a conference of women was held in Petrovac. Mitra was its originator and leader. In a speech delivered on that occasion, Tito declared that he was proud to have so many women in the ranks of his army. This was indeed an advantage for the Partisans. It confirmed the quest of women for emancipation, and ennobled the men and inspired them to heroism. I observed that among the Partisans the women were braver than the men, perhaps because the very act of joining the army and the revolution constituted a greater turning point for them.

In late December the First Congress of Antifascist Youth was held in Bihać. As if by some higher law, the activities and hopes of the Communists had spread over all areas, into all strata of society. Lola Ribar conducted the proceedings and talked without notes. After reading the stenographic record of his speech, I remarked to him that he wrote exactly as he spoke. "Yes," he agreed, "we just have to set a few new lines and make some corrections." Žujović had the presence and the voice of an orator, but only Lola Ribar was the complete orator. No wonder; being an outlaw like Žujović was hardly conducive to oratory, while Lola began his career in the public eye at student meetings.

But the youth congress was not the chief reason for my presence in Bihać. Tito had summoned me to discuss my forthcoming trip to Dalmatia, where I was to hold talks with representatives of the Croatian Peasant party. Tito attached special significance to the talks as a means of gaining recognition abroad. I departed in haste; the roads were covered with snow. Olga Dedijer and Mitar Bakić were with me, each on a separate assignment. On the road from Glamoč to Livno we almost froze, and kept losing one another in the blizzard. From Livno I went on to Imotsko, accompanied by Vice Buljan. We reached the outskirts of the town and there we waited all day. But the delegates of the Croatian Peasant party never showed up; they were frightened. Still, I will never forget my association with Buljan, who was charming and witty. Dalmatia itself was half-free and passage through it safe. Only the cities were unsafe because of the Italians, and the Serbian villages because of the Chetniks.

In Livno I held a meeting with the Second Division's party organization. The division had taken Livno only recently after fierce battles with the Ustashi, and it subsequently suffered heavy losses—in vain—at Kupres. The building in which the meeting was held throbbed with the strength of warriors who even here didn't lay down their arms. I made a great effort to control my feelings of concern for them, particularly those of the Second Brigade; many of my acquaintances were no longer alive, while the survivors resembled their fathers on the Salonica front of 1918—caps, long overcoats, mustaches. They listened carefully to every word I said; it was all the harder to find the right words, because I had to inspire them to further sacrifices.

Back at Bihać I had two pleasant encounters—with Mika "the stocking king" and with Kusić. I met Miko as I was crossing the market square. He looked strange in civvies; when he saw me he unbuttoned his cap to reveal a five-pointed star, then lifted his pants to show me his boots and wide leggings. "I dress according to the occasion," he laughed. "And you Partisans are losing millions. I've set up a business. They don't have tobacco here, and in Glamoč and Livno they don't have salt—a profit both ways. I've made about a million lire and kunas, and I've turned the money over to the Partisans."

I also met Kusić on the street. We were happy to see each other. Partly through my influence, Ranković made it possible for Kusić to be made commissar of the Fourth Krajina Division after he had been relieved of his duties as political commissar of the Second Brigade. I asked him if he was pleased. With a knowing smile Kusić replied, "Yes indeed, I am." But Kusić never had a chance to command; he came down with typhus and committed suicide while in a fever—as did many, until hospital personnel began to take weapons away from such patients.

About two hundred Ustashi had been captured in Bihać. The Supreme Staff created investigating commissions. Evidence kept coming in from Serbian villages about thousands who had been killed with pickaxes and thrown into ditches, and from the towns about the horrors of Ustashi jails. Young Ustashi had fun with the girls in this manner: when they shook hands with them during walks, they would place human ears, fingers, or noses in their hands, just like village toughs who get a kick out of offering tobacco pouches with snakes in them. But once a government is formed, it must act on its own; therefore our investigators decided to check on the peasant reports. They disguised several Partisans as Ustashi and had them led through the massacred villages. They were almost lynched, especially by women who swore that they recognized them as the murderers of their kinfolk.

At Tito's—some fifteen kilometers to the north of Bihać, in a manor overlooking the Una Valley—we found the renowned writer Vladimir Nazor, who had recently fled from Zagreb to liberated territory. Nazor was past seventy. When the Zagreb underground discovered that Nazor was an opponent of the Ustashi regime, they proposed to him that he go over to the Partisans. He replied, "I will—to see how our Serbian brothers fight!" His decision to join the Partisans was prompted by the Yugoslav idea and Slavic sentiment. He was thin and frail, but tough. All through the war he suffered from diarrhea and ingested huge quantities of food. This was obvious at the dinner with Tito, as was an inexplicable pettiness on Nazor's part, side by side with courage and idealism. We all recognized the significance of Nazor's joining us, all the more so since he was a Croat and had never shown any leaning toward the Communists. And Tito liked associating with a writer who was of a high repute and well known.

Ranković and I went over our business with Tito easily enough. But we were worried by the situation; the Supreme Staff had received reports of German divisions concentrated north of the free territories, and of the first encounters with them. As panic spread through the population, the will to fight increased among the soldiers. Tito grew noticeably pensive, as though before something fateful. He speeded up the anti-Chetnik declarations of the Executive Council for use on Radio Free Yugoslavia, and also the redeployment of troops.

We had reached an informal agreement in the Central Committee that in the spring we would move considerable forces to Montenegro and southern Serbia in the direction of Kopaonik and Toplica, and the border of Kosovo and Macedonia. This was also prompted by the developments abroad—the Soviet victory at Stalingrad and the Allied landings in Africa. The possibility of an Allied front in the Balkans was also contemplated; Arso Jovanović wrote an article about it which Tito approved, though he did not firmly share this opinion. To judge by their propaganda, there was no assurance that the British would stay out of our conflict with the Chetniks. Serbia, along with Montenegro, was of decisive importance: the Chetniks had to be beaten on their own breeding grounds.

That same day Ranković and I returned to our duties with new cares. His duties were more important and numerous, but our cares were the same.

In the Cauldron

1

We did not know in November that Hitler had received Pavelić at Vinnitsa, and decided on far-reaching and radical measures against the Yugoslav Partisans. Nor did we know of the German-Italian conference in Görlitz (East Prussia) of December 18–20, 1942—attended by Hitler, Ribbentrop, and Keitel on the German side, and by Galeazzo Ciano and Ugo Cavallero, the Chief of the General Staff, on the Italian side—at which the Führer's decision to crush the Yugoslav resistance movement before the end of winter was announced. Likewise we did not know that on January 3, 1943, in Rome, a plan for the destruction of the National Liberation Army had been elaborated by a conference that included the German commander for the Balkans, General Alexander Löhr; the Italian Chief of Staff, Cavallero; the Commander of the Second Army, Mario Roatta; Pavelić's generals; and the Chetnik leader for Dalmatia and Hercegovina, Jevdjević.

But this lack of information had no essential effect on our conduct or on the course of the fighting. We knew that we had five German divisions against us—among them the famous Seventh SS Prinz Eugen Division, made up of Yugoslav Germans. Several Italian divisions were also set in motion, Chetnik bands were transported to starting points, while the desperate Ustashi attached the incompetent Home Guards to the German commands.

We were unaware that the enemy plan—Operation Weiss, known as the Fourth Offensive in Yugoslav historiography—called for the separation, encirclement, and destruction of our forces; or that the least enviable role in that plan had been assigned to the royal "army," the Chetniks. Fearful that in case of an Allied landing in the Balkans the Chetniks might join the Allies, the Germans had agreed to the participation of the Chetniks on condition that the Italians disarm them upon the completion of the operation. But the engagement of so many German forces, combined with the simultaneous movement of all our enemies, told our common soldiers and peasants that a time of unequaled mortal struggle was before them. This message was all the more clear in that the offensive began after the campaigns at Stalingrad and in Africa, and the setting up of our army and regime in Hitler's rear.

Also, the fierceness of the offensive from the day it began in Banija on January 16, and in Kordun on January 20, left no doubt about Hitler's determination to squash the resistance movement in Yugoslavia. The Germans crushed our defenses with tanks and artillery, set our villages afire, and shot hostages and prisoners. From morning till night their aviation pounded everything in sight; from the very first days our bases at Bihać, Petrovac, etc., were constant targets. I wrote in *Borba* that the Germans were not as well prepared or as militant as in 1941. That was not so, except insofar as it concerned us directly. Our experience was greater, and our army larger and better prepared, than during the German attack on Užice in late 1941. And the people, the Serbs of Krajina and the Croats of Dalmatia, stood firmly and fearlessly—and not merely because the only alternative for the Serbs was to be massacred by the Ustashi, and for the Croats to be denationalized and decimated by the Italians. Through our warfare and the populist government, the Communists and the people were joined in their destinies as never before or again. Thousands of new fighters flocked into our units in a general mobilization that was not compulsory or decreed. It was then that the Ninth Dalmatian Division was formed. And when it fell apart in the course of the offensive, it was because of ineptness, inexperience, hunger, and typhus, and not irresolution or a change of heart.

The offensive found two of our elite divisions, the First and the Second, moving across central and southeastern Bosnia in the direction of our intended spring breakthrough into Montenegro and southern Serbia. Our intended breakthrough operation had begun in haste, under pressure from the enemy. Tito was determined about this and really had no choice, if he wanted to escape the encirclement of the greater part of his forces within a small area. But the German plans did not develop as scheduled either: our resistance was so stiff, our destruction of bridges and roads so thorough, and our units so experienced, that the German command had to rescue its own surrounded troops. The winter weather was hard on us, but even harder on the Germans; it limited their use of aircraft and increased their supply problems.

Borba had a telephone link with the Supreme Staff, and the acting chief, Terzić, was extremely co-operative in keeping us informed of the course of operations. But it was Dedijer who informed me on January 30, around two o'clock in the morning, that we had to evacuate in a hurry. The dismantling and loading of the printing press on a truck took place quickly and in orderly fashion. Miraculously, the truck was under way before dawn, carrying a group of the workers as well. I set out on foot for Oštrelj with the guards and several members of the editorial staff. The whole village gathered in front of the building that had housed the press to see us off. Women were crying and everyone was upset, more over our withdrawal into uncertainty than over their own misfortune. I said farewell in the name of us all, with a few words of advice in case the Germans should break through. I told them to take refuge in the hills: the Ustashi would go with the Germans, and the Germans would go after us.

We were still in the plain when the morning was heralded by a distant roar of aircraft. However, their target was the road from Petrovac to Oštrelj, so we got to the forest safely. When we arrived at Oštrelj around ten o'clock, we found the road covered with wrecked carts, dead horses, and dismembered children, while little groups of refugees huddled around roadside fires. Thousands of refugees had come all the way from Banija and Kordun under a sky which showered them with bombs and bullets, through blizzards, and across a hungry land. We too took refuge in the woods, on the spot where our Supreme Staff had been located earlier, and we were given a message to carry on with our evacuation.

It was instantly clear to me that this retreat of the population was senseless, even had it not been a burden for the army. I suggested to the Supreme Staff and local commands—and I was surely not alone in this—that the refugees be distributed among the Serbian villages and eventually sent back to their native regions. And in point of fact, the wave of refugees subsided as our units progressed further eastward. But we could not do the same with the wounded, who hobbled along on crutches, on horseback, or in carts, or were supported by women, old men, or other soldiers. They were a part of the army; we had to take care of them. And their number grew as fierce fighting continued to be waged by the units attached to the Supreme Staff, and as the rear-guard commands sent their wounded to "safety" with us at the Staff, as if to an imaginary future free territory. And so—in spite of everything, and without approval from the Supreme Staff or the Central Committee—a central hospital was set up which became a terrible burden on our conscience and our maneuverability, and a most attractive target for our enemies.

At Oštrelj the *Borba* personnel separated: some continued on foot, while I went off by truck with the press in the direction of Livno. We settled in the village of Priluke, some ten kilometers from Livno, in a requisitioned house above the road. We reassembled the press that same day. Two days later the editors arrived who had set out from Oštrelj on foot, and *Borba* resumed its work.

Priluke was an Ustashi village, but only the defenseless were left. We kept a strict guard anyway; even the members of our editorial staff took part. The peasants just barely spoke to us: "yes" and "no"—a "yes" that most often meant "no," and a "no" that most often meant "yes." Even so, they collaborated with the people's committee which they had elected at a gathering in the presence of a political commissar. The committee carried out requisitions according to what each household could contribute, and co-operated honestly in the matter of food supplies, fuel, and transportation. We were an occupation force that imposed certain duties on them, but not identification with our cause.

We were convinced that someone would inform the Ustashi and the Germans of our presence in Priluke, but that they would not bomb us in an Ustashi village. And indeed, at one point German planes flew so low and slowly that they must have seen us running to the cliffs. But they zeroed in on the road, and riddled a soldier and a child.

One day I went to Livno on business. The little town was full of soldiers, refugees, and the wounded, but was also very active culturally and politically. One night, in a room reeking with smoke, iodine, and sweat, I listened to the reading of "The Pit," a poem by Goran Kovačić. It had a stirring effect, being a tale of recently dug up victims and of a horrible imagined world. It was natural then for a Partisan poet to condemn the Ustashi massacres of the Serbs, and of course, morbidly symbolic: Goran was a Croat who was later butchered by Serbian Chetniks.

We were able to publish only one issue of *Borba,* because the Germans were advancing on us from Bihać and grouping in Bugojno to cut off our breakthrough across the Neretva River. At sunset on February 15, 1943, Ranković and Terzić rode into Priluke on horseback. They told us that the next day our side would attack Prozor, where the Italians had established themselves so as to block our crossing of the Neretva. Ranković also told us that the publication of *Borba* had to be stopped and its personnel reassigned; he needed experienced Communists to organize the wounded. Yes, to organize the wounded! There were over three thousand of them, and they had to be organized like an army to enable them to accompany our units. We had to work among them to raise their morale. We agreed to let Dedijer go right away. As for the press, the workers, and the editorial staff, we were to carry on as long as we could. The Germans had surrendered at Stalingrad, and the Russians continued to advance. In the flaming circle which was tightening around us, we recognized that event as very important. Thus we were all the more embittered by Eden's senseless, malicious declarations regarding British efforts to unite all the resistance forces in Yugoslavia. At that very moment the Chetniks were rallying under the Italian command to finish us off, bloodied and exhausted as we were, and driven into the Neretva canyon by the Germans. We were pleased by the *New York Times*'s reports about our struggle because, though sparse, they revealed Draža's collaboration with the invaders. When the BBC finally commented on the heroic struggle of the Partisans, I said with bitter irony, "They think the Germans are going to destroy us now, so they're chanting our requiem."

On February 17, again at night, I set out with our truck and printing press for Prozor. In the pitch-black darkness, at Kupreško Polje we skidded into a ditch. There soldiers emerged from frost-covered wilderness, and so did uncertainty: Prozor had not fallen the day before, as planned, so the attack was to be repeated that night. Silent figures passed, somber in the cold. These were the wounded, the typhus victims, and the refugees. When I told them that we were *Borba* and that I was a member of the Supreme Staff, they paused, glanced at the truck, blew on their frozen hands, and pulled their scarves over their ears. Finally we got the truck out, by lifting it with our bare hands.

When I arrived in Šćit in the cold predawn light, I found Tito and his staff alert and excited: Prozor had fallen. The day before, the Italians had bombed the monastery where the Staff was lodged. Only one bomb, a dud, hit the mark: it broke a wall under its weight, and lay in the courtyard like a singed sow. Tito thought it harmless, but I disagreed: "It's a big bomb. If it explodes, it'll bury you in your shelter."

Everyone was full of details about the fighting for Prozor. We heard at the time that the Russians had offered terms of surrender to the Germans they had surrounded in several towns, and when the Germans refused, they destroyed them. We offered the same terms to the Italians, but at noon, instead of an answer, they opened fire on our positions. That evening, February 16, our attack was repulsed. Thousands of inadequately protected wounded were inching ahead of the advancing German divisions. The army was unable to secure a crossing for them over the Neretva. Tito ordered that Prozor must be taken. On the evening of the next day, February 17, Sava Kovačević dragged an artillery piece over and destroyed an Italian bunker at a distance of 150 yards. The army poured through the breach. Gligo Mandić threw some planks across the barbed wire on the other side, got a group over, and cleared a passage. The Italians in the little town surrendered, and we took over the outer fortifications. All the Italian troops—the entire Third Battalion of the 259th Regiment of the Murge Division—were put to death. We put into effect the conditions they had rejected, and vented our bitterness. Only the drivers were spared—to help transport the munitions and the wounded. Many corpses were tossed into the Rama River. Several got caught among the logs, and I shared with our officers a malicious joy at the thought of Italian officers on the bridges and embankments of Mostar stricken with horror at the sight of the Neretva choked with the corpses of their soldiers. The booty was enormous—carloads of munitions. The Partisans were as happy as children to be able now to fire as much as they liked. Several mortars were also captured which the former royal officer Branko Obradović, later an artillery commander, seized joyfully and, with his exceptional gift for improvisation, turned against the Germans.

2

With the fall of Prozor, the way to the Neretva was opened, on whose banks the Italians and the Chetniks had entrenched themselves. Dapčević's units also reached the Neretva the day before the fall of Prozor, on February 16, after taking Drežnice. Even though the passage toward Hercegovina and Montenegro seemed secure, heavy fighting was obviously still ahead of us.

At first I tried to stay in Prozor, but the little town was being intensively bombed throughout the day, so I joined Tito and the Supreme Staff. The truck with the printing press was ordered to forge ahead—as far as the road would take it. We camped in Gračanica, which was safe only at night when the air raids stopped. At daybreak we withdrew into the narrow, winding hollows—those miniature canyons in which this terrain abounds. To be as inconspicuous as possible, the Central Committee and Supreme Staff separated from other personnel. We were crowded into narrow valleys over an area some thirty kilometers long. Whenever the weather permitted, the planes pounded us with clusters of bombs which dotted the ground with explosions.

The Executive Council of AVNOJ was with us, and Nazor had been co-opted by it in the course of the offensive. But Nazor didn't have much use for the civilians—or, as he called them, the "politicians." Though he was eating special food and plenty of it, he claimed he was dying of hunger. "Yesterday," he said, "I ate a chicken and they all envied me! What is one chicken?" And so he was temporarily assigned to the Supreme Staff. One morning, when everyone was in a hurry to get moving, he got it into his head to stay in a friar's apartment next to the church: "Saint Vitus will watch over me and the friar," he said. "The church is dedicated to him." And there he remained. It was not a problem for Saint Vitus to protect two of them in view of the fact that here, on the road and beside a prominent target, there was no one else to protect. But Nazor was exceptionally brave, so brave that, as I watched his behavior under aerial attack during our return march from Montenegro, it crossed my mind that perhaps he was really not quite aware of the danger.

We Central Committee members stayed together except when we were sent on a special task, most often one involving the wounded. This was the only time during the entire war that Tito didn't shave every day. He always shaved himself, on arising, whereas the battalion barber shaved the rest of us twice a week. All day the enemy aircraft roared overhead; all through the night couriers came and went with reports. Wrapped in tent cloth, we slept on the ground next to small fires—or on rocks which radiated warmth from the sun, since we had now gotten out of the mountain blizzards and frosts, and into valleys with a mild Mediterranean climate.

Tito exhibited nervousness, even rashness, in issuing commands. While he was confident in determining strategy that was more political than military in character, as a commander he reacted too quickly to the changes so inevitable in war, and as a result frequently changed his orders. Temperamental by nature, with an exceptional sense of danger and a keen, quick intelligence, in battle he didn't have the necesssary detachment, and often moved large units to protect himself and the Staff. At first the Staff missed Arso Jovanović, whose remarks and counterproposals at least prompted Tito to reflect. But before long the commanders learned how to cope with this weakness by adapting orders to circumstances, or by citing new facts to persuade Tito to correct his decision. But Tito was not petty as a commander, nor did he stifle the initiative of his officers, so agreements were reached without serious problems.

The perilous difficulties multiplied with our victories, so to speak, and resulted from the number and immobility of our wounded. They couldn't move, and we had no means of transporting them with our units. Their immobility necessitated a constant regrouping, which gave the enemy time and opportunity. We didn't dare think about abandoning them, because of the kind of war that had been forced on us—a war of massacre and slaughter. Thus the very conditions of war imposed rashness. On receiving a radio report that the offensive in Croatia had spent itself, Tito issued an order, on February 28, to return the wounded to

Croatia—only to withdraw that order upon receiving a radio report that the communication lines to our rear were in the hands of the Germans. A moment of uncertainty arose: Acting Chief of Staff Terzić told the wounded on March 5 that we would make a drive on eastern Bosnia and not on Montenegro and southern Serbia, though later that same day the decision was made to force a crossing of the Neretva and push ahead as originally planned.

It was Tito's idea to take Konjic and thus make possible a withdrawal of the wounded into the inaccessible hill country, instead of over the rocky slopes looming before us. No one suspected that Konjic would hold out, because it was held by the Italians and we planned to attack in force.

The entire territory from the mouth of the Rama to Konjic was cleared by February 20, 1943; as for Konjic's rear—that is, the approach from Sarajevo—the First Proletarian Brigade took this over, occupying Ivan Sedlo. Units of the Third Division had been designated for the taking of Konjic, but the commander of the First Proletarian Brigade, Danilo Lekić—an experienced fighter from Spain, brave and enterprising—attacked Konjic with his two battalions. Obviously Lekić did not expect a German attack from Sarajevo, and everyone had enjoyed victories over the Italians. The attack took place on the night of February 19–20 and a part of the town was taken, but the Italians hung on tenaciously. By the next day, February 21, a unit of the German 717th Division at Sarajevo regained Ivan Sedlo with the help of tanks and planes, then pushed forward to Konjic.

Certainly Lekić had been overly enterprising this time, and Tito was angry when he heard of the German advance. Yet it was unlikely that Lekić could have stopped the Germans for long with his two auxiliary battalions, even if he hadn't rushed on to Konjic. Tito's anger appeared to be only momentary. Since the Germans were not yet at Konjic in numbers, and Sava Kovačević's Fifth Brigade and other units were arriving there, Konjic's fall still seemed likely. Besides, we were joyous over another success: the taking of Jablanica on February 22, where an entire Italian battalion had been destroyed by the Fourth Brigade. Our approaches to the Neretva River were now widened, and the Prozor-Rama-Jablanica road also freed. The captured Italian commander held himself proudly, and admitted that he had fought in Spain on Franco's side. Our "Spaniards" could not forgive him, so our couriers were ordered to shoot him, despite his request that this be done by one of the Spanish Republic volunteers. The rest of the Italians were simply drafted to transport the wounded, or assigned to our supply unit. The same was done with the Italians from the other strongholds. And so a kind of lower class was created among the Partisans, consisting of Italians who were more poorly dressed and fed and yet performed the hard jobs.

The battle for Konjic was still in progress and going well for us when Tito, convinced of success there, ordered the bridge at Jablanica destroyed. This was carried out on February 28 or March 1, 1943. Vladimir Smirnov, who had charge of destroying the bridge, had observed that Tito held up the attack just so Smirnov could execute this order. Before the bridge was blown up, the Second Proletarian Brigade, which had taken the bridgehead on the left bank, was transferred to the right bank.

But the Germans received reinforcements in Konjic—from the Chetniks, among others—while at the same time the Germans from Gornji Vakuf threatened to cut off the wounded. We had all the conditions for a breakthrough, yet we had to put it off. Lekić's "error" now assumed fateful dimensions, and Tito was overcome by irrational fury. It was obvious that the wounded couldn't have made it, even if Konjic had been taken. Tito was also angry over the premature destruction of the Jablanica and other bridges. But something had to be done, so anger was quickly set aside. Finally, making the best of it, Tito said, "Well, maybe we can turn that demolition into a stratagem of war." Later Tito wrote this up, and I confirmed it in Moscow in 1944, in my article on Tito, because the Russians insisted that I describe Tito as commander, and I couldn't think of anything else. This "stratagem" was celebrated in the film *Neretva,* even though some Yugoslav military specialists have since described the demolition job as an error, and many were aware of this at the time. I don't think the Germans were deceived, because we had no other avenue of retreat. The film *Neretva* would have been more dramatic and convincing, I believe, had it shown Tito's human failings and doubts, rather than the infallibility and omniscience of the supreme commander.

Yet it was Tito who, in that complex and dangerous situation, came up with a risky but brilliant maneuver: first, to pin down the German 717th Division, which was descending on our wounded, and then carry out the breakthrough on the Neretva—through the Chetniks, since the Italians had already been beaten. In making these grave decisions, Tito consulted with the Central Committee. The meeting was short, for everything was clear; nobody opposed anything, and Tito's initiative prevailed.

Our units were drawn back in haste, though they were already in the valley of the Neretva, ready for the crossing and withdrawal. The Fourth Montenegrin Brigade was the first to arrive at the battle area. On March 2, in combat with the Germans at Vilića Guvno, it stopped their advance on our wounded. Soon the other brigades arrived as well and a three-day battle ensued. In that battle we surpassed the Germans in artillery, thanks to Partisan improvisation and the skill of Branko Obradović. We pinned down the 717th Division and auxiliary Ustashi units, so that the 369th Division—the so-called Devil's Division, made up of Croats under German command—had to come to its rescue. Our scouts reported that, in an effort to escape us, the Germans had scrambled frantically into trucks, so that this battle accomplished an important transformation in our minds: the mythical fear of the German army vanished.

On March 2, the day the Montenegrins beat the Germans at Vilića Guvno, we decided to bury the printing press. We chose a spot not far from the water mill on the left bank of the Rama, where Tito had spent the night. It was obvious that we couldn't get through Konjic by road, and because of its weight we couldn't carry the press off the road. Someone later told the Germans where the press was buried, and they destroyed it.

Pijade also found a place for himself in the vicinity of the mill, in a sandy cave safe from machine-gun fire, and settled there with his sacks of printed forms, papers, typewriters, dishes, food, clothing, and all kinds of odds and ends which as a prisoner he had gathered by force of habit over many years. Now he too had to get rid of everything that he couldn't take along in his saddlebags or his knapsack. It wasn't easy for Pijade to abandon anything, so I helped him make up his mind, jokingly eliminating one item after another. He ended up with less than he was able to take along, so he began the reverse process of salvaging what was more important. In any case, he couldn't abandon his attractive little secretary Drinka. A few days later, during the breakthrough, she went over to the Germans. It was not established, at least not at the time, whether she had been planted by the enemy.

Late in the evening of March 4, Tito took Ranković and me in a car to inspect the battle area. We drove slowly. On the narrow winding road we passed empty vehicles and overtook trucks carrying munition. It was around nine o'clock when we reached the plateau at Gornji Vakuf, amid campfires and the roar of artillery. We walked among the soldiers, who were eating their supper around the fires. They recognized us and saluted with an easygoing seriousness. The commanders were stiffer, and terse and direct in their reports. The immediate surroundings of the battle area were as I had imagined them from novels and pictures. The battle had already been won, in the sense that the Germans had been pinned down, thus ensuring the wounded a successful retreat. Only the advance units, the rear guards, and the artillery had not yet completed their task.

The next day Tito ordered the breakthrough to the Neretva through the Chetnik detachments. The breakthrough was carried out on the night of March 6–7, 1943, with astonishingly small forces—three battalions from each of our two brigades. First our bombardiers and gunners clambered up the skeleton of the wrecked bridge over the Jablanica and, with losses, drove the Chetniks out of their redoubt. The penetration deepened and spread rapidly. On March 7 a wooden walk had been constructed over the bridge scaffolding, and on the night of March 8–9 the crossing of the wounded began.

Though we didn't expect high morale among the Chetniks, we were nevertheless surprised at their disarray and lack of spirit; only small groups of officers and gendarmes showed vigor. This was the biggest battle thus far with the Chetniks. Later we engaged them in bigger and sharper encounters. The Chetniks fell into hopeless, senseless contradictions militarily as well: lacking political organization, without which there can be no real army in a civil war, they carried out a general mobilization, relying on legality, though in the minds of many Serbs legality had ceased to exist. Moreover, individual leaders continued to undermine this legality: during the Neretva campaign alone they massacred over a thousand defenseless Moslems in the Plevlja region. There were some capable commanders among them, but they lacked either military or political leadership. In London their government was being unhinged by feuding Serbs and Croats, while at home its war minister and supreme commander, Draža Mihailović, sat out the Neretva battle in Lipovo near Kolašin, presumably to keep out of reach of Germans and Partisans alike.

At that very time the royal government approved collaboration with the Pavelić forces, and the Chetnik officers forbade listening to the BBC because it gave reports of Partisan fighting. What was left of Serbian nationalism and Orthodoxy, of the alliance with Britain, in the eyes of the ordinary Chetnik? On the other hand Tito and the Communist leadership, united in a single aim, were in the thick of the battle, with an army continually engaged in vital military operations. The German propaganda—leaflets often dropped by planes—was not literate or effective; its themes were great German victories, and amnesty for deserters.

But the Chetnik propaganda was incomparably worse: the King, myths, folklore, and omission or distortion of the facts. We Communists also made use of folklore and myths, but not the old, worn-out ones: we composed new poems about the Partisans in the traditional style, and emphasized our ties with the centuries-old struggle of our freedom-loving peoples. As for facts, sometimes we exaggerated and sometimes we passed over things in silence, but always from our viewpoint and for our cause.

The Chetnik commanders—and particularly Draža's delegate, Zaharije Ostojić, who landed with Captain Hudson on my territory in Montenegro in 1941—gave strict orders to their units: "Prisoners are to be executed after a brief interrogation."* But the fighting did not afford them many such possibilities—not so much because this policy stimulated desperate Partisan resistance, but because the Partisans were on the upswing. Yet the Chetnik leaders, possessed by a mythical fear of German power, were convinced that we couldn't survive, and that the role of dealing the final *coup de grâce* would be theirs.

We, on the other hand, urged our men immediately to free Chetnik prisoners—except, of course, the officers, of whom very few were captured. So hundreds of prisoners returned to the Chetnik rear full of praise for the Partisans' treatment. On my way to Jablanica, where I was to supervise the transfer of the wounded, I met a Chetnik accompanied by one of our men.

"Where are you taking that man?" I asked the soldier.

"That's my own brother," replied the soldier. "I'm going to finish him off in the ravine. He had no mind for our sufferings. He betrayed us." The soldier was a Montenegrin, as I could tell by his speech. He had apparently recognized me and seemed anxious to vent his feelings to a fellow Montenegrin.

"Don't you know our policy regarding the captured Chetniks?"

"I do, but I tell you, he's my brother. I have the right to judge him."

I ordered him to take his brother into his own unit. He agreed instantly, though he would have killed him if I had said otherwise—contrary to my desire but in accordance with my phraseology.

That desperate, fateful battle also exposed to the world the collaboration of the Royal Army with the occupation forces. With this aim in mind Tito sent a telegram to Free Yugoslavia, our radio station under Soviet control, that twelve thousand Chetniks were also fighting against the National Liberation Army alongside the Germans and Italians. After he

* Vladimir Dedijer, *Dnevnik* [Diary] (Belgrade, 1946), p. 141.

had read the telegram to us, I said, "I think there are about twenty thousand."

"I think so, too," Tito said with a grin, "but it's awkward admitting they were able to get so many together."

I may have grinned, too, while agreeing with Tito.

Once he had learned of our battles with the invaders and of the Chetniks' defeat (the British mission of Colonel S. W. Bailey was with Draža), Churchill decided to send his own military missions to the Yugoslav Partisans. At the time of our breakthrough we didn't know this, but we were not surprised when we learned of it soon thereafter. In war only the victorious gain a right to hope.

3

Some dozen Germans were captured in the Gornji Vakuf battles, among them a high-ranking officer by the name of Stoecker, a short man of dignified bearing. The idea came up in a conversation involving Velebit, Rankovič, Tito, and myself that a letter be sent to the Germans through the captured Major Stoecker, offering the captured Germans in exchange for our arrested comrades, especially since the Germans had agreed to such an arrangement in 1942. It was Tito who developed the idea—or rather, immediately sought ways of putting it into effect. He brought together the Central Committee members—Ranković, Pijade, and me—in his water mill by the Rama River, and suggested that we send a letter to the Germans through Major Stoecker proposing, in addition to an exchange of prisoners, that the wounded and prisoners be treated according to international conventions, and demanding specifically that the Germans recognize us as a "belligerent force." We had been briefed before in detail on the issues of the "belligerent force" by Vlatko Velebit, who was a good lawyer. The covering letter bore the seal of the Supreme Staff, but Terzić's signature, not Tito's. However, it was clear to the Germans that the offer had been made with the knowledge and approval of the supreme command: they knew that our movement was centralized. Our assumption was that the Germans would not easily agree to our proposal, and we phrased the proposal in a way that left room for negotiation.

This Central Committee meeting was held the day after the Chetniks had been defeated on the Neretva, and a passage had been cleared across the Prenj highlands. But this hardly brought our difficulties to an end. What if the Germans attacked us after the breakthrough, and tied up our forces while they were protecting the passage of our wounded? The number of wounded in our central hospital alone had risen to four thousand, while typhus fever had hit some of our units hard, despite the vigilance and care of our medical corps: the regions along the way were infected, as were the houses we stayed in and the clearings where we bivouacked. What was to become of our basic plan to rally in southern Serbia? We would disintegrate before we even got started. Besides, the Chetniks were a problem for us—more of a political than a military one. What if the Allies landed and came upon broad areas of Chetniks, most of whom would join them? Even if no landing took place, the Chetniks could hold on for a long time in league with the occupation forces, perhaps to the war's end, and then get support from the British, as from the Royal Army.

Surprisingly, we received an answer from the Germans within two or three days: the message that we could immediately send our negotiators was signed by an officer and sealed with an eagle. On the day the German reply came—March 9, 1943—another meeting was held, attended only by Tito, Ranković, and myself, to appoint a delegation and work out tactics to deal with a hypothetical German offer. Tito regarded the matter as so delicate and important that he proposed that I be appointed to the delegation as a member of the Politburo. No one raised any objection, and I did not demur. I knew enough German to follow a conversation and get along somehow or other. After all, we didn't intend to discuss Goethe and Kant. Tito also felt that a senior commander should go; Koča Popović was designated; he knew German fairly well. Vlatko Velebit's participation in the delegation was taken for granted; he had shown adroitness in handling the exchange with the Germans in Livno, and he knew German so well—he had studied it in Vienna—that the Germans thought he was Viennese.

The tactics to be followed in the negotiations could only be formulated generally, especially since Tito did not get into hypothetical situations and strategies. The Germans were not to know that our chief objective was to penetrate into Serbia, or that we intended to occupy northern Montenegro, Sandžak, and parts of Kosovo and southern Serbia. We were aware of German sensitivity with regard to Serbia as a central Balkan region with a strongly anti-German population and a sense of national identity. But we had to offer them something convincing: Sandžak was the most expendable, being our poorest and most backward territory, while the Chetniks were an enemy of ours of whom the Germans were also apprehensive—though they had not fought against one another in some time, but on the contrary were collaborating, as on the Neretva. In short, we were to name Sandžak as the future Partisan territory, and the Chetniks as our main enemy. The question of our position with regard to the Italians was to be avoided, but in extremity it had to be defined sharply: as long as the Italians were arming the Chetniks and collaborating with them, clashes could not be ruled out. There was not a word about the cessation of fighting between the Germans and ourselves, but this too was understood. Tito's views and reactions could be summed up as follows: while we were spreading and growing stronger, the Germans would grow weaker, and then we would see. This was never clearly outlined, but it followed in part from the actual situation, and in part from Tito's thoughts—sufficiently, so that everything took shape in my consciousness. I would not make public the essence of these negotiations with the Germans, if they had not already been made widely known abroad.* Moreover, some Yugoslav historians as well—among them, Vladimir Dedijer in 1967—have tried to explore the course and content of these negotiations. The official Yugoslav silence has only enhanced the rumor. There is no reason for this silence today, except to preserve the idealized image of the leaders of the Yugoslav revolution, as if it weren't sufficient for them to have carried out an original revolution.

We were in agreement on the course of the negotiations, though Tito was the least skeptical of all. I raised the question, "What will the Russians say?"

Tito replied almost angrily—in anger at the Russians, not me—"Well, they also think first of their own people and their own army!"

That was the first time that a Politburo member—and it was Tito—expressed so vehemently any difference with the Soviets: a difference not in ideology, but in life. I was very pleased with Tito's reaction: yes, it was clear to me that we were beginning to differ with the Soviets over a very sensitive question—the most sensitive of all—and one that was vital to us. Had someone asked me then if this divergence from the Soviets agreed with our ideology, I would have replied, "Well, our struggle is also a contribution to the Marxist-Leninist teaching." In other words, as long as life fits into the ideology—as long as the ideology makes possible a productive orientation—the ideology is alive.

* In greatest detail in Roberts, *Tito, Mihailović and the Allies,* pp. 106–12.

Besides, Tito had already received Moscow's reply. At the same time as the letter to the Germans, a dispatch had been sent to Moscow which mentioned only an exchange of prisoners. But this time Moscow was quick and discerning, and we received an immediate and angry reply, true to style: Is it possible that you who were an example to all of enslaved Europe—you who have until now shown such heroism—will cease the struggle against the worst enemy of mankind and of your people?

The next day Koča Popović joined us, and the delegation met with Tito; this time there was less attention to detail than at the meetings of the Central Committee. On the evening of March 10 Tito left for the Neretva. That same day we were informed that the Germans had entered Prozor: our defensive units were crossing the Rama while our heavy weapons—howitzers and mortars captured from the Italians—were being put out of commission and dumped in the river.

The three of us spent the night in the little mill. At the break of day on March 11, 1943, we set out for Prozor with a little white flag on a stick. The last of our units had crossed the Rama and the wooden bridge over the river was in flames; meanwhile the Second Proletarian Brigade was dispersing the Chetniks around Lake Borak.

About a dozen kilometers of highway along a terraced hill separated the little mill from Prozor. We made our way without haste, but we were tense, expecting to come across the Germans at any moment. However, there was no sign of them, nor any noise or movement. We felt extremely anxious: what if we came across the Ustashi first? We decided we would show them the German letter, bluff them into thinking that hundreds of captured Germans were involved, and tell them that the Germans were expecting us. Yet we weren't sure that the Ustashi would give a damn about all this; they might simply slaughter us on their own. In our desperation, we joked and insisted that Velebit should be the one to carry the white flag because he had the lowest rank.

We reached Prozor without meeting a soul along the way. In Prozor we spotted a German sentry in front of an old fortress. He stepped into the house next to his post; we understood that he was telephoning the command of our arrival, which reassured us. Going on, we reached the first few cottages, which were demolished and charred, where some ten Germans sprang forward with pointed tommy-guns; one of them, an officer, demanded sharply—much more sharply than we had expected—that we identify ourselves. Velebit handed him the letter, while the soldiers took away our pistols—in accordance with international law.

With the German escort on all sides of us, we proceeded down the main street. To the left, along the walls of a razed house, several Ustashi, in tatters and half-uniforms, were doling out breakfast. We stopped for a moment—I no longer remember why—and as the Ustashi gazed on us in astonishment, one said, pointing at me, "That one sure has nice shoes!" The officer escorted us into a little house on the right and set up a guard. We waited for about an hour, then another officer let us out and put us in an automobile blindfolded. We were still tense, but now we were curious as well. Trucks frequently roared by, and occasionally a tank rumbled past. Since we couldn't see them, they sounded all the louder—like heavy iron rods hurtled down a chute. We spoke sparingly, even reluctantly. Koča could not resist his own cynicism: "That's nice! We've just moved out of the Rama Valley and they're pouring two divisions into it. That valley is so narrow that they'll need a lot of time and a lot of thought to get them out again."

We were not surprised when our car stopped in Gornji Vakuf. We knew by the incline where we were going. Our blindfolds were removed and immediately we were taken into a small building to the left of the road. We were led up a flight of wooden stairs and into a little room. Behind a table sat a German lieutenant-general, the commander of the 717th Division which we had just recently pinned down. He was a ruddy man, graying, in his fifties, with cultivated manners and very neat. I now know from published documents that his name was Benignus Dippold. He didn't shake hands with us, but with cold courtesy invited us to sit opposite him, on previously arranged chairs.

The general immediately pointed out that he was not empowered to discuss the matters at issue, but would listen to us and report. Our conversation with the general lasted for a little over an hour, and was terse and tense, for the simple reason that he wanted to gain as much as possible while giving nothing. We concluded our talks around eleven o'clock, when the general told us that we would have to await the reply of his superiors. Again he didn't shake hands, though he was now less reserved. They put us up some fifty yards away, in a small one-story building, in a room with three plain beds covered with blankets. In Gornji Vakuf one could hardly find better lodgings; not even the general's was much better.

We were now under the jurisdiction of the military intelligence service of the Abwehr. We had not expected anything of the sort, and would have preferred to remain under the army. Immediately two intelligence officers joined us who talked with us more out of curiosity than to ferret out secrets. They behaved properly and unobtrusively, and did not ask questions regarding our mission. This was their first contact with high Partisan officers, and they were interested in our mentality. We in turn were interested in them, all the more so in that they departed from our preconception of German officers as narrow chauvinists or fanatical racists.

One of them was a captain, slender and very handsome, and the other, I believe, a major, rather heavy and with a coarse face. They took care not to linger with us, and we not to provoke them by attacks on Nazism, but neither side concealed its views. Moreover, they were rather defensive, more concerned with Germany and the German army than with Hitler and Nazism. What surprised me more than anything during all these negotiations was how little of the Nazi ideology and mentality was evident in the German army, which did not seem at all like an unthinking automated machine. Officer-soldier relations seemed less disciplined and more cordial than in other armies. The junior officers ate out of the soldiers' kettle, at least here on the battlefield. Moreover, their army did not appear particularly organized or blindly obedient. Its militancy and homogeneity sprang from vital national sources rather than from Nazi discipline. Like any other men, they were unhappy that events had embroiled them in a war, but once embroiled they were resolved to win, to avoid a new and worse defeat and shame.

We had agreed that only Koča Popović would give his real name: because he had introduced himself as the commander of the First Division, it made no sense to conceal it, and the Germans probably knew of him through prisoners. Velebit changed his surname to Petrović, for fear of reprisals against his family, while I assumed a common name—one borne by a Montenegrin hero of long ago: Miloš Marković. I was too prominent a figure to reveal myself, and too tempting a prisoner for the Gestapo in case the Germans reneged on their bargain. Later, when Velebit and I went to Zagreb to negotiate, I permitted Velebit to give his real name and to visit his family. The Germans in Gornji Vakuf took photographs of us by surprise, but I covered my face. Later, when Velebit and I arrived in Sarajevo, I caught sight of one of our deserters in the corridor of the German command, and tried to cover up. I am convinced that he recognized me, but I don't think that he reported me. Thus my pseudonym remained unidentified until the publication of Roberts's book.* Prior to that, only Vladimir Dedijer had mentioned—in the third

* Walter R. Roberts, *Tito, Mihailović and the Allies, 1941–1945* (Rutgers University Press: New Brunswick, N.J., 1973).

edition of his *Diary,* published in 1972—that I was with Velebit in Zagreb.

Yet the German officers in Gornji Vakuf were not deceived by our secrecy. When I told them that I was the quartermaster of a division, the coarse major remarked with irony, "This one is their commissar!" On the morning of March 14 both officers wished Koča a happy birthday with cordially ironic expressions. Koča wasn't at all taken aback; he thanked them and added, "That was easy enough for you to find out: the Belgrade police have had a file on me for a long time."

The German officers spoke with contempt of the Chetniks and the Home Guards. And the Ustashi, they said, were cruel, and had no experience or real military organization. We agreed with this analysis, adding that the Ustashi fought like desperadoes. They asked us where we got all those mortars. We replied that we took them away from the Italians. To which they retorted with devilish irony, "Of course! Who else? Our dear allies!"

Both officers had fought on the Eastern front. They did not think very highly of the Red Army. The coarse major said, "They surrendered in droves. In 1941 only small isolated units put up any real resistance."

But the handsome captain added, making a clucking sound, "In front of Moscow, the Siberian divisions were wonderful. They fought to the last man." He wore a ribbon on his lapel, for a wound he received at Moscow.

The defeat at Stalingrad bothered them, though they belittled its importance: "So what? We lost two to three hundred thousand men, but in some operations the Russians lost a million!" We argued that this wasn't at all the same, that Stalingrad had come after a year and a half of war, on foreign soil, when there were no longer any surprises. They replied that no one had a monopoly on the fortunes of war. They regarded Africa as a secondary theater of war, though one could feel their dread of the American potential.

When we commented on the unprovoked nature of their attack on the Soviet Union, they replied, "It had to come. Russia would have attacked us the minute we became engaged in the West." We did not deny such a possibility, but held that just as there was no victory in the East, there would be no compromise in the West. Though they spoke of a German victory, we were under the impression that they were simply hoping for an outcome which wouldn't bring ruin to Germany. When we tried to prove to them the hopelessness of Germany's position, they replied, "Others are giving thought to this." They never once said, "The Führer is thinking for us!"

They thought highly of the fighting qualities and discipline of the Partisans, but they were horrified at our warfare. The handsome one said, "Look what you have done to your own country! A wasteland, cinders. Women are begging in the streets, typhus is raging, children are dying of hunger. And we wish to bring you roads, electricity, hospitals."

We did not deny Yugoslavia's backwardness, but we added, "Still, one could get along here until the Germans attacked. Perhaps you would give us roads and electricity, but in exchange for our ores and our food!"

We demonstrated the senselessness of their warfare against us. "We can't be beaten," I insisted. "Even if you destroyed our organization, we would regroup somewhere else, our other units would stay intact."

Koča added, "I can slip my division through your ranks whenever I want—tonight, if you like!"

The coarse major conceded: "Yes, it's easy to make war that way—a piece of bread and some bullets in a bag, and off you go into the mountains!"

"We can make an exchange if you like; we get the tanks and motorized equipment, and you . . ."

"You'll never get your hands on that—never!" the coarse one shouted.

The food was good—soldier's fare. They gave us cigarettes, then offered us liquor, but we declined it. We were served by a Croat, a young man from the Devil's Division, who was ashamed to look us in the eye, let alone talk to us. Was this chance, or did the Germans wish to confront us with the bad blood among our own people?

We stayed in Gornji Vakuf until March 14, waiting for the negotiations to begin. But no answer came, so we decided that Koča should go back, which he did that day. Later that same day an order came through directing us to Sarajevo. The officers returned our ammunition, on condition that we not load our guns. Velebit and I took an almost cordial leave of them, before we drove off in a military car.

In Sarajevo the Germans put us up in the apartment of a dark-skinned, pretty woman of thirty—a Serb and, if I'm not mistaken, the wife of a captured officer. She received us graciously—almost as if we were relatives—though she was on intimate terms with the German officers staying in her apartment. She hated the Ustashi, and saw in us Serbs who weren't very smart, but Serbs nevertheless. Her apartment was on the bank of the Miljacka River, luxuriously and tastefully furnished. And like all Serbian women she was an excellent cook.

The next day we were escorted to the German command, which was located in the building known as the Konak. This was the headquarters of General (later Field Marshal) Alexander Löhr. We were taken to Section 1C, that is, to the Abwehr, where we were received by several high-ranking officers. The conversation was informative and restrained on both sides. They mostly asked questions, and we tried to say as little as possible. We stayed within the limits of set positions. The Germans made no commitment other than that they would exchange prisoners. The atmosphere was not menacing or even unpleasant. The Germans then demanded that, prior to any negotiations, sabotage on the Zagreb-Belgrade railroad be stopped immediately. However, we did not commit ourselves to this, but tied everything to the recognition of our rights as a belligerent.

It was our impression that this was only the beginning of our talks. But then the Germans came to a standstill, which led us to believe that they didn't know when the negotiations might continue. I decided to go back, but to leave Velebit in Sarajevo so he could let me know if the negotiations were to continue; if not, he was to come back, too. Since the Germans insisted that we send back Major Stoecker and the other prisoners as soon as they could be rounded up, I asked to have one German come with me so we would give them a definite answer.

I left Sarajevo for Konjic in a truck with a semi-trailer driven by a noncommissioned officer, which was to bring back Major Stoecker. The noncom was rather short, apparently obtuse, and spoke reluctantly in some dialect; he may have been a Yugoslav "Schwab."

The road was bumpy and the land devastated, but we traveled unhindered. At Konjic the noncom received information concerning the sector where Ranković and I had agreed that I would cross over. This was some four to five kilometers south of Konjic, where the Germans held the highway while our side held the heights. The noncom and I immediately drove there.

On the highway the Germans took cover behind tanks, and brought several wounded men there for protection. From a jagged crest the Partisans were firing in short bursts, the bullets whistling all about us, so we stopped below an overhanging cliff. The noncom quickly explained our mission to an officer, and while the soldiers gazed at us with curiosity, he and I ran across the meadow overlooking the road.

Near the top of a hill small groups of Germans were making their way through the bushes. Several Stukas swooped down, circled around the crest with great noise, and then dove headlong to plaster the heights with volleys of bombs. A battery on the right bank of the Neretva also began to pound the crest, which seemed to be swaying in the gleaming light. The Germans rushed forward under a cloud of smoke. I was almost indifferent, although it never left my mind that my comrades were dying on the crest of the hill. But when the Stukas descended into the valley and, one by one, greeted their troops by dipping their wings, I was overcome by a bitter envy.

Apparently the Germans had taken the hill, for the fire had subsided. The escort and I had to climb up the slope. We came upon two Germans, one of whom was unwinding a telephone cable from his back. On the hillside a German was resting, screened by a crag. On the ground below lay a young Partisan, in a brand-new Italian overcoat, woolen peasant stockings, and peasant shoes; the image of our fervent hopes, our victories, and our backwardness. Driven back by the attack, the Partisans hadn't managed to take him away.

We rested and smoked. The Germans received us calmly, as if they hadn't just been through a skirmish. "They didn't get us. If it hadn't been for the Stukas, we would have gotten ours!" They spoke about their experience as if it had been a quarrel in a tavern.

Carefully sticking out my head, I surveyed the scene: a rocky valley with sparse underbrush, ending with a ridge on the horizon. "Where are the Partisans?" I asked a German officer, trying to find our units with his binoculars.

"Over there!" He waved toward the ridge.

It was around three in the afternoon. I began to call out, "Comrades, don't shoot! I'm a Partisan! I'm returning from a parley!" I shouted it several times loud and clear. Not a sound or sign of life from the other side.

I told my escort to stay under cover: the Partisans might be suspicious on seeing me with a German. I emerged from my shelter and started to run, but after fifteen paces a volley of bullets rained down on my left. I shouted, "Don't shoot! I'm a Partisan!" Two short volleys answered me, ricocheting on the rocks all around. I returned at a fast pace, but not at a run, to my shelter. Again I called, again I went out, and once more was turned back by volleys showering on the crags nearby. While I was returning to my shelter a German soldier in a helmet, and with a martial air just like in the posters, snapped at me, "You should be more careful. They're good shots!" I was probably never closer to death, but not even for a moment did I feel danger, at least not the reflexive kind to which one reacts with a quick jump or a fall. Ranković later explained what had happened: there had been a change of units, and the outgoing commander hadn't passed the word on that I would be crossing over in that sector.

There was nothing for me to do but to make my way at night, when at least they couldn't aim with such precision. I waited for nightfall and set out with my escort. I began to climb toward the ridge, shouting now and then, "Comrades, don't shoot! We are Partisans!" At the top, I was stopped by a sharp shout, unexpectedly near: "Halt!" I said I was a Supreme Staff delegate. The voice continued, unchanged: "Hands up! One come forward, the rest halt!"

It was a junior officer of the Fifth Brigade. He recognized me on shining his flashlight into my face. I explained that I was exchanging prisoners, and I summoned the sergeant out of darkness. "You almost killed me today," I told him. He replied laconically, "That's our machine gunner. How could he know who it was?" At the battalion headquarters I was told that Sava Kovačević, the commander of the brigade, was up front not far away.

Sava was sprawled out by a fire, eating supper. He said to me, "The Supreme Staff must be somewhere around Lake Borak." And suddenly, with a sly smile, he added, "Don't you go making peace between us and the Germans!"

I felt trapped and confused, nevertheless I was on the offensive: "Don't be a wise guy! Don't you have any confidence in the Central Committee? This is an exchange of prisoners. And to protect the wounded from being killed."

"I *do* trust them!" Sava said, "but the army has just barely gotten started against the Germans. They're our worst enemies."

Even so, Sava gave my German escort a good supper, and we continued on our way, overtaking long columns of soldiers, many of them wounded. We traveled through the night. Morning found us on the slope between Bijela and Lake Borak. We overtook a small group of young men—typhus cases. One of them staggered and sat down, propped against a tree. I walked over to him and encouraged him to go on. He looked at me—and died. I resented having the German witness our suffering but was proud of our determination, which amazed him.

As we were descending to Lake Borak, several Stukas flew overhead. We got off the road and hid under a pine, though their target proved to be elsewhere. The roar of the Stukas impressed the German more than us; we had already grown accustomed to it. At the tip of the lake I ran into Major Stoecker and left the noncom with him. Stoecker did not know where I was going or to whom: Tito and the Supreme Staff were up there by a cliff, deep in the forest.

Tito was overjoyed to see me. I recounted the negotiations briefly: no firm commitments, except over the exchange of prisoners; the Germans didn't seem averse to further talks, but as a condition demanded an end to the fighting in Slavonia. Tito was dubious: "Ha, I knew that's where it hurts them! We can't agree to this until they stop attacking us." Of the Central Committee members, only Ranković was there. We decided to go on with the talks—but nothing more specific than that, as the Germans hadn't presented a stand on a single political issue. I was convinced that the Germans would turn over their prisoners, and that the release of Major Stoecker and the other prisoners by us would be looked upon as a token of good will, so Tito approved this immediately. We sent Major Stoecker and my escort on their way to Konjic within the next twenty-four hours. As for the remaining dozen or so prisoners—there were no more—we couldn't send them yet because they weren't rounded up.

"And how did the Germans treat you?" Tito inquired.

"Correctly, very correctly."

"Yes, it seems that the German army has kept something of the spirit of chivalry," Tito commented.

I told Tito and the others of my impressions and experiences, talking until the afternoon, when Tito had to go on to Glavatičevo and I had to return to Bijela, to await word from Velebit.

Tito's wife Herta, with whom he lived before Zdenka and who had borne him a son on the eve of the war, was on the list of prisoners whom we were seeking from the Germans. During our conversation Tito returned to the subject three or four times. Obviously her situation pressed upon him, and he said feelingly, "The most important thing now is to get our comrade back! Do all you can to have her released."

I waited several days in Bijela while the German prisoners were assembled. The fighting went on, though not as intensively. The German troops had nowhere else to deploy, while our smaller units were engaged in protecting the evacuation of the wounded. Even so, the Germans bombed Bijela while I was there. It was there, too, that I saw Mitra again. I heard that the Germans had come upon our wounded in several villages and, instead of killing them as they used to, had distributed cigarettes and chocolate. Apparently, some good had come from our negotiations.

One evening while supper was being dished out—it must have been on March 23, 1943—there was an uproar in Bijela: "The Germans!" Someone reported that they were bearing a white flag and shouting, "Don't shoot! A parley!" Realizing that these were the Germans who had come for the prisoners, I restored order and went out on the road to meet them. The Germans were already standing before the nearby houses. Koča Popović was also there, commenting cynically, "That's how my army stands guard! It's a good thing the Germans don't know who they're dealing with!"

A German officer brought a message requesting that I go to Sarajevo. The prisoners were waiting. One of them was gravely wounded; lifting himself up on the stretcher, he stammered, beaming with joy, "How good to see you!"

The officer knelt down, stroked the man's head, and said with emotion, "Now everything will be fine. Don't worry, everything will be just fine."

The Germans took over the wounded and we set out for Konjic. In Konjic, in a tavern dimly lit by a gas lantern, I saw bearded faces around the tables—Chetniks drinking raki. But the Germans paid them no heed. Then we got into a truck. They sat next to the driver, and we set out for Sarajevo. There I found Velebit and Hans Ott, the engineer who had arranged the first exchange between the Germans and the Partisans in the fall of 1942. The next day we set out with Ott for Zagreb: we traveled as far as Brod by car, and from there proceeded in a guarded train compartment.

Having renewed their acquaintance, Ott and Velebit spoke fairly openly, though not without tact. On the other hand, I was quite tactless: I stated that the whole world, except for the Germans who had been *gleichgeschaltet* by Goebbels's propaganda, were already aware of Germany's impending downfall. Hans Ott strained to deny this, but in doing so just confirmed our impression that not even he believed in Germany's victory. Moreover, from our lengthy talk with Ott on the train, we gathered that among the upper classes in Germany, particularly among the officers, uncertainty and discontent were beginning to emerge. At one point Ott even let slip, "Hitler is a maniac!" Did Ott belong to the active opposition? And was he close to Edmund Glaise von Horstenau, the German general commanding in Croatia, who was said after the war to have been against Hitler? I don't know, and all that was incidental at the time. Ott was a mere engineer, involved by an accident of fate in German-Partisan relations. Very likely he reflected the critical thinking of the

German leaders in Croatia, and the discontent among the higher strata in Austria and Germany. At the end of the war our intelligence people hauled him back to Yugoslavia, not for any wrongdoing, but on the assumption that he might know German intelligence agents who were close to the Partisan leaders. Ranković told me that they never found out anything from Ott, even though he stayed in prison until his sad end.

Ott was the most convenient person through whom to get Tito's former wife Herta Has released. Ott complained that the Ustashi were giving false information about the prisoners we were seeking, so that the Germans had to check through their own agents. Such was the case with Herta as well. Velebit had already spoken with him about her, and I was so adamant that he finally asked me, "Why are you so insistent over her release, if she's no longer a high official?"

I thought it best to be frank to a degree: "She's the girl friend of one of our commanders."

"Oh, then I understand," said Ott.

The German service knew that on her father's side Herta was a Volksdeutscher, but since she was a Communist this made no difference to them. Ott promised that Herta would be found and exchanged, and so she was.

The Germans put us up in a hotel-like room in one of their institutions on or near Zrinjevac. The negotiations were conducted in the same building. They must have begun on March 26, the day after our arrival. As there was nothing for us to do, the Germans offered to take us to the movies. We went, while Velebit went to visit his parents.

If I remember correctly, there were two meetings. The German team was headed by a colonel with fine blond hair who acted more like a diplomat than a military officer. He continued to keep the negotiations on a low level; this was unexpected, and in my eyes diminished their significance. Even so, progress was made toward a truce: the Germans indicated that they would cease operations as soon as we stopped our raids on the railroad line in Slavonia. But no agreement was ever signed, nor was there talk of our getting any weapons or help from the Germans.

No cessation of our struggle against the Ustashi was ever discussed. We acted as if we had special rights with the respect to the Ustashi, as well as the Chetniks, since they were our internal enemies. But we didn't place emphasis on them as our main enemies since the Germans regarded them as allies. If the Germans brought up our stand toward the Ustashi, we were to argue that the Ustashi were killing our people, massacring the

Serbs, and so on, and therefore we couldn't cease fighting them. The Germans, in turn, avoided the Italian issue. Or else they simply didn't get around to raising it, since the negotiations never got beyond a discussion of general positions.

We didn't shrink from declarations that we would fight the British if they landed. Such declarations didn't commit us, since the British hadn't yet landed, and we really believed that we would have to fight them if—as could still be concluded from their propaganda and official pronouncements—they subverted our power, that is, if they supported the Chetnik establishment.

Again, there was a delay on the part of the Germans. They told us that they had gathered together the required prisoners in Sarajevo, which was a good reason for us to go to the Supreme Staff. We arrived in Sarajevo on the morning of March 30. I believe that Velebit was also with me. The Germans told us that some of our prisoners taken in recent battles—their names weren't on our list—were in the military prison, and would be turned over to us. I went there—to the barracks of the former officers' school. While I was talking to a German in the corridor, I suddenly recognized that the girl scrubbing the floor was Lenka Jurišević, a close friend of my sister Dobrana. I called to her to get ready, because she was about to be exchanged. With her wet arms she clasped me about the neck and wailed on my chest. The German officer was moved to tears as I comforted her. Later, on the way back, she told me that when she heard my voice, she thought that I too had been captured and that the Partisans were done for. "That was the worst moment in my life," she said. "I couldn't believe it, though I saw, I heard." Lenka was extremely brave. The Germans had come upon her while she was asleep and captured her. She lost her life toward the end of the war. Perhaps that is why the memory of her today so moves me.

After lunch we set out by truck for Trnovo with our exchangees—about fifteen of them, a few more than the exchanged Germans. In Trnovo we stopped to get information about our positions. An officer of the Croatian Home Guard gave us the information. I tried to persuade the Home Guard officer to join our side. At first he was startled, then he said, "That's not so easy. We're under the control of the Ustashi."

We continued by truck for another ten to fifteen kilometers, and then proceeded on foot. Among the exchanged prisoners was Ivo Frol, the author, whom I knew personally. He told me in great excitement that the Ustashi in Jasenovac had turned him over to the Germans in Zagreb, in the belief that he was being summoned for interrogation. When the Ustashi learned what was going on, one of their officers remarked, "We'll see each other again, over a rifle sight!" On the way to Sarajevo the German officer kept reassuring him, "You have no need to worry, Herr Professor, the Partisans are not bandits. They have printing presses and schools, and there is order among them."

I found Tito and the Supreme Staff in a village not far from Kalinovik. I made my report to Tito, but he didn't seem quite as interested as before: the Germans had, in fact, already called a halt to their drive, while our units had won a hard-fought victory over Pavle Djurišić's Chetniks, and were penetrating into Hercegovina toward Montenegro and Sandžak. Tito immediately approved Velebit's return to Zagreb, and stopped the operations of the Slavonian Partisans, particularly on the Zagreb-Belgrade railroad. Velebit carried out this assignment, taking quite a bit of time. He also brought Herta back. He told me that he had trouble in Slavonia: the Partisans suspected him of being a provocateur, and the supreme command in Croatia had to intervene.

At the beginning of our stay among the Germans, it occurred to me that they could turn us over to the Gestapo for torture and execution. But the Germans gave no reason for such misgivings, and eventually the misgivings vanished.

Neither I nor the other Central Committee members had any pangs of conscience that by negotiating with the Germans we might have betrayed the Soviets, internationalism, or our ultimate aims. Military necessity compelled us. The history of Bolshevism—even without the Brest-Litovsk Treaty and the Hitler-Stalin Pact—offered us an abundance of precedents. The negotiations were held in great secrecy. There were no differences among the top leaders, except that Ranković and I were more dubious of the outcome than Tito. As for a more permanent truce or broader agreement, no one really believed in that.

The negotiations with the Germans could not have produced any more significant results. This was because we essentially sought a respite, while the Germans were setting a trap for us. The Germans couldn't permit our stabilization and expansion, and we couldn't permit them to gain strength with the help of pro-German elements and to continue the war in the Balkans. Among the Germans in the army, and among German representatives in Croatia, there was an element that favored a truce. But this element didn't even have the courage to make its desires known. Hitler cut off the negotiations—to be sure, at a moment he regarded as propitious—with one of his categorical fallacies: "One does not negotiate with rebels—rebels must be shot."* With this, Operation Weiss passed over into Operation Schwarz. In official Yugoslav historiography these two operations are called the Fourth and Fifth Offensives, though they constitute a whole—both in the analysis of historians and in the experience of the participants.

* Roberts, *Tito, Mihailović and the Allies*, p. 110, citing the work by Walter Hagen, *Die Geheime Front*, p. 268.

4

The Supreme Staff spent the greater part of April 1943 in the village of Govza. I was there too, idle, clambering among the cliffs in search of chamois and roe deer, sometimes all by myself, even though the Moslem militiamen had fled from the village to the woods. The only people who were worried by the militia, and with good reason, were the sick and the unarmed. One day I talked Tito into going hunting with me. A buck ran at me and I shot him. Tito hadn't done much hunting before the war.

At the beginning of April our units had reached the Drina River, which was defended by the Italians and the Chetniks. Swollen by spring rains, the Drina was a seemingly insurmountable obstacle. But we had to cross it in order to get to Sandžak, and Tito was quite adamant about it.

I was present when Pavle Ilić, a captain in the engineers corps and later a general, explained to Tito and Terzić that the Drina had such and such a velocity per second, and that it was impossible to cross it without adequate boats and pontoons. Nevertheless Tito stuck to his resolve. After Ilić left, I asked Tito, "What do you really think? Will we cross the Drina?"

"You know," Tito answered, "The experts as a rule don't take into account the human will. Humans who are determined to do something achieve it, even if it's considered impossible according to calculations."

And so it came about. Through their courage, enterprise, and improvisation, the Partisans conquered the Drina, then again beat the

Chetniks and Italians. The roads were strewn with Italian tents, and the crossroads with the hastily shaven beards of Chetniks. The Italian commander in Foča abandoned the town and posted himself on a bare hill in the vicinity, while his colleague fled from Čajniče. Finally, the Vasojević Chetniks also suffered a serious defeat.

In Govza, beside a campfire, I gave a lecture to our escort battalion and Staff personnel on the occasion of May Day. Tito was also present. At the end he rose to speak. He recited routinely that heavy fighting awaited us, then announced unexpectedly, "I'm convinced that we'll observe the next May Day in Belgrade." It sounded overoptimistic in this Bosnia backwater, with the Allied fronts thousands of kilometers away. But he was not far wrong: instead of on May 1, 1944, we entered Belgrade on October 20.

We also had to prepare political cadres for the areas we planned to take over. We selected political workers with exceptional skills mostly from the army. I conducted the course, with the help of Krsto Popivoda, in the village of Lijećevina. Our basic text was Stalin's little work, *The Foundations of Leninism.* Our lectures and discussions extended into imperialism, the Marxist view of the world, and basic political economy. But *The Foundations of Leninism* was not only the most concise source, but also the most trustworthy and convenient, for mastering Leninism. To be sure, this work didn't have the significance of Marx and Engels's *Communist Manifesto.* But for the tempering and the indoctrination of the party bureaucracy as a new ruling force, it couldn't be surpassed, precisely by virtue of its mathematically bare, total, totalitarian "truths." Later Žujović joined us as a lecturer. His girl friend Mileva Planojević also attended the course. Their relationship seemed stable, and there was no need to be puritanical, so we assigned them a separate room for the enjoyment of love in bloom and Leninism.

The course was to have lasted about a month, but we had to complete it in haste because of the German advance from Foča. This was a bad sign, offering little hope that it might be a local defensive action. The Fifth Offensive—Operation Schwarz—had in fact already begun on May 15, 1943.

On May 22 I set out with a group of comrades for the Supreme Staff, then at Crno Jezero, by the little town of Žabljak, on Mount Durmitor. In Rudine our hospital was bombarded. The next day, as we were descending Durmitor, German planes—about thirty of them—whirred above the peaks and dove down on Žabljak below us, on the huts and cabins scattered over the wasteland and tucked among the evergreens. There was no longer any doubt that the Germans had begun a broad operation.

Our units were on the verge of realizing our objectives: they had reached the Tara and Lim rivers, and were about to take Kolašin, Bijelo Polje, and the Chetnik center at Berane, while their right wing was to unite with the Albanian partisans and their left to take Nova Varoš and Sjenica. The Italians and Chetniks had been beaten at Podgorica and Nikšić, and were obviously unable to counter our units, which were being replenished with fresh manpower.

On May 11, apparently under British influence, the royal government-in-exile sent Draža Mihailović instructions to cease collaboration with the invaders and quislings, and to establish peaceful relations with the Partisans. The instructions remained unfulfilled, because of bad blood and the insubordination of the Chetnik leadership. On May 15, the day Moscow decided to dissolve the Comintern, the United States government decided to send a mission to Draža Mihailović, thereby linking itself to forces whose collapse the discerning Churchill had already sensed.

Reports concerning the German offensive were sparse until the Germans appeared before us in the valley of the Lim and the Tara. The only reliable report came from Arso Jovanović; on his way from Slovenia to the Supreme Staff by way of western Bosnia, he sent word that we would be attacked by the German forces which had taken part in the Fourth Offensive, and that he was coming to our aid with the Fifth Division.

On the day after my arrival at the Supreme Staff, a report arrived concerning the concentration of significant German forces in the valley of the Lim and the Tara, and their offensive operations. Waving the report, Tito shouted at Ranković and me, "The Germans are lying! We have never been in greater danger! We have to go back to western Bosnia. There is no other way out!"

On leaving Tito, I remarked to Ranković—with sarcasm aimed at myself as much as at Tito—"So much for our negotiations!"

Ranković replied, with less resentment against Tito than myself, "This is no time to talk about that!"

It seemed to me then that the Germans had lulled us with the negotiations, at least insofar as they prevented us from carrying out a better planned and more timely disposition of our units. But under no circumstances could we conceal our forces in the area under their occupation. Today I would amplify this thinking: had we known (as we did not)—had we seen through German intentions—we could have dispersed some of our units and services, and thus avoided their encirclement.

Truly, there was no better way out than the one which Tito had resolved to take. Tito had previously given the First Division an order to strike the Germans on the right bank of the Tara and break through to the north, into eastern Bosnia. The division broke up some German units in the hills, but it couldn't break through because of the density of the Germans, and the advantages which the Drina and communications along the river offered them. Moreover, Tito had already effected a hasty regrouping of units for their return—that is to say, for the breakthrough into western Bosnia—though the formal decision was made only on May 26.

On May 15, Moscow had decided to dissolve the Comintern, because its existence aroused misgivings over the worldwide aims of the Soviet state as a revolutionary center. However, before the formal dissolution, Moscow sought the opinion of fraternal parties. Tito did not hurry with his answer. This was not because he felt strongly about the Comintern's survival; rather he was preoccupied by the offensive, had to consult Kardelj in Slovenia, and knew well that our opinion would not influence Moscow's intentions.

All the Central Committee members approved of the dissolution. Not even Pijade, a veteran Communist, mourned over it. Tito's explanations made one feel that he was personally liberating himself from an unnatural burden. I too looked upon the dissolution as a relief and a way to greater opportunities: we would not have hanging over us the discipline of an abstract worldwide agency; we would conduct direct, open relations with the Soviet government, and thereby gain both respect and rights in our relations with them. I reasoned aloud: Communists are united by the same ideology, but they count in Moscow only to the degree that they are strong in their own countries.

However, Moscow was in a hurry—as always, when its own needs were concerned. It again requested an answer, and Tito sent it after reading it to Ranković and me. He had all the Politburo members sign—I don't know why, unless to stress our unanimity. When I was in Moscow in 1944, Veljko Vlahović told me that—next to the Soviet party, of course—the Yugoslav party was the most categorical in its acceptance of the Comintern's dissolution. The representatives of the other parties—for example, the Czechs led by Gottwald—had demurred, perhaps for fear of falling directly under Moscow. In contrast, our party—because of its greater strength and different position—found a direct tie with the Soviet government to be advantageous.

Meanwhile Tito gave me a new assignment: the organization and direction of a Commission for the Suppression of the Fifth Column and Terrorism, as a separate service of the Supreme Staff. Anticipating our consolidation in Montenegro and southern Serbia, Tito was setting up an agency which would secure our power and maintain our control. This was in fact to be a secret police and counter-intelligence service. I had begun the selection of staff and the organization of services, but the offensive and the loss of stable territories made the idea obsolete until the Supreme Staff was settled and expanded at Drvar, in the spring of 1944. In May 1944 Tito entrusted that task to Ranković; I was in Moscow at the time. That doesn't mean that, had I not been in Moscow, I would have been chosen again for the task. Radovan Zogović told me that he had played godfather in naming the agency. They settled on OZNA (*Odeljenje zaštite naroda*): Department for the Security of the People. They thought how powerful and unforgettable the Soviet Cheka sounded: in politics, names should be memorable, suggestive, and—for such organizations—mysterious. I didn't envy Ranković, but I wasn't overjoyed either that this sword of the revolution didn't fall into my hands. Though I am glad that this duty bypassed me, I don't doubt to this day that I would have carried it out conscientiously and perseveringly, even though not with Ranković's calm equanimity.

A year had gone by since we had left Montenegro in bitterness. Now we were leaving it again, in even greater haste, but with the realization that this was only a temporary retreat before an alien force. We were no longer confounded by our own inadequacies, while our foes had been exposed through their collaboration and their acts of violence. We were sure, especially after the battle on the Neretva, that the Chetniks would be weakened by their way of governing—the restoration of the old police regime without any rights, let alone laws. And yet the people's abandonment of the Chetnik movement surpassed our expectations. New and hardened fighters were joining us voluntarily; only the sudden and violent offensive hindered the replenishing of our units and the creation of new ones. Also, our behavior contributed to the strengthening of our movement. We did not arrest, let alone execute. True, we had few potential victims, since the Chetnik officers and bullies fled to the towns and the Italians. But as if by agreement the peasants, when reproached for having sided with the Chetniks, replied, "Well, you're not the way you used to be either!" If Chetnik outrages—dungeons, raids, burnings, executions; the handing over of Partisans to the Italians for money; the inflicting of twenty-five lashes for minor offenses—didn't push into oblivion the earlier Partisan terror, they did raise a comparison favorable to us, all the more so in that we showed ourselves now to be not only a stronger force, but one with better prospects and more statelike.

There was no indication, other than Chetnik stories, that the Chetniks would ever wriggle out of the invader's nets. London was extolling the Partisans, while defeats had forced the Chetniks into the towns, into fortified places and under the direct control of the invaders. To the end, the Italians kept the name for the Chetniks which they gave to the original road guards: *milizia voluntaria anticomunista.* The Partisans were the boast and pride of enthusiastic young people who yearned for something new, but who also saw in the movement the chance to improve their lot. And that wasn't all. The Partisans brought with them a party and a government whose concern for all didn't yet reveal a desire to control everyone. At meetings, and in talks with peasants, the response to slogans was more thoughtful than emotional: whatever happened had to happen, the new order had to be accepted in the form in which it was being created.

We learned that Draža Mihailović had been staying in Lipovo, near Kolašin, since the fall of 1942, and that he had actually launched the campaign on the Neretva from there. I would have accepted his sojourn in my homeland as my own personal shame, had it not been politically explicable: in hiding from the Germans, Draža moved to where he felt the safest and where his movement seemed the strongest. We also got wind of a report that during the campaign on the Neretva—I know now that it was on February 28, 1943—Draža had delivered a speech in Lipovo such as not even we Partisans could have written for him: he said that he didn't need the Western Allies since they weren't helping him anyway; that he wouldn't stop collaborating with the Italians as long as they helped him; and that the Croats, Moslems, and Partisans were the enemies of the Serbs—that is, of the Chetniks—and that only after settling scores with them would he turn against the Germans and Italians.

However, there were politicians among the Chetniks who didn't approve of collaboration with the invader, and who assessed their own destiny with pessimism. At that time we could only sense this from the confusion and disunity within Chetnik ranks. With ensuing defeats and dissolution, Draža was abandoned by all the politicians—though the outstanding among them were not in the country, except for Dragiša

Vasić and Živko Topalović. But Topalović and Vasić had belonged to small, uninfluential parties before the war: the Republican party and the Social Democratic party, respectively. Vasić joined Draža as early as 1941 during joint Partisan-Chetnik fighting in Serbia. Topalović, however, placed himself at the head of Draža's political movement in early 1944, then flew off to Italy to fight lost battles with memoranda and articles. At the time of the Fifth Offensive, we knew nothing of the differences between Vasić and Mihailović. However, I recall that our secret service had shown a special interest in Vasić. These differences were ignored, if not even hushed up: Vasić was the most important political figure in Draža's movement, and an opponent of co-operation with the Communists.

Recently I learned from the author Dobrica Ćosić that Vasić resolutely, if in vain, opposed collaboration with the invader. While gathering material for his novel *Deobe,* in the archives Ćosić had come across Vasić's letters in which the latter claimed that the Chetnik movement was doomed to destruction with the first gun it received from the invader for use against the Partisans. After the Chetnik collapse in Montenegro, Vasić wrote to Draža from Serbia that Serbia was politically lost to the Chetniks, because the movement had been taken over by sergeants and had instituted a senseless reign of terror. Ćosić gave seven such letters to the author Miroslav Krleža, who had been on close terms with Vasić, particularly after World War I. Both were gifted writers and against war, although Vasić quickly lost his fire. All things considered, Vasić didn't wield any major influence on Draža and his officer-ridden movement, except that he shook them somewhat in the beginning. Although I know that these assertions may provoke the most disturbing themes of "What might have been," I maintain that the Chetniks wouldn't have won even with Vasić's orientation: the Communists were more militant and visionary, and most important of all, more Yugoslav. Everything that fails does so in the ugliest way.

During Operation Schwarz the Germans disarmed and interned the Chetniks, who were herded together in the towns. This was done against the will of the Italians, and sometimes without their knowledge. To be sure, this occurred only in the towns of northern Montenegro, which was as far as the Germans penetrated. The Germans no doubt did this because they distrusted the Chetniks in the event of an Allied landing—a reason which we were reluctant to recognize and gladly hushed up. Draža Mihailović barely got away into Serbia. Pavle Djurišić was captured and interned in Germany. But I have read somewhere that, on the intervention of Nedić's government, after Italy's capitulation the Germans brought Djurišić back to Montenegro to fight the Partisans, this time in league with them.

Even if we had reason to delve into German intolerance of the Chetniks, our own fate on the battlefields was of greater concern. The Supreme Staff at Crno Jezero was expecting a British military mission. One mission had already been parachuted into eastern Bosnia, and another into Croatia. But we attached special importance to a mission to the Supreme Staff. Its activity wouldn't be limited to the gathering of information, and its very coming signaled the end of British support of Mihailović.

On the night of May 28, 1943, near the village of Negobudje on Mount Durmitor, the mission descended by parachute on the very spot where, in the beginning of 1942, Pijade had awaited Soviet planes. The following morning the mission arrived at the Supreme Staff, at Tito's little tent by Crno Jezero. The mission was led by Captain F. W. Deakin, who was outstandingly intelligent despite his general reserve. We found out that he was a secretary of a sort to Churchill and this impressed us, as much for the consideration thus shown to us as for the lack of favoritism among the British top circles when it came to the dangers of war.

The mission was greeted by all the Staff members present. The introductions were followed by a conversation in which Tito was the most loquacious, though he was ordinarily not overly talkative. Tito informed the mission of the position the Supreme Staff was in, while saying nothing, to be sure, about the black forebodings that assailed him. "We are constantly surrounded," someone observed. Giving us a mischievous look, Tito remarked at one point, "And you will have the opportunity to convince yourself whether it is we or Draža Mihailović who is fighting the Germans and Italians." But Deakin and the mission members remained silent and expressionless—exactly as we had pictured the English. They only pointed out that they wished to help us and that they were interested in the enemy forces.

This was a festive occasion, so to speak, so we ate together on the rocks, and on tent flaps under the firs. We gave the mission horses to carry its radio transmitter and belongings, as well as its own escort—mostly selected and instructed party members. Then, having waited till nightfall because of enemy air raids, we set out across Durmitor. The British mission stayed with the Supreme Staff throughout the offensive.

It was a hard march, one of the hardest. The mists and rains made the night impenetrable and the mountain path muddy and slippery. Though

I grew up in the mountains, I was astounded at the thickness of the mud and the chasms our imprints made. The column kept breaking and getting lost in the fog. There were a few flashlights around, but they were of little use even to their owners. Nor could we ride our horses, since they stumbled and their loads and saddles were overturned. Moša Pijade's sight was poor at night, and as he was exhausted and wearing uncleated boots, he kept falling all the time. I tied his horse to mine, lifted him, and held him up for a good part of the way.

It was morning by the time we reached the canyon of the Sušica, by a lake. A journey that would have taken three hours by day took us nine at night. The British were muddy and tired but they did not complain, nor react to our inquiring glances. They shared their cigarettes with their escorts and learned from them, amid laughter, how to roll their own.

The following night we continued our withdrawal from the canyon of the Sušica. There was no rain and everything went well except for a mix-up over Tito's horse, as a result of which an angry Tito started out on foot. I felt awkward riding beside him, and so I dismounted and offered him my horse. He refused, but when he realized that I wouldn't ride either, he mounted. But my horse had the bad habit of lying down when he felt like resting. He gave in to this urge with Tito on top of him—luckily, not until after we had gotten out of the canyon; Tito landed on his feet. At least we had a laugh in the midst of all our troubles. "Suddenly," Tito recounted, "the horse sank under me."

The Staff settled down in the forest for the day. Tito sent me ahead to find a suitable spot for the Staff to rest after the next evening's march. I took along some ten soldiers, because there were runaway Chetniks about, and set out toward Barni Do, above the canyon of the Piva River. The morning was clear, and I found the sun and the dew exhilarating, despite the exploding of bombs in the gorges and the yapping of machine guns along the roads and over the bends. We moved from grove to grove amid the hollows and mounds.

In the woods next to a village we came across a hollow full of sheep, where we surprised two men. They broke into a run. I ran after them, shouting with the rest, "Stop! We'll shoot!" They kept on running. I took out my revolver, but was forestalled by two rifle shots: a soldier behind me had fired. One of the fugtives staggered, rested his arm against a tree, and sank into the grass. We ran to him. He was a youth with a thin, unkempt beard, the kind the Chetniks wore. The bullet had pierced his chest. He didn't reply to our questions, he couldn't. Back by the sheep we found two rifles with cartridge belts, leaning against a beech tree. The Chetniks had been milking the ewes; taken by surprise, they didn't have time to grab their weapons. On the youth we found a little notebook which contained peasant-style verses in a scrawling hand, dedicated to his shepherd sweetheart. We sent a patrol into the village with the news. Meanwhile the youth expired and our soldiers, who were thirsty and hungry, drank the milk and took a rest. The patrol came back and reported that the two men were a well-known local Chetnik and his nephew.

We continued to Barni Do, and found caves suitable for shelter in the cliff under the edge of the canyon. In one of the caves we came upon some smoked meat and dairy products: the peasants had a premonition of the violence and scope of the offensive, and had hidden away food, cattle, and belongings. We didn't touch a thing, despite our hunger and the tempting aroma of the supplies.

On the next evening, May 31, a group of Chetniks attacked a column of the Supreme Staff. The bullets buzzed around Tito's head and a soldier was killed, though the Chetniks were beaten as soon as our side counterattacked. The Supreme Staff arrived in the morning with its escort. They didn't reproach me or my squad, yet by their silent bitterness I could tell what they were thinking: that the night attack wouldn't have taken place had we not killed that youth. I thought so, too, feeling sorry for the young victim felled by the lightning stroke of alien forces here on his mountain, next to his sheep. Even so, this was a good opportunity to show the British mission that while the Germans were showering us with steel from the slopes and skies by day, the Chetniks ambushed us by night.

Since speed was a factor in escaping destruction, our units were given permission to confiscate livestock at will. That decision was made with a heavy heart, but also in the conviction that our people would understand: the Durmitor region was loyal to the Partisans. Still, Tito forbade anyone to touch the supplies in the cave, even though the Staff and escorting battalion were in dire need. When I visited this region in 1948, the peasants showed us receipts made out by us for requisitioned livestock, but not one dated back to those days. The peasants couldn't remember how much livestock had been confiscated or from whom. They said, "The poorest suffered the most, and the families that gave you your leaders, because they didn't hide their livestock. When are you going to honor these receipts?"

"Well, we've given you UNRRA," I replied.

"That's American and for everybody," they retorted mischievously.

I was a member of the government and knew that individuals wouldn't be paid war indemnities. I replied with a good-natured smile, "Those receipts are for you to frame—so your descendants can proudly remember that you gave all you had for such a glorious struggle."

The peasants smiled, too—at my quick thinking, and at the fickleness of wartime promises.

The planes patrolled throughout the day with such consistency that we were unable to stir from the caves. They attacked horses no less zealously than humans. The supplies of the Supreme Staff were depleted. But that day the news from the battle areas was good. Until the previous two or three days, the Germans had controlled the Sutjeska River. But the Second Proletarian Brigade first repulsed the regiment that tried to block our exit from the Piva Valley, and then took Suha, the entrance to the canyon of the Sutjeska River, and established a beachhead on it. But what was on the other side, in the hills overlooking the Sutjeska? How great were the enemy forces there? In any case, the exit from the Piva canyon and the road to the Sutjeska were secured.

On that day the generally argumentative Zdenka topped each discharge of spleen with another. Having quarreled with the guards and cooks and everyone in sight, finally she lit into Tito, perhaps all the more furiously because Herta, Tito's former wife, wasn't there; Herta's presence usually restrained her. Though Tito also had a fiery temperament, before his staff he always restrained himself with Zdenka—for the sake of his reputation and also, I would say, because of Zdenka's emotional influence over him. (She did not, and could not, have any other influence over him.) At that point I boiled over. I yelled at Zdenka to shut up or I would grab her by the hair and throw her over the cliff, because the Central Committee had other worries than to put up with her hysteria. Tito kept silent; Zdenka got scared and shut up. That was how the quarrel ended. I then resumed correct relations with her, but we were never cordial.

The next evening we set out for the Piva canyon. The canyon was over three thousand feet deep, and the narrow path wound amid the sharp rocks. We moved slowly because it was impossible to pass a loaded horse, or stretchers bearing the wounded. Even so, the going was easier than it had been across Mount Durmitor. We were worn out by the stumbling, the turns, the halts. In the morning someone commented that this descent bore the name of Zlostup (Bad Step), at which an exhausted Pijade retorted in a lively voice, "It has certainly earned it!"

On the morning of June 3 it began to rain, and mists covered the canyon with gloom and travail. The roaring of artillery assailed the mountain from the plain, and the mountain responded with the feeble barking of machine guns. There were no planes at least, and the Supreme Staff continued its march—up to the cavern below the village of Mratinje. We took refuge in it because of the rain, but didn't stay long. We ate something. I don't remember if we received any reports. But all of us, and particularly Tito, were worried over the slow movement of the gravely wounded and of the units protecting them. No one said it, but it was as if we had gotten out but they had stayed behind, without much prospect of escape.

Suddenly Tito said, as if after much hard thinking, "One of the Central Committee members has to be with them." We were silent. Milutinović and Žujović were not present. "It's best that you go, Djido," Tito proposed rather than ordered. Ranković was the only other man who could be considered for the job. Pijade was worn out, and anyway lacked the authority of a Politburo member. I believed then, as I believe now, that Tito picked me and not Ranković because he worked with Ranković more closely and had grown used to him: Ranković selected and proposed officers and officials, he took impeccable care of the day-to-day details regarding the Staff and the Central Committee. On observing that I didn't accept his decision with enthusiasm, Tito added—more to justify his decision to himself than to arouse my vanity—"After all, the Bolsheviks also sent their leaders into the toughest spots!"

However, I had already made up my mind to go, though I had a premonition of failure. "How would it be," I proposed, "if we tried to cross the Tara into Sandžak?" The enemy concentration there was not nearly as dense, and there were greater possibilities of maneuvering.

Tito reflected and said, "Good, take a look yourself. We'll be in touch by radio."

On the terrain entrusted to me there were two divisions, the Third and the Seventh Divisions. Since these two divisions had to be joined into one operational unit, it was decided that Radovan Vukanović, commander of the Third Division, should take over both divisions, and that Sava Kovačević, commander of the Fifth Brigade, should take over the Third Division. An order to that effect was issued later.

I had no reason to wait, especially since Tito and the Staff were about to move on. They saw me off with anxious, almost mournful expressions. I mounted my horse and set out all alone back again to the jerrybuilt little bridge over the Piva. The cable bridge further down, at Kruševo, had already been demolished. I pondered as I rode, but couldn't come up with a solution for the army or its several hundred wounded. I tried to console myself: perhaps it isn't as bad as all that, given our determination and resourcefulness . . . As I was emerging from the canyon, the weather began to clear. Several planes darted through an opening in the clouds, and I had to get off the road until the clouds closed in or the planes went away.

5

In the Fifth Offensive—Operation Schwarz—the German command chose a favorable terrain, the most favorable they could find for surrounding and destroying our troops: a deep canyon with few steep passages, swift and unfordable rivers, and bare pasture heights. A giant fortress, with all the advantages and drawbacks of a fortress: an army with enough food and ammunition could easily defend itself here for a long time, but, being surrounded, would find it extremely difficult to get out. We had experienced the many disadvantages of this terrain in 1942. But only now were we faced with all the horrors of this otherwise wonderfully scenic landscape.

The Germans gathered greater forces now than in the Fourth Offensive: seven German and several Italian divisions, some Ustashi units, and a Bulgarian regiment—all in all, 120,000 men against 18,000 Partisans. The Chetniks had been disarmed by the Germans or else "discharged" by the Italians, so they didn't take part in this offensive, except to attack small peasant groups and the wounded. The Germans had carried out special preparations for mountain warfare: they had gathered hundreds of horses and thousands of bearers—largely Moslems from Bosnia, to carry supplies under German guard; they even had specially trained units with ropes for mountain climbing, and units with bloodhounds. (I have also read somewhere that the Germans had poisoned many springs. That is not true; we leaders would have known about this, even without drinking from those springs.)

The entire region was ringed with roads that were narrow and winding, but with a stone foundation and usable; even though we had the advantage of being concentrated in choice spots, the Germans were able to bring in reinforcements by truck. The Germans treated the whole operation as exclusively their own. Their troops occupied the crucial positions, and it was with them that we waged a life-and-death struggle. I would say that our only advantage was the springtime, even though it favored the use of their aircraft; at least we could rest a bit on the ground, and take cover and maneuver more easily, than if those crags had been covered with snow.

My military conclusions here are in fact thoughts which at the time obsessed not only me but, in one way or another, every soldier down to the lowest ranks. The leaders did not hide, nor could they, the mortal peril into which we had fallen—a ring of some thirty to fifty kilometers, a circle of fire which was drawn around us tighter every day, and which was penetrated by lethal jets of flame first in one spot, then in another. We leaders and soldiers were a brotherhood condemned to die, and which could be saved from utter destruction only by our collective heroism and personal sacrifice.

This heroism and sacrifice was what I felt and counted on as I took leave of the Supreme Staff in the cavern below Mratinje. Our confidence and solidarity were such that every least bit of both continued to live, to fight, and to take root in the people: there is no revolution without a solid, adequate organization, and none without faith and brotherhood. I knew the Germans couldn't grasp this special supermilitary strength of ours. I believed that the Germans couldn't destroy us, that we would get through piecemeal and again regroup our units. I only feared for the gravely wounded and for the newly recruited cadres.

As for Tito's decision concerning the splitting into two groups, and my joining one of them, I believe to this day that it was justified militarily and certainly politically—though it hit me hard then and galls me to this day, because of the failure. Militarily: had we withdrawn all our forces simultaneously and along the same route, the enemy would have been able to deploy the greater part of their forces along that route. Politically: all units could have withdrawn along with the main force, only if the gravely wounded had been abandoned—which was incompatible with the conscience and tradition of our army. We had to fight for our wounded regardless of the odds in battle. And indeed, as I was making my way to Sava Kovačević's headquarters, my determination grew: if we are unable to save the wounded, we must confirm by our deeds and lives that it was not possible to do so. Moral considerations before tactical ones.

I came upon Sava Kovačević in a peasant house in a tiny village in the Piva Range. He was extremely glad to see me. Soon Ivan Milutinović joined us as well, more by force of circumstances than through any Central Committee decision. The Executive Committee of AVNOJ, of which Milutinović was a member, had stayed behind the Supreme Staff; but when the decision was reached in that cavern below Mratinje that I should return, it was also decided that the Executive Committee should stay with me. Since Milutinović was a member of both the Politburo and the Supreme Staff, and thereby my equal, I was obliged, though formally in charge, to conduct all matters jointly with him, and I did this the entire time.

Our most urgent and vital problem was where and how to cross over into Sandžak. No other routes came into consideration, since they were all closed off by major forces and unsuited to maneuvering. But even the passage into Sandžak was complicated and difficult. We had to negotiate the swift and impassable gorge of the Tara River. The enemy had mined all the bridges along its lower course. The only free and accessible spot was the confluence of the Sušica and the Tara; there was no bridge there, which may explain why the enemy had not mined that route. This was the spot we chose. We informed the Supreme Staff of our choice, and sent a company along with responsible comrades to clear a path. There was no time to build a bridge, so on my recommendation we decided to throw ropes across some calm spot and build a pontoon bridge over inflated sheep hides, or rafts made of logs left from prewar days.

The company set out a couple of days after my arrival. However, one could feel a certain lethargy, even apathy, among them. To be abandoned and sacrified for the wounded—who could not be gotten out, even if every one of us were to become a stretcher-bearer—was more than human strength and spirit could bear, even among men who had endured six months of marches, want, disease, and battle. Consequently, I personally took a detail of five or six men and went to the confluence of the rivers to see what could be done about crossing the Tara. To shorten our journey, we made our way across a pathless stretch into the canyon of the Sušica.

We got there faster but with greater difficulty, because we had to get around crags and bluffs, and slide down ravines. When we arrived at the mouth of the Tara, we found that almost nothing had been done. I studied the possibilities with the commanders, though the complement lacked adequate tools and means. Displeased by the negligence of Sava's staff and the assigned men, I returned to the Piva Range—again across a pathless land, this time with even greater effort. I arrived at headquarters before dark, terribly exhausted. This was most likely on June 6.

I was greeted by a mute consternation which not even Sava Kovačević's bravado could cover up: the enemy had occupied the right bank of the Tara, at precisely the spot—the only possible one—where we had intended to cross. We could clearly see small clusters of soldiers in blue-green uniforms on the meadows above the canyon. According to the report received by Sava Kovačević's staff, enemy soldiers had descended to the banks of the Tara—apparently, while I was climbing from the Sušica up the Piva Range—and had killed or wounded several of our men working on the crossing.

All my anger over any lack of initiative became senseless now: we had to make new decisions. That same day, June 6, I informed the Supreme Staff that the enemy had occupied a stretch of the Tara, including the best spot for our crossing. The next day we investigated other crossings over the Tara. But by that time the enemy had crossed over to the right bank, at its lower course. Our canyon fortress had been invaded, and I informed the Supreme Staff of this development.

While I was observing the enemy on the plateau above the Sušica, I was assailed by the suspicion that the Germans were intercepting our radio messages. That suspicion weighed on me all the more heavily during the offensive. There is no more terrible thought, at least in battle, than that the enemy knows your intentions. That suspicion never left me, and was confirmed after the war. Ranković was the first to tell me that the Germans had indeed been reading our messages at the time. Most probably the Germans gained our secrets by deciphering our extremely simple code, rather than through an agent at the Supreme Staff—though there were hypotheses concerning this as well.

Radovan Vukanović, who had been appointed commander of the combined Seventh Banija and Third Divisions, had arrived. But even before the enemy had blocked the crossing at the Tara, Milutinović and I discussed the possibility of attaching the Seventh Banija Division to the troops accompanying the Supreme Staff. With Vukanović's consent, we decided to do this before the Supreme Staff's approval arrived. Consequently, Vukanović was a commander without units to command—a commander who was occasionally consulted by Sava Kovačević, commander of the remaining Third Division, though he also issued and signed orders.

We decided to attach the Seventh Banija Division to the Supreme Staff units because it had long been unfit for battle. This was not for lack of determination or militancy; rather, typhus and exhaustion had incapacitated it. The January offensive had driven it from its native soil of Banija, and so it was attached to the Supreme Staff units. At the beginning of the January offensive, it had numbered 4,000 men—young men, largely Serbs from the villages and an occasional Croat from the towns; by the end of the offensive, it was reduced to 1,500. Even though it had participated in a large number of battles, its greatest losses were brought on by typhus fever, aggravated by the fact that the stricken also starved and had to keep moving.

These men came from mild and prosperous regions, and were less used to privation, hardship, and sickness than those who lived in the mountains and in poverty; consequently, discouragement and dejection had spread throughout all the division. Both their commander Jakšić and their commissar Kladarina were obviously suffering, as if they were to blame for the misfortunes that had befallen their devoted and courageous men. Like skeletons to which only tawny skins still clung, these young men of Banija trailed behind the army, through villages and mountain hamlets, tattered, listless, spent. They died without a murmur, lost and nameless. The division had no deserters; even if they could have made it back to their homes, this never occurred to the men from Banija. Not one voice among them was raised in protest: they had pledged their lives to saving their people from slaughter, and to freeing them from the invader. The division's manpower simply wasted away. Those who still made up the division continued to go on their military assignments obediently and devotedly, but they were physically unable to carry them out, and were unhappy and ashamed because of it. And because that deterioration was irreversible, this racked us in the leadership and evoked the unspoken sympathy of the soldiers in the other units. Seemingly unheroic, this was one of the fullest and most consummate of tragedies.

We decided that along with the Seventh Banija Division we would send off the walking wounded, some six hundred of them. We also sent along with it the poet Vladimir Nazor and the president of AVNOJ, Dr. Ivan Ribar, not only because we were obliged to protect them, but also because their advanced years would not permit them to endure the hardships of the breakthrough. We did not send Nurija Pozderac, a member of the Executive Committee of AVNOJ and a Moslem leader. Marko Vujačić also stayed behind. Openly voicing his regret over not having withdrawn into Bosnia in 1942, Vujačić insisted on joining the staff of the Third Division, with whom Milutinović and I remained.

The Seventh Banija Division left the Piva Range on the evening of June 8. It did not go into the unknown: we knew that the First Proletarian Brigade had ensured its crossing of the Sutjeska River as early as June 5. Thus a bridgehead was established for the next and decisive breakthrough at Balinovac, which was carried out on June 10.

As for the Third Division, including Milutinović and myself, it did not withdraw with the Seventh, though it could have done so. I had proposed this to the Supreme Staff, after a council with the officers which Milutinović and Vukanović also attended. That council was held, I think, on the day I returned from the confluence of the Sušica and the Tara, which is to say, as soon as we knew that the enemy had occupied the crossings over the Tara. In that same message I also recommended that the seriously wounded, whom we obviously couldn't take along or defend, be hidden in the canyons and woods. I don't remember the date of my request. I couldn't even determine it on the basis of published documents. Most probably this was in the evening of June 7 or the morning of June 8. But my memory is certain that a reply didn't arrive in time, that we waited two days for an answer, and that we got one only after I had pressed for it.

What is fixed in my memory—and not in mine alone—is the experience of waiting for that reply. It was clear to me that Tito was having a hard time making up his mind to approve leaving the wounded behind. It was certainly not easy for Tito; he was faced with a most terrible decision, unacceptable to himself or the army. We had been faced before with the possibility that the enemy might capture and kill our wounded, but never before had we consciously accepted such a possibility. I took Tito's reluctance to act as shifting responsibility to me for the fate of the wounded. This I couldn't consent to. This was not for the sake of my prestige, or very little so, but out of a sense of duty and responsibility: we were in communication with the Central Committee, with the supreme commander, and they were in a position—they alone, in fact, had the right—to make such a decision.

At the same time I was no less concerned over the opinion and mood of the officers, and even of the common soldiers, around me, who were aware of the hopelessness of our situation, and by the same token, of the senselessness of our struggle over the wounded. They all looked to me and awaited my decision. And that decision was not forthcoming from the Supreme Staff. This only worsened the prospects for the wounded, since if the enemy engaged us near their hiding places, they would surely be discovered. Meanwhile the enemy kept advancing, with each day taking peak after peak and ridge after ridge on Durmitor, the only mountain range that we still held. It was this situation and how I felt about it that caused me, on June 8, to send the following message to Tito: "Situation very grave. Enemy occupied both banks of Tara. If our penetration fails, we shall fight to the last man."*

But Tito could have had, and did have, other dilemmas besides abandoning the wounded. His own group had not yet succeeded in breaking through, though it had better prospects by virtue of its numbers. The dangers and difficulties enveloping the Supreme Staff groups were confirmed by Tito's order to bury the artillery and archives at Vučevo, which he issued on the same day—if not before—when he sent me to the Third Division and the wounded: nothing was permitted to hinder the speed and mobility of the units. Moreover, as early as June 5 the Germans had cut off, with great losses to us, our exit at the most convenient spot for a withdrawal, through the village of Vrbnica. And when a breakthrough was finally accomplished, it was not in the desired direction of western Bosnia but toward eastern Bosnia—in line with the progress of the fighting, and the discovery of the enemy's weak spots. From later accounts and documents it may be seen that Tito was upset about the fate of my group, and constantly made inquiries concerning it. He paced up and down and exclaimed in agitation, "We cannot abandon them!" Yet this two-day delay of a reply to my proposal that we leave the wounded in hiding places, and withdraw the Third Division with the Supreme Staff groups, was the only reason that a gap developed between the Third and the Seventh Divisions, into which the Germans hurled themselves.

Finally Tito sent us his approval to leave the wounded in the canyons and retreat with the Supreme Staff units. By my reckoning Tito sent that approval on June 9, the day he was wounded. In anticipation of Tito's approval, and under constant pressure from the enemy, I and the officers around me had already undertaken to hide the wounded and to deploy our units adequately. There was nothing else we could do, even if the approval hadn't come. The village of Mratinje and the bridge on the Piva over which the Supreme Staff had crossed, and which I used to

* Cited in Dedijer, *Dnevnik,* Vol. II (1946), p. 206.

return to the wounded, were occupied by the Germans as soon as they had been abandoned by the retreating units of the Supreme Staff. Thus we had to build a new bridge downstream. But this one was so narrow and shaky that the horses couldn't cross on it; we had to drive them into the river and drag them to the other side, using ropes to keep them from being swept away by the current.

Meanwhile the Germans were having their own troubles, even though they were on the offensive. They didn't come out from Mratinje onto the plateau of Vučevo, and so failed to cut off our only exit from the Piva Valley. However, once the Supreme Staff units had gotten through, the exit from Mratinje was left undefended. We took that exit before they did—right behind the Seventh Division.

On the day we got permission to move out and leave the wounded behind—that is, on June 9—we pulled in our flanks and worked intensively at hiding our wounded. Late that night the staff of the Third Division began its descent into the canyon of the Piva River, toward the little bridge at Gornje Kruševo. Milutinović and I went with the staff members, as did the party leadership of Montenegro and a large number of functionaries and patriots.

It was a clear night. The path was relatively wide, and not as steep as the one at Zlostup, along which the Supreme Staff had descended to the Piva a few days earlier. Yet we didn't cross the bridge over the Piva by night, as planned: most movements in war aren't carried out according to plan, and this was especially true of us, since practically everything we did was under pressure. The road was clogged with the gravely wounded, who were being carried by soldiers and peasants. Several hundred lightly wounded, weary soldiers, and refugees trailed behind, in little clusters no longer held together by organization, but by a common misfortune and ideal, moving at their speed on the same path and in the common stream of withdrawal. We were held up by the corpses, too, since we never knew but that they were merely exhausted soldiers who needed to be roused from sleep.

So we were caught in our descent by the morning, which coated the rocks with dew and cast a bluish mist over the river. We came across some fifty gravely wounded men among a heap of boulders by the path: a sour stench and the muffled cries overcame the dewy freshness, as well as my determination as a commander. It was then that I discovered how much I dreaded facing the gravely wounded. What could one say to them? How could one explain why we were abandoning them, why we couldn't die with them? Why this selfishness toward them without whose sacrifice we couldn't retreat or fight? Should we tell them the truth—senseless but nonetheless true: that the wounded enjoy every consideration, but only those who can fight have the undeniable right to live? The wounded asked for nothing, made no reproaches. We left our medical personnel with them, those who volunteered to stay—and there were more of them than duty or the spirit of sacrifice required.

Only one ruddy, handsome youth without a leg—the political commissar of his battalion, a native of Kragujevac—turned to me and said, "Comrade Djido, I beg you to give orders not to have our weapons taken away." I had no idea that anyone was doing such a thing. But I immediately understood that the medics were doing it to keep the wounded from committing suicide. It occurred to me: those weapons are their only means of defense, to escape torture, if the enemy should come. I ordered that they be allowed to keep their weapons. I also spoke a few words to them: I said we would do something for them, if we broke through. And the commissar from Kragujevac shot back, "Just you get through. Never mind about us!"

Dejected and downcast, we resumed our descent through the grove. But as soon as a log slide appeared through the thinned-out trees in front of us, we were stopped by our scouts. They explained that the slide clearing was being shelled by a battery of howitzers and that it was impossible to get around it: there were cliffs above and below. We knew from reports that German scouting patrols were giving our positions to batteries at Šćepan-Polje, and that we had to run that gauntlet, as did our whole army. Sava Kovačević told us to run, and if the shells began to sound, to fall flat, then start again during the next lull.

We took this as a signal. I started running and the rest followed me—along a little path amid the boulders, over corpses of men and horses. The slide was about 150 yards wide and we ran across without stopping; our scouts had led our horses across the night before. We were in luck: the enemy reconnaissance was asleep, or else a patch of fog had hidden us, so the howitzers didn't open fire.

Not far from there we were met by the roaring azure of the Piva, and the little bridge built over the rapids from cliff to cliff. The howitzers didn't shell the bridge, but the bombers more than made up for it. Impassable gorges protected the passage from a German attack along the river. Though the suspension bridge three or four kilometers downstream at Lower Kruševo had been damaged, this one, at Upper Kruševo, had survived, as if defying all the skill and technology of the Germans. Nevertheless, the surrounding banks were freshly gouged, and covered with corpses. Beyond the bridge—on the left bank, among the rocks in the river—there were some fifteen casualties: from Banija, to judge by their youth, their horribly emaciated appearance, and their tattered clothing. The steep meadows and little fields beyond the bridge were also strewn with corpses. We no longer had the men or time to bury the dead, and preoccupation with the dead seemed senseless anyway, after we had abandoned our wounded.

As soon as we crossed the bridge, the artillery began to resound, planes roared, and the plateaus rumbled with explosions. Still we had to go on, making our way across the clearing in between attacks, and taking advantage of the sparse trees and bushes. Thus we crossed a steep incline of over a thousand yards, perhaps more quickly than if we had proceeded unmolested. By nine o'clock we emerged on the plateau of Vučevo. Close behind us, the enemy had already appeared on the Piva Range. Smoke rose from burning villages and machine-gun fire echoed across the mountains, while from the north, where the Supreme Staff was, came the all-too-frequent roar of bombs. The planes didn't let us alone all day; at great peril, we moved from crevice to crevice. The cooks, as always, found a way of dealing with the situation: they lit their fires with dry, sapless branches which made no smoke. It was easier to cook a meal than to get to one.

It was there, on Vučevo, that Nurija Pozderac was wounded. He was with his son Hamdija, a dusky cross-eyed youth, lying among some beech trees. A piece of shrapnel had blown off his heel. He was immediately bandaged, and a squad of soldiers was assigned to carry him.

On that day, too, on Vučevo, the Supreme Staff order was carried out to execute the Italian prisoners which our units were escorting and using in special services since February, when the Murge Division had been defeated in the Neretva Valley. These prisoners had been decimated by disease, hunger, desertion, and bombing. I don't know how many of them there were altogether—maybe two or three hundred, maybe more, but no more than ten were attached to the Third Division. Though there was a stifled opposition inside of me to that order, I accepted its necessity, all the more so since I couldn't change it.

What had given rise to that order was the desertion to the Germans of a prisoner who had been serving Nazor. That Italian had been most obliging, and being with Nazor, he had also received better rations. He had pleased Nazor by his solicitude—a trait which hardly distinguished the Partisans—and because Nazor got along with him well in Italian. But this didn't keep the Italian from taking one of Nazor's manuscripts. Also, it was suspected that he gave welcome information to the German air force concerning the location of the Supreme Staff. Of course we realized at the time the information could be of only limited value, because the Supreme Staff moved about so much and was ahead of Nazor and his Italian orderly, but the enemy would be glad to learn of our condition and particularly the direction of our breakthrough. In that mortal struggle in which the Supreme Staff and all our units were engaged, the giving away of information became a grave matter, even if such information didn't reveal anything that the enemy didn't know already. Even so, amid the death and destruction on Vučevo, moving stories reached us: there were Partisans who wept as they executed the Italian prisoners, with whom they had grown close in suffering and travail, even fondly giving them Yugoslav nicknames.

It was on the Vučevo plateau also that we decided to take along with the staff of the Third Division a group of party workers who had been helping with the wounded. Their job—to move the gravely wounded to safe places and to organize the less seriously wounded—was ended, insofar as anything but death could be the end. We foresaw a final mortal encounter with the Germans, one in which the party functionaries could only perish, and it would have done no good to anyone to leave them with the wounded. Among these workers was Mitra, who had been brought to this assignment a few days before, from the Third Sandžak Brigade, where she had been working to strengthen the party organization. Certainly the decision to move out those party functionaries was also influenced by Mitra's presence among them. But my concern over Mitra was not decisive, nor was it I alone who made the decision. Thus, once the reorganization was completed and the wounded were directed to the units of the Supreme Staff, several hundred wounded gathered and set out with us, some of them—such as Danilo Janković, today a general—on crutches. More kept coming, from ever fiercer battles in progress. All these wounded had their own leaders, and were under the constant care and protection of the staff, Milutinović, and myself. It was not necessary or in consonance with previous practice that party workers temporarily assigned to hospitals as assistants be needlessly exposed to danger during the march. And when Mitra joined us the following morning, she was overjoyed, not so much from gaining greater safety—who was safe, in those days?—but from rejoining familiar and respected comrades with whom she could forge ahead confidently into mortal danger.

Soon after we got to the top of Vučevo, we realized that we couldn't catch up with the Seventh Banija Division and the units of the Supreme Staff. A gap had developed into which our enterprising enemy would zealously throw himself. The party leaders would have to return to Montenegro: every moment of hesitation meant greater danger for them and ever gloomier prospects. I sat down with them to discuss the situation, as well as their return to Montenegro—a decision which I had already made in my mind.

We agreed that the Chetniks were falling apart and that their authority in the villages, even if restored with Italian help, would no longer be stable. This offered prospects for open marches and semilegal activity, whereas after the retreat of the Partisans into eastern Bosnia in 1942, a small number of party members and sympathizers had barely escaped Chetnik dragnets by hiding in the woods and in dugout shelters. Yet the Chetniks in Montenegro, although in disarray, were still a danger to the party; we had no military organization there, and not a single military unit could break through. The party leadership was going into a land devastated by invasion and civil war—without communication lines, without bases, without protection.

There was no opposition, but no enthusiasm either. This was not out of fear for one's fate; rather, despair oppressed us all because of the rising death toll before our very eyes, and the fate of our gravely wounded—a fate so terrible we didn't dare speak of it, though we most certainly felt it.

Only Blažo Jovanović inquired, "And how do you propose that we get through?"

Blažo was very experienced, more so than I, in moving through occupied territory. Thus I took his question to be an attempt to cover himself for the losses they might suffer, and I answered pointedly and impatiently, "You know better than I do! Lose yourselves in the woods, in the canyon, anywhere you can—so long as you keep as far as possible from the army. The Germans and Italians are after the army."

And so it came about. We embraced one another and they made their way into the woods, and from there to the canyon of the Piva River. But as Blažo was to tell me later, they ran into enormous difficulties. Three or four enemy divisions were combing the area, searching behind every branch and rock, so that the comrades in the Montenegrin leadership saved themselves by sheer evasion. This dodging lasted some fifteen days, so they were practically dying of hunger. On top of that, Chetnik groups were in the area also.

That afternoon we set out across the Vučevo plateau toward the Sutjeska canyon, using the brush as protection against the planes. But there was one clearing we couldn't avoid, even though it was within range of German machine gunners who had clambered up the rocky peak of Maglić. They shot at us from a cluster of pines as we scurried across one by one. We suffered hardly any losses, but the march was slowed down and the column broken.

Nevertheless, by evening we reached the heights overlooking the deep murky valley of the Sutjeska and faced the ruddy peaks of Zelengora—the image of a lovely romantic painting. The battle had subsided, as if the armies had agreed on supper and a rest, but we were neither glad nor able to rest. The wounded were catching up with us, the uncertainties were multiplying, our fighting ranks were thinning. To the left of us the bare ridge of Prijevor was occupied: we could see the Germans—so close were they—in little groups around improvised shelters; there was no getting through over there. Still less could we proceed on the right—toward the Drina, amid clearings and strong enemy communication lines. We didn't know for certain what lay in front of us—on the Sutjeska and in the mountains on the other side—or behind us either, since our last units had left the Piva Range that night, as the fires of burning villages and shepherd camps flickered from the peaks of Durmitor. There the Italians were advancing, always eager to burn, and this time more zealous than the Germans in tracking down and massacring disabled soldiers and unprotected civilians.

We spent the night above the gorge of the Sutjeska. At dawn, under cover of the mist, we proceeded to Dragaš Sedlo, a crest overlooking a high cliff, which we reached after a three-hour march over steep slopes. We settled down at the edge of a small meadow, up against Dragaš Sedlo, the most commanding point of our narrow territory. Our forces seemed sparse even in such a small space. Before us was the First Dalmatian Brigade, with us the Fifth Montenegrin, and behind us the Third Sandžak Brigade. There was also the excellent Mostar Battalion, now reduced by half. At that time the Fifth Montenegrin numbered about six hundred men, the First Dalmatian as many if not more, the Third Sandžak over four hundred; there couldn't have been more than a hundred men from Mostar.

We anticipated that the morale would drop, particularly after the abandonment of the gravely wounded. It did, but in a special and unpredictable way: the men continued to carry out their duties, perhaps with less grumbling than before, but were fearful of getting wounded—yes, wounded, and not killed. Relations between officers and men grew closer than ever—so close, in fact, that the Communist form of address "comrade" disappeared from usage as something official and superimposed. It did not occur to a single man to give himself up, not because he knew he would be automatically executed, but because we were all bound to one another in suffering and death. It was as if life had lost its value; and we continued to fight automatically almost; unwilling to give in to an inhuman, inanimate, abstract force.

That morning somebody told us that the stretcher-bearers wanted to leave AVNOJ member Nurija Pozderac behind. The bearers were afraid that they might miss out on the breakthrough, or perhaps felt that Pozderac could be left like so many others. We summoned the squad leader among the stretcher-bearers and I threatened to have them shot if they abandoned Pozderac. "He is not a party member," I said to him, "and we must take special care of him." The squad leader looked at me with frightened comprehension. They carried Pozderac up alongside the leadership. But Pozderac was a constant burden to us, less so because of his misfortune than because of our obligation toward him. What would become of him, if we failed to break through? What if the Germans hit us, massacred us, and threw us into ravines? A doctor told us that Pozderac was breathing his last, that there was no hope for him. He proposed, not only because of Pozderac's mortal agony but because of our own concern over what would become of him, that he give him an injection to finish him off. I rejected this without the slightest hesitation. As it turned out, Nurija died that same day, on the afternoon of June 11, thus ending our torment and his own. We gave him a decent funeral, though we simply left our own dead behind. His son, who grieved more than was appropriate under the circumstances, was made a member of our staff personnel. That he survived was more the result of chance than any privileged position of the leadership. Everybody had become equal, even in dying.

There weren't many battles that day, at least not severe ones. We came across corpses and injured horses, the remnants of units which had gone ahead of us. The artillery systematically pounded the forest path along which our men and wounded were descending. From the woods overlooking the meadow at Dragaš Sedlo, we caught sight of a cliff the size of a house, and thought we would camp behind it. However, when we got closer, we saw a dozen dead soldiers around what had once been a fire, lacerated by a howitzer shell. The corpses were already decomposing. They belonged to the Supreme Staff units which were also on the run—so preoccupied with breaking out of the death-dealing pincers, that they had no time for anything else. Nauseated by the stench and shocked, we chose a high cliff lower down as a refuge. We didn't bury those dead, just as we didn't many of our own; that would have been senseless, when we couldn't even get out the living.

Our cliff was useful as a lookout as well, being quite steep and high. On top there was a clear ledge on which we could doze. Before us stretched the Sutjeska Valley, with narrow, deep gorges in the upper part, and tiny fields and wooded knolls in the lower.

Suddenly, from out of the Suha River gorge, an extended column of Germans emerged which was making its way down the Sutjeska toward the plain of Tjentište. It stunned us. We had anticipated that, after the crossing by the Seventh Banija Division, the Germans would take the narrows of the Suha. But that they would ply the river, as if we weren't in the nearby woods, was inconceivable. Such behavior could not be explained in any other way except that the German commander knew nothing about the location of our forces. That became obvious when the Germans encamped in the middle of the plain of Tjentište, pitching their tents and unloading their horses.

It was at that very spot that Gligo Mandić had to pass with his First Dalmatian Brigade. They were already on the march through the woods, from where they couldn't see what we saw from the cliff above Dragaš Sedlo. In his haste Gligo had been unable to gather together the entire brigade, but instead took along a battalion from the Fifth Montenegrin Brigade, while two of his own battalions lagged behind. Mandić's assignment was to cross the river and occupy the hills over the left bank of the Sutjeska, and thus secure a passage for the remaining units and our brigade-sized medical corps. The movements of all the other brigades hinged on his success.

Sava Kovačević quickly sent Mandić a courier with the news of the German movement and encampment. Then for hours we waited impatiently for Mandić to attack the Germans and secure the crossing. The fickle mountain rain began to fall. We knew that Mandić needed time to deploy his battalions and prepare for the attack. Drenched, we continued to keep watch on our cliff.

Late in the afternoon, as the sun began to shine and the mists to rise from the lowlands, the sound of machine guns reached us from the Sutjeska. The Germans ran so fast that they abandoned their horses and tents. "Just look how we have them on the run!" Sava Kovačević exclaimed enthusiastically. It did seem to us, from some distance away, that Mandić had surprised the Germans and smashed them. The truth turned out to be quite different. While Mandić was conferring with his officers, a

German had discovered them, so the Germans who had dug in along the bank opened fire. Having no choice, Mandić without hesitation had ordered an attack, forded the Sutjeska, and with a loss of over thirty men, scattered the Germans at Tjentište. It was beginning to get dark and the men, starved as they were, fell upon the food in the tents. The Germans regrouped and chased out the Partisans, and Mandić lost another thirty men. Though disorganized, his units continued their retreat along the passage cleared by the Supreme Staff, which the Germans had not closed off, probably in the belief that they would block the crossing at the Sutjeska.

We were overjoyed, all the more since we believed that there were no Germans ahead of us: during the day we hadn't noticed any in the hills the other side of the Sutjeska, and the Seventh Banija Division had passed through two days before without a struggle. Night fell quickly, which only increased our joy and optimism. We received news that the Supreme Staff units had made a major breakthrough at Balinovac on the previous day, June 10; also, we no longer heard sounds of fighting from Mandić's brigade, which we interpreted to mean that the Germans hadn't yet closed off the breach made by the units of the Supreme Staff, and that Mandić would occupy the undefended hills and hold them through the following day, or until the remaining units and the medical corps got there.

They went to sleep—couriers, patriots, party workers, and the entire army—around the fires in our refuge, in ravines or wherever they found ourselves. We leaders also dozed awhile. But we kept waking up and rousing one another, anxious because there was no word from Mandić. Our anxiety, which turned to anger, grew as the night passed, and plunged to irrational depths at daybreak, when we finally received a report from him. The message, delivered by a single courier on a scrap of paper, was confused: in one sentence Mandić explained that he had not secured a passage through the hills above the Sutjeska because the brigade had gone on ahead, while he stayed behind with only two couriers.

Sava Kovačević had worked with Gligo from the start of the uprising, and they were on close terms. I had also got to know Mandić during my stay in Montenegro in 1942. He had the reputation of being extremely bold and venturesome, but by the same token a willful commander. The brigade he led, though not fully manned, was regarded as among the best. We wondered what had happened to that brigade. What of Mandić? This didn't sound like his willfulness, which found expression in great daring and exploits, but never in cowardice or lack of solidarity.

Sava Kovačević was upset, personally offended: "To be betrayed by Gligo Mandić—this never would have crossed my mind!"

Milutinović kept muttering bitterly, "This calls for a court-martial!"

I myself did not explode in anger, yet I was terribly worried and bitter at this complex and unexpected development.

By leaving us without a secure passage, which it had gained with little effort, the First Dalmatian Brigade had confronted us with an unexpected and difficult problem. Our forces thereby decreased: we were left with two halved brigades and two halved battalions—a total of only 1,200 men, who had to defend themselves on all sides as well as make a breakthrough. As for the Germans, who far outnumbered us—by how much we didn't know—they were not sleeping any more than we did, nor were they idle after the surprise attack and breakthrough by the First Dalmatian Brigade at Tjentište.

On the morning of June 12, behind the cliff at Dragaš Sedlo, we held a council. The suggestion was made—which Sava Kovačević also favored—that we set out for the Sutjeska immediately. Similar suggestions had been made the day before, when Gligo Mandić had scattered the Germans at Tjentište. Such a decision seemed militarily justified. However, I opposed it, on the grounds that the medical unit and our protective units were still on Vučevo or just beginning to descend, so between us and them a gap would develop which the Germans could enter. I myself was tempted by our only exit, but I didn't want to leave the wounded without protection or exclude them from the fighting column; at least those who were able to move had the right to share our fate. My view prevailed, but with the proviso that the movements of the medical unit and protective troops be accelerated, so that we could cross the Sutjeska at the latest by evening.

However, our situation grew considerably worse in the course of the day. That very morning the Germans had begun to attack our right flank, with the obvious intent of occupying Dragaš Sedlo and thus splitting our forces. The immediate move of the local units to the Sutjeska would have resulted in the separation and piecemeal encirclement of our units. Therefore we threw in major forces—"major" in view of our small numbers and the complexity of the battle—in order to block that penetration from Borovno. Through the afternoon we were engaged in bitter combats, which the Germans supported with artillery and we with sparse mortar fire, from Dragaš Sedlo. The day was cloudy and rainy, so that at least the planes didn't bother us.

Once again, that afternoon, we climbed the cliff to observe the

Sutjeska. We could detect Germans only on one bluff above the left bank, downstream. But not far from Tjentište we caught sight of a flag with the swastika which the Germans had unfurled to indicate their position to their air force. It was an evil omen, because it confirmed that the Germans had occupied the left bank, and probably also the hills which Mandić had reached the night before, hoping to pass through without a struggle. Though it had no effect on the battle, Sava Kovačević could not resist firing an artillery piece at that unfurled flag. After the second shot the Germans lowered the flag, Sava let out a whoop like a child who had just scored in a game, and the rest of us felt unwarranted joy.

Our cares and duties quickly reabsorbed us. I had such a strong presentiment of a murderous encounter with the Germans that I took my little leather bag with some notes and *The Mountain Wreath*—the only book I had carried with me—and stuffed it inside a crevice, so the enemy could find nothing on my body to identify me or give him something to boast of. After the war, while visiting that area in 1948, I found the crevice, but not my bag; too much time had passed for someone not to have found it.

At one point in the day the comrade who monitored news on the radio ran to me joyfully: the Allies had taken Pantelleria! Nobody knew what that was, or where. "Now that's a big help!" I commented, then explained that Pantelleria was a tiny island between Sicily and Africa. Our radio man withdrew, crestfallen, while Pantelleria's fall and the relief it could bring us became the butt of bitter jokes among us leaders. But the news spread fast and gave courage to the fighting men; they grasped at straws—all the more readily because they knew nothing about Pantelleria except that the name sounded unusual.

That afternoon, as our units descended toward the Sutjeska, the Germans were pounding the road from Dragaš Sedlo to the river. While behind the cliff under a beech tree, I heard an unusual, cautious rustle: a startled roebuck glanced at me, then scampered away into the forest.

In the afternoon we received a dispatch, the last one, from the Supreme Staff, which according to Radovan Vukanović read as follows: "Our situation grave. Turn to the source of the Neretva. Will send you scouting units from Kalinovik. If you encounter bad situation, get through in smaller units. . . ."* The Supreme Staff deviated from the principle—for the first and only time, to the best of our knowledge—of

* Dedijer, *Dnevnik*, Vol. II (1946), p. 342.

not dividing up units. We too anticipated such a solution, though we did not mention it even among ourselves, aware how close such a solution was to chaos and disorder.

By evening all our units were on their way to the Sutjeska. The leadership set out with them, along the muddy road and into the wet forest. However, no sooner did we cross the little meadow of Dragaš Sedlo and enter the forest, when we came across dead bodies scattered by the roadside and the thunder of shells below us. Sava Kovačević dashed off the road and down the slope in the direction of the booming shells, with the rest of us all bunched up behind him. Fearing more shells, I shouted, "Get down!" But Sava grabbed me by the arm, shouted "Forward, forward!" and sped downhill with me in tow. At that point another volley came, further to the right and higher up. The forest roared, branches broke, leaves and dirt fell over us. Again we dashed downhill, and then to the right along the road, while shells thundered over Dragaš Sedlo and around it. "They're combing the road and the forest with their howitzers," Sava Kovačević explained to me. "Always run toward the explosion, then the next round of shells will overshoot you."

Shaken, we continued our descent toward the Sutjeska very slowly. We were waiting for reports, the medical corps, and the units which had not been able to escape the enemy. Besides, the troops had to eat, however miserably; all day they had fought and marched on empty stomachs. By the time we gathered on the clearings overlooking the Sutjeska, it was long past midnight. The troops were silent, subdued. One sensed tension in the curtness of speech and spareness of movement. The sky cleared, but the moon had already gone down over the mountain.

Finally we were on the move again—the Mostar Battalion up front, and the leadership behind. As soon as we had left the little meadow, we found ourselves on the steep banks of the Sutjeska. I crossed the river on my horse, whereas the troops waded across, holding on to a cable thrown across the river. Many did not bother to hang on: the Sutjeska is a small river, and at the time it was hardly above our knees. Thus the stories about the drowning of multitudes are not at all true, unless it was someone who was also gravely wounded. Like the other leaders, Sava Kovačević rode across on his horse, while repeating the order, "Troops will cross in uniform!"

As soon as we reached the left bank, Sava observed, "There are no Germans, or they must be very weak. They would have been waiting for us at the river."

Everyone else was silent; only I voiced the doubt: "Maybe they're waiting for us up ahead."

Our mute, tense march was broken by the weak, pleading voice of a wounded man in the bushes above the road, just as we turned toward Tjentište: "Comrades, help me! Save me, comrades! Don't leave me, comrades!" Probably having fallen asleep, the poor fellow had been left behind by the units ahead of us, and had hid himself from the Germans, who had been crossing the area and bivouacking there the day before. But not one of us said a word, nor did we break step or slow down our horses. We had no men to carry him, we couldn't ask it of troops that were facing combat. Besides, each of us was repeating to himself the excuse that the medical corps was following behind us, and that they were responsible for the wounded.

We began the ascent of Krekovi. We had been climbing for about fifteen minutes, maybe twenty, when directly above us—near the top of Krekovi, beneath Ozren Mountain—machine guns crackled from the bushes. We jumped off our horses and stopped dead in our tracks. At least in that first instant, our position didn't strike me as serious, though it did seem unwise for the leadership to blunder into a German ambush at the head of a column of undersized battalions unfit for battle.

Sava Kovačević and several officers went ahead. Handing over my horse to a courier, I followed them. When I caught up with them, Sava said: "The Mostar Battalion is done for. They got trapped in the middle of some bunkers!" The battalion was lying on the ground, firing; though almost a third of them had already been cut down, the men of Mostar didn't run.

This encounter took place at the crack of dawn on June 13. Sava Kovačević and the rest of the officers recognized the awkwardness, indeed the senselessness, of our situation. They immediately ordered the advance battalions from the Third Sandžak Brigade, which was marching behind the leadership, to go forward and fan out. By the time they arrived, the sun was up and the leadership and the Mostar men were on the road. The soldiers hugged the ground, taking advantage of hollows and gullies. We leaders remained standing, with the commanders, even though the men around us were being wounded and killed. Dr. Simo Milošević had curled himself up in a little hollow, but he was wounded anyway.

At first I kept close to Sava Kovačević: he was a renowned hero and I knew that I wouldn't bring shame on myself beside him. But he pushed me away several times, and once grabbed me by the shoulders and forced

Tito, after being wounded on Mount Ozren, June 1943

A Partisan brigade marching toward Prozor, March 1943

Ivo-Lola Ribar (center) with F. W. Deakin to his left, and members of the British military mission attached to the Partisan army, November 1943

Randolph Churchill with the leadership of the Eighth Partisan Division, June 1944

Review of the Second Proletarian Brigade and the Third Dalmatian Division in the Bosnian mountains, 1943

Sava Kovačević, leader of the Partisan uprising in Montenegro

Ivan Goran Kovačić, Croatian poet

A Partisan crossing the Piva, May 1943

Mitra Mitrović, with Soviet officers, October 1944

Blažo Jovanović (left) and Milovan Djilas (second from left) in Montenegro, May 1945

The Partisan leadership in Vis, summer 1944. Left to right: Vladimir Bakarić, Ivan Milutinović, Edvard Kardelj, Josip Broz Tito, Alexander-Leka Ranković, Svetozar Vukmanović-Tempo, and Milovan Djilas

me to the ground. "No, that's dangerous! Get down!" he yelled. "Now we'll get them!"

Milutinović too was moving close beside Sava or me. At one point our artillery commander turned to Sava: "Comrade commander, the artillery is ready! Give us the target!"

"Shoot at the mountain," Sava replied with a broad grin. "The important thing is for our guns to be heard."

The commander went away somewhat bewildered. But soon our guns began to thunder behind our backs.

When the battalions arrived from the rear, Sava roared, "Forward, comrades! Your commander is at your head. Forward!" And he rushed on down the road.

At that moment Savo Burić, the newly named commander of the Fifth Montenegrin Brigade, was next to me. Under his left arm he had some Italian offensive grenades which didn't kill when they exploded. With his right arm he motioned me to lie down. He too was one of our most renowned heroes, and a lively, unconventional spirit besides. It was no shame for me to be beside him either. "Why are you throwing grenades when you don't see the enemy?" I asked.

"The important thing is that they go off. That clears the path." And he hurled a grenade.

Burić's grenade exploded in a thicket. We went around the thicket to the left, and saw a bunker consisting of a wattle fence lined with earth. Inside the bunker, all alone, lay a stout German soldier with closely cropped hair. I thought for a moment that Burić's grenade had landed in the bunker and, exploding in a tight spot, killed him. "It wasn't my grenade," said Burić, who from my expression guessed my conclusion.

At that moment a man from Sandžak emerged from the thicket, a Moslem whom I knew slightly. His expression was one of ferocity, but also of joy. "I got him, that German bastard!" he cried, then over his shoulder slung the machine gun—a favorite weapon of the Partisans. Unlike the Moslem, Burić was pale and stiff, but also much in control of himself. I imagine that I was, too.

Burić and I returned to the spot from where Burić had thrown the grenades. Milutinović ran up to me and told me in disbelief, "Sava's been killed!"

"Where? How?"

All of us felt, and someone expressed it—I think it was Milutinović—that Sava's death would have a disastrous effect on our troops' waning morale and on our relations with the commanders. We had officers who were in no way inferior to Sava, but not one was so closely and vitally linked with the soldiers in a legendary way. Not even the popularity and reputation of the political leaders could replace Sava's. Sava was a man of war, who waged battle by inspiration rather than by methodical command, who knew every soldier individually, especially in his own Fifth Montenegrin Brigade.

I also felt that Sava Kovačević's death was not only a threat to our prospects, but to my own personal safety and confidence. I was surprised to find him no more than ten or fifteen yards away, lying on his back, on the left shoulder of the steep road. He was covered with a tent cloth, though the tips of his artillery boots, which he loved to wear more than anything else, were sticking out. I called out to one of Sava's couriers to bring me Sava's revolver; I wanted to save it. The courier ran off, poked around Sava, and came back. "Somebody took the gun!" he reported in disappointment. We leaders stared at one another and shuddered: under machine-gun fire someone had actually pilfered the gun, while we were mourning and holding council. This was taken as a sign of demoralization and dissolution—perhaps not so much of demoralization, but no less horrible than Sava's death.

During the war, as in my illegal activities and in prison, I often felt fear, not only of torture and death, but at the thought that as a writer I might not have a chance to prove myself. If duty and the accomplishment of our aims required it, I faced danger. But I also avoided it, if it seemed wise or not shameful to do so. Yet in the midst of this particular skirmish, just a few yards from Sava Kovačević's body, dying did not seem terrible or unjust. This was the most extraordinary, the most exalted moment of my life: death did not seem strange or undesirable. That I restrained myself from charging blindly into the fray and death, was perhaps due to my sense of obligation to the troops, or to some comrade's reminder concerning the tasks at hand. In my memory I returned to those moments many times, with that same feeling of intimacy with death and desire for it, while I was in prison, particularly during my first incarceration, from 1956 to 1961.

But the battle went on.

Meanwhile the men of Sandžak arrived, furious at some wavering and even shirking in their brigade. The attack was led by the staff of the Third Sandžak Brigade. I took note of the acting commander of the brigade, Moma Stanojlović, an ex-officer from Šumadija. His pierced, loosely bandaged shoulder was bleeding, but he laughed it off: "That's nothing!" There was also the political commissar of the brigade, Božo Miletić, brother of Petko Miletić, leader of the leftist faction and proclaimed enemy of the Yugoslav party, who had gone to Moscow on the eve of the war to seek party justice—and had landed in a Soviet prison. From the very beginning of the uprising, Božo's courage and self-sacrifice were such that he not only dispelled all suspicions of deviation which clung to him because of his older brother, but gained special recognition. In the revolutionary movement those who are in "error"—though Božo was "guilty" only of being on his brother's side as a brother, and not as a factionalist—are gladly accepted when they "see the light," just as they are done away with unhesitatingly if they are not "sincere." Also present were the two Bultović brothers, the sons of a cousin of mine; Veselin, a political officer in the brigade; and his younger brother Vojo, or "Strunjo" as he was known, who had not yet recuperated from a wound and who had tuberculosis.

Božo Miletić tried to assure us—or us and himself—that, to judge by the firing, there weren't many bunkers ahead. And Strunjo smiled as he twitted me, with grenades in his hands, "Now you CCs [Central Committee members] can see how we fight!"

Crouching, Miletić went forward; Strunjo and the rest followed. Some twenty yards beyond Sava, Miletić was cut down. He was still alive, writhing in the ditch beside the road, trying to get up, when the medics came to his rescue. But they also were cut down, whereupon he took out his gun and shot himself. I did not witness this. Strunjo also lost his life, as did Moma Stanojlović, who was cut to pieces in a charge.

There were other charges, equally in vain. It isn't true that Milutinović and I and the rest of the officials also led charges. However, it's quite true that we neither took cover nor fell behind the military officers; the fact is, there was no place in which to do either.

Our area was dominated by the grassy and rocky Ozren Mountain. Frequently I glanced at it through the brush from the road. Its bareness and steep incline made it impossible for us to occupy, nor did the units of the Supreme Staff occupy it either. While we were coming to terms with that fact—each of us to himself and for the hundredth time—a rocket flashed from Ozren in our direction. Soon thereafter, mortars began to tear up the ground and bushes around us. We also received a report, delivered orally by a frantic courier, that the Germans were advancing on our unprotected rear, that is, on Tjentište and the heights above it.

We had no units left; the firing and losses had driven them off and in part scattered them. We held a brief council, right there next to Sava's dead body. We decided to move to the left, in the direction of the forests and cliffs of Zelengora, and to seek a new spot for a breakthrough.

It must have been around eight o'clock. There was no dew left on the leaves or in the grasses. The heat was already baking the hillsides. We crossed the road and set out along a sheep path up the slope, leaving behind us the field kitchen, the radio transmitter, the dead, and the wounded. Before long we came across Dragiša Ivanović, the political commissar of the Fifth Montenegrin Brigade, who after the war became a professor and the rector of Belgrade University—a brilliant student, model soldier, and articulate political worker. We held a short meeting with him and decided that, if the penetration did not succeed, we would make our way back in groups across the Sutjeska, that is, through Perućica Forest and on to Vučevo, into Montenegro and Sandžak. I believe I proposed this plan. No one opposed it because no one could think of any other way out. We also agreed to maintain communications, to send patrols and couriers to one another.

So we got going in search of a way out to the northwest, around Tisovo Brdo. There were about a hundred troops with us, from various units and staffs. We entered the forest and felt safer, at least for a while. Deep down I felt ashamed, not so much because we had abandoned our breakthrough, but because I knew that whenever volleys echoed from the Sutjeska, our wounded and refugees were being killed off; not only could we not help them, but we didn't even have the courage to talk about it.

Some five hundred yards ahead—in the woods beneath the cliff, by a spring—we stumbled onto two Germans. Our advance guard—if one can so designate the group at the head of our disjointed, disorganized crowd—had already disarmed them.

"Where are the German soldiers?" I asked in German.

One of them replied in good Serbian, making a circular motion with his hand, "All around."

This heightened my outrage, both at the carelessness of our own troops and at the insolence of the Germans. Apparently these two were from the Seventh SS Prinz Eugen Division, which had relieved the German 118th Division after the latter had been penetrated by our Supreme Staff units, and which was now being regrouped in a wider circle. The Seventh SS Prinz Eugen Division was free to do this once our side left the Suha gorge, thus clearing the way down the Sutjeska. Mandić's brigade had clashed with the division at the Sutjeska. The Prinz Eugen Division had learned a lesson from that defeat and occupied covertly the positions which the German 118th Division had held.

I unslung my rifle. Since I didn't dare fire, because the Germans were some forty yards above—we could hear them shouting—I hit the German over the head. The rifle butt broke, and the German fell on his back. I pulled out my knife and with one motion slit his throat. I then handed the knife to Raja Nedeljković, a political worker whom I had known since before the war, and whose village the Germans had massacred in 1941. Nedeljković stabbed the second German, who writhed but soon was still. This later gave rise to the story that I had slaughtered a German in hand-to-hand combat. Actually, like most prisoners, the Germans were as if paralyzed, and didn't defend themselves or try to flee.

I then took the rifle of one of the Germans. It was well cared for and precise, similar to the one whose butt was broken. I was to carry that rifle until I was sent to the U.S.S.R. in the spring of 1944, at which time I left it with the battalion escorting the Supreme Staff, where it got lost—probably during the German raid on Drvar on May 25, 1944.

We continued on our way. After a while we paused on a wooded ridge. Scattered troops and stragglers had joined us along the way, so that we looked more and more like a mob. I called a halt and ordered the party members to come forward. There were about forty of them. "You cannot be a mob—you must not be disorganized under any circumstances!" I said to them. Milutinović and Vukanović joined me. We assigned officers and formed them into a company. Then the nonparty people joined them to form platoons and squads. And so we became a unit of sorts.

We did some reconnoitering from there: the Germans were above us, all around. In the course of the day we moved a bit further. We established communications with some of the Sandžak troops. But by evening our communications were broken off because they had moved on. We stationed guards around us, and could hear the Germans killing our wounded and noncombatants along the Sutjeska, and on the road which we had traveled that morning. It was then that a husband answered the plea of his gravely wounded wife to kill her, seizing his opportunity while she was dozing. I had known them since the Montenegrin uprising in 1941. It was then also that a father fulfilled the same request by his daughter. I knew that father, too. He survived the war, withered and somber, and his friends regarded him as a living saint. We counted the shots the Germans above fired as signals. We were unable to decipher them, but knew that they spoke of our annihilation. Planes also flew about, but we no longer seemed worthy of their attention. They unloaded their bombs far away, where the units of the Supreme Staff were fighting.

In our uncertainty we decided to move on and continue to look for a break in the encirclement or a forest the following morning. As we were crossing a road a German patrol opened fire, more likely at some other group than at us. That same night Dragiša Ivanović, the political commissar of the Fifth Montenegrin Brigade, found a way out not far from the place where we had first attempted a breakthrough, and led two battalions of his brigade through, along with some 350 wounded. They passed without a fight: the bunkers were scattered and the Germans had abandoned their ambushes.

Although Ivanović didn't get his whole battalion through—his communications with them were broken—he had shown greater courage and enterprise than we leaders, perhaps because he was closer to the men and had more military experience. It is interesting that Ivanović did not run into the encirclements at Balinovac and on the Foča-Kalinovik road which the units of the Supreme Staff had to penetrate. Was there panic and confusion in my group—among us leaders, in myself—so that we weren't as resourceful as Ivanović? There was tension and agitation, but no fear. We were beyond fear and didn't think of death. There was disorientation. The worst of all, however, was the lack of communication: everybody fought in his own circle, caring little for the rest. Every individual fought desperately, but the hierarchical organization and unity were shattered.

We stopped, faced by cliffs above, to the left, and below us. It was dark in the forest, with only an occasional star among the branches. We were exhausted by anxiety, fighting, and hunger. Twenty hours must already have gone by. We judged that the place wasn't bad for defense and for our renewed efforts the next day, so we stationed guards and lay down in a heap like logs.

After I fell fast asleep, something seemed to awaken me. Suddenly in my mind Christ appeared; the one from the frescoes and icons, with a silky beard and a look of pity. I knew that this image was synthesized from the stories and impressions of my childhood, but his presence was pleasing to me, as if I found myself in some safe and glowing warmth. I tried consciously to dispel that image, but in vain: it only melted into a still sadder gentleness, firm in its contours. I opened my eyes. Around me, the trees and my slumbering comrades. And silence, endless and lasting, as if there had never been any firing or screaming. I closed my eyes, and there was Christ again—tangible, close enough to touch. I began to speak to him: If you came into the world and suffered for goodness and truth, you must see that our cause is just and noble. We are, in fact, carrying on what you began. And you have not forgotten us, nor can you abandon us. You live and endure in us. As I was saying this, I knew that I was not ceasing to be a Communist, and kept telling myself that this was in fact brought on by nervous tension and exhaustion.

I don't remember how long this lasted, or when and how it stopped. But I'm quite sure that I was awake, and that the image appeared only when I closed my eyes. It never even crossed my mind that this was a miracle or that miracles occur, though the apparition inspired calm and courage. Even today I am of two minds about setting this down. Yet it seems to me that without it my own personality and the circumstances would remain unclarified. Thus I include it in this work, though my atheism and the dogmatic purity of the revolution may suffer for it.

Then, unobtrusively yet insistently, various thoughts came to my mind concerning the Germans, the Partisans, and ideology. Why were doctors from Berlin and professors from Heidelberg killing off Balkan peasants and students in these ravines? Hatred for Communism was not sufficient. Some other terrible and implacable force was driving them to insane death and shame. And driving us, too, to resist them and pay them back. Perhaps Russia and Communism could account for this and justify it to some extent. Yet this passion, this endurance which lost sight of suffering and death, this struggle for one's manhood and nationality in the face of one's own death, with one's death as a clarion call and an inspiration—this had nothing to do with ideology or with Marx and Lenin. When the sun rose, I suppressed these abysmal thoughts, for I sensed how destructive they were for the ideas and organization to which I had given myself. But I never forgot those thoughts. They cropped up whenever any violence threatened me, and became a reality of my suppressed and hidden spiritual self.

The morning of June 15 spread its bright warmth through the forest. It was even more unpleasant than the chill, hard night. The Germans were on the cliff above us. Two Partisans tried to slip through the crags on their own, but the Germans opened fire and we heard screams some fifty yards away as their bodies plunged into the ravine. About four or five hundred yards below us, on a tiny meadow along the Suha, the Germans were preparing breakfast, after which they walked over casually to the river.

By now there were around 150 of us. Even so, nobody dislodged a pebble or snapped a twig. We also took care not to smoke too much. We found a spring nearby, so we didn't lack for water, which volunteer water boys brought us with the adroitness of cats. We prepared to defend ourselves. Almost all of us were loaded down with ammunition, but hardly anyone had brought a bit of bread or meat. This was, in fact, our second day without food.

We sent out our most responsible comrades, including Vukanović, to scout around. They came so close to the Germans that they saw them lighting their cigarettes. But as for a safe and secure passage, they could find none. The afternoon frowned darkly on us. I proposed that we all destroy any papers and documents by which the Germans might identify us. This was done. Vukanović destroyed his war diary. Masleša ripped up some papers, probably notes for essays. Milutinović tore up the decisions of the Executive Committee of AVNOJ. I considered throwing away my watch, but changed my mind: it was war booty and it didn't matter who got it.

We held council in a fine drizzle of rain. We decided upon the exact place for a breakthrough across the Sutjeska into the virgin forest of Perućica, and put together a shock brigade of some thirty comrades to secure that crossing. It was easy to get lost at night, particularly in the forest and in battle, and so we designated the village of Gornje Izgore as a meeting place.

When the night fell we turned back, to gain a less rocky approach to the Sutjeska. Everything went quietly and smoothly. We descended a good part of the way along the road. As we turned off the road, we suddenly found ourselves just above the bank of the Sutjeska. Our advance guard and main force had already crossed the river; it looked as if the Germans had not anticipated our taking cover in Perućica, and that we would get by without a fight. But just as the leader got down to the river bank, machine guns opened fire from the opposite shore. Raja Nedeljković was standing next to me. "Shall I shoot?" he asked in a whisper.

"Shoot!" I whispered back. He fired four bursts, singly, with his submachine gun.

"Hang on tight to my belt!" I said to Mitra. I was afraid that if we were separated she would get lost. It also occurred to me that, if we had to die, it was better to die together.

I scurried down the bank, with Mitra clinging to my belt. The river sparkled and everything was more visible by its edge. Stopped by the firing, a group of comrades had gathered there. I caught sight of Milutinović and Vukanović to the left, and Masleša to the right. The machine gun clattered once again. The bullets splattered across the river and splashed my face. To my right a body plopped into the water. I am convinced to this day that it was Masleša's. Our group was thrown into confusion. Milutinović and Vukanović went off to the left.

We were threatened with disintegration, with flight into nowhere. But that was not what I had in mind. I knew I had to assume the greatest responsibility—who else but me, since there were no military leaders left? Crying out, "Forward, comrades!" I waded into the water.

I expected the water to be no higher than my hips. To avoid the later discomfort of soggy boots, I tied them together and slung them over my shoulder. Actually, the water wasn't deep, but I came across a smooth spot in the middle of the river, slipped, and was completely submerged. I extricated myself immediately and hastened on, with Mitra in tow. But the water carried off my shoes. It would have made no sense to look for them even under normal circumstances, let alone at night when in a rush, though the loss of shoes was awkward and even a little humiliating. I still had thick woolen socks which would protect my feet from the soft forest floor for a day, but not two days.

I hurried on, without stopping. I got to the other bank, which was steep—steeper than I expected. But there were bushes and roots, so I pulled myself up and I clambered on some ten yards more. The machine-gun fire continued in brief spurts. I saw sparks, some ten yards away, to the right—exactly where I suspected they were.

I unscrewed one of the two bombs I had, and took aim: I was afraid that the bomb might hit a tree trunk or branch and roll back at us. The sparks from the machine gun were clearly in view; it looked as if the bomb had a clear path. I lighted it, hurled it, and pulled Mitra down with me to the ground.

The explosion silenced the machine gun, though I don't know whether the bomb killed those Germans or just frightened them off. To try to find out in the dark would have held me up and put me in danger. In any event the curtain of machine-gun fire had been penetrated, and we continued our climb more slowly, more calmly. The newspaperman Mirko Ćuković, who was lame in one foot from childhood, caught up with us and held on to Mitra's belt. The thicket was now less dense. We even rested a bit, as small groups moved past.

We turned the first bend and entered a thick forest; utter darkness. The branches cracked and the twigs snapped as groups made their way. I called to them to stop and gave out the password, but no one answered. Ferid Čengić, a party worker, joined us. I knew where we had to go: up to the ridge, then down into the valley of the Perućica River. I found my way easily in the mountains.

Again there was firing to one side. We were slowly arriving at the last bend when suddenly, out of nowhere, we were stopped by the cry, "Halt!"

The shout sounded foreign to me. "Lie down!" I whispered, and gave the password. No answer came. "Answer!" I cried, still in the hope it wasn't a German ambush. "Don't fool around! Partisans here. Who's there?"

The answer came from a machine gun. I got down. The gouging bullets splattered dirt into my face. Then a firm command in German—"Two to the right, two to the left!"—followed by two bursts of a rifle one after the other. The Germans were some ten to twelve yards away. We could clearly hear the clicking of their bolts. I also fired, but my rifle only clicked, probably because the breech was still full of water and the spring impeded.

"Let's run to the left, around the hill!" I whispered.

We got up. Suddenly the dawn touched the white of the beech bark and blanched the mist. We ran as fast as we could to get ahead of those two Germans who were to block our way to the right. Quickly we got around the hill and, walking along the side of it, caught up with some fifty Partisans who were rushing in disorder toward the Perućica River. They were mostly soldiers from the group that had spent the day with the leadership at the foot of the cliff, as well as some wounded. The idea of pushing back into the wilderness of Perućica had come up before we had been scattered, but it also occurred to many independently. Yet Milutinović and Vukanović were not among them. They had crossed the Sutjeska at the same time as I, but had stumbled on an ambush, veered to the left, and got lost among the cliffs, where they spent the day. Then they proceeded to Perućica.

I stopped a group of men, stepped up on a rock overhanging the bank, and delivered a speech: I stressed the need for organized squads and platoons. Only as firmly knit units, however small and separated, I said, could we survive and again be the army that we once were. They stopped and listened, even though the Germans must have been somewhere nearby.

I hadn't even finished my exhortation when the rock collapsed under me. I went crashing down with the rubble onto the grassy and sandy bank, tumbling some twenty yards—almost all the way to the Perućica. Raja Nedeljković rushed over to help me. Everyone thought I was broken in pieces. But I was only slightly bruised, and continued on my way to the stream with Nedeljković, then on through the reeds and across the fallen branches, back to the crowd that was descending diagonally to the stream.

Just as we were about to overtake them at a small grove, some firing broke out not far behind. The bullets deafened us as we grasped at one another and stumbled into a thicket. Miraculously not one of us was even grazed. The Germans were also nervous and exhausted, so they were shooting badly. But the group disintegrated—all except Mitra and three or four officers.

We continued through the woods. It occurred to me that we ought to be wading through the stream, to get any bloodhounds off our scent.

Suddenly there appeared before us, as in a dream, the waterfall of the Perućica, some thirty yards high. We climbed up above it and rested on the moss, on a little mound under the beeches. At that point Raja Nedeljković's submachine gun went off by mistake. The bullet whizzed by just above my head. Nedeljković turned pale and began to stammer—after all the tribulations and perils through which he had handled himself so calmly and fearlessly.

It was mid-morning when we came across some ten soldiers in a small clearing. They were skinning a horse beside a crackling fire. We had a hard time getting anyone to stand guard. Some women succumbed to hysteria and began to scream, so we had the job of quieting them as well. But we roasted the horse meat and ate our fill. Someone even had some salt wrapped in a rag. We took along a good deal of the meat. I also took a piece of the hide. I dried the hide in the sun, and Pavle Pekić's wife made me some sandals from it. There were many abandoned, unfit horses. When Milutinović and Vukanović finally reached the Perućica a day later, they also came across a horse and fed themselves.

Large groups were more easily detectable. They were composed of people who didn't know one another, and therefore weren't as suitable for defense. So we decided to break up into squads led by political and military officers. There was no end to the resistance, to organizational rallying and renewal. About ten officials and fighting men gathered around me, among them Mitra, Pavle Pekić's wife, Raja Nedeljković, and Nikola Gažević, who was to become a general after the war.

I found a clearing and took a look at the ridge of Prijevor, along the route of our retreat. The Germans were strolling or standing in clusters around improvised rock breastworks. I assumed that they intended to stay put until they finished mopping up. That evening one of our small groups attempted to get out by way of Prijevor; the Germans let loose a hurricane of fire along the entire length of the clearing.

We found a dead German inside a cabin. We couldn't guess how he died, because the body was already beginning to rot away. He had been killed long before we had come that way. At any rate, his pants came in handy for Pavle Pekić's wife, who was half naked, but not until after she cleaned them of lice and washed them in the stream.

We spent three days in the forest of Perućica. My experience with Montenegrin poverty and prison hunger strikes served me well in picking edible forest vegetation and conserving my strength. I wasn't afraid that we would die of starvation as long as we used our energy rationally. Nor did we fear the Germans: Perućica was so vast and untouched that to search it thoroughly would have required more manpower and time than the Germans had. Besides, the Germans didn't even venture into Perućica, except for patrols which routinely inspected the forest roads by day. Not even these SS men from the Seventh SS Prinz Eugen Division were eager to become the targets of captives who were used to death.*

On the fourth day, June 18, the ridge of Prijevor appeared a pure green in the morning light, clear of Germans. Gažević went ahead to reconnoiter, just in case. In our impatience the rest of us pushed on, so he wouldn't have to come all the way back. We caught up with him before he was able to get out of the forest, and we set out together in the direction of Trnovačko Jezero, a lake under Mount Maglić.

We rested beside the lake, beneath the mist and mountain evergreens. We tried eating snails, but without enthusiasm, even though we had heard that the French, the upper classes, considered them to be delicacies. Then, out of the rock-hewn mountain above the muddy stone-covered road, came rain and fog. Only now, on meeting the first shepherds, did we become aware of our inhuman condition—emaciation, exhaustion, and dishevelment. Our weapons alone gave us the appearance of men.

* A phrase from the poem "The Death of Smäil-aga Čengić" by the Croatian poet Ivan Mažuranić.

6

It was late afternoon when my group arrived at the rendezvous, the village of Gornji Izgori. Milutinović and Vukanović had arrived two hours earlier. They were waiting for us along with over one hundred Partisans, among them some wounded as well. They were mostly soldiers from the Fifth Montenegrin Brigade—all too few of them. We wondered what had happened to the rest: we didn't know at the time that Dragiša Ivanović had succeeded in getting through with two battalions. We were less concerned over the fate of the Third Sandžak Brigade, having heard that its elements were making their way toward their native region.

Resourceful both as a provider and as a Communist, Milutinović had already ensured a milk supply from the peasants and had begun the reorganization of military units. Since Gornji Izgori was not able to accommodate so many people, we sent some groups on toward Golija and Župa Piva. The villagers of Gornji Izgori secretly marveled at the way the German soldiers had climbed up the cliffs of Mount Maglić. They also praised the Germans: they had passed out chocolate to the children, and then paid for their milk. That was in Gornji Izgori. As for the neighboring village at Donji Izgori, the Ustashi killed the whole population, some 120 souls. And the Germans were not so good everywhere: in some villages they acted like the Ustashi. An order was found against abusive treatment of the population, the destruction of houses, and confiscation of property: the differences in behavior apparently arose from individual differences among the officers.

Our soldiers lent themselves to organization without resentment. As for us leaders, we avoided talking about the horrors we had experienced. Each had reason to reproach the others as well as to praise them, but that was neither here nor there. Our moral unity and our determination to begin anew remained intact. Only one battalion commander, or acting commander, rejected an assignment, though he was known for his bravery: "I'm going home first," he said, "to eat and to rest, and then I'll come back. Everyone knows that I belong to the army, and everyone knows which army." His obstinacy bordered on a breakdown. Milutinović and I agreed that, though he was a party member, we should let him go "on leave."

We spent the night there. The next morning we held a council on how to rally our forces and where to go. Our council was interrupted by brief scattered firing from the woods at Donji Izgori. Yet even such firing was enough for our forces—still unhinged, and insecure in their units—to start running. Nevertheless, a core remained with us and took up positions. Those who took flight grew ashamed and came back. We soon received a report: it was a patrol of ours, which had clashed with an Ustashi patrol from Gacko.

We decided that Milutinović should go to Montenegro, to organize the survivors and mobilize new forces, and that I should go to the Supreme Staff to report, though we had no idea where it was or what had become of it. I had about thirty officers and party workers with me. We had more functionaries than soldiers or space. Milutinović set out that same afternoon, June 19, while I and my group spent the night in Gornji Izgori.

The next day I also started out—for the villages around Čemerno (Gacko), to investigate the situation and look for survivors. We found no survivors, but the peasants told us that small groups of the Mostar and Dalmatian Battalions had succeeded in getting through that way.

The situation had changed since the spring of 1942, when the villagers in that region went to bed at night as Partisans and got up in the morning as Chetniks. The villagers were now frightened. They opened their doors to us, but with caution. They marveled at the Partisans, who stood up to the German power. In their minds the Partisans were growing into an invincible, victorious force, all the more because the Chetniks had fallen apart without a struggle before that German power. Even so, the local Chetnik leader, Milorad Popović, was still gathering a little group to attack treacherously. In every village there were people who favored us. But without a large unit for a nucleus, there was no immediate prospect of growing into a larger one.

We decided not to expose ourselves in any way because we had to make contact with the Supreme Staff, which was our most urgent, imperative task. Three or four days later, probably on June 23, we started down the Sutjeska, back to the scene of death and destruction which our memory had not yet absorbed, though we already knew that it would never fade from our mind. Following the Sutjeska up to the Suha, we were reminded of the recent battles only by the occasional German graves on prominent plateaus—in neat order, as on picture post cards. But at the gorge of the Suha we were accosted by the stench of putrefaction, billowing and inescapable as if the river itself were festering. That foul, heavy stench did not let up until three or four hours later, when we emerged from the canyon. For days I kept smelling that noxious, unbearable stench, as if it had permeated my clothing, my skin, and my mind. Inevitably, I thought of that rebel song from Krajina, in all its monotonously sad melody:

> O thou Glamoč, level plain,
> Hear the people moan in pain!
> Grave on grave upon thee lies.
> For her son a mother cries. . . .

Above all, I was beset by cares concerning the fate of the Supreme Staff, Tito, and Ranković. I thought to myself that even if they had all perished—though this was not likely—the most urgent task would be to reconstitute the Central Committee. This did not seem impractical to me, with Kardelj in Slovenia, Milutinović in Montenegro, and deputies in Croatia and Serbia. Such thoughts and plans even renewed impulses within me: the uprising, the party, the revolutionary army—all these became overpowering and indestructible, once they entered one's life and spirit.

Very soon—down below the Suha, on a patch of wasteland by the side of the road—we came across our first group of executed victims. There were about fifteen of them—bloated and blackened, in miserable dress and sticky with rot. Their nostrils and eyes swarmed with white worms. There didn't seem to be any soldiers among the victims, who were dressed partly in peasant clothing. There were several women in black kerchiefs—the sisters and mothers of those who had joined our army, or women simply caught in the encirclement.

We went by the place of our nocturnal breakthrough in the Perućica forest in the hope of finding Masleša, but he wasn't there, it was never determined whether the Sutjeska River swallowed him up or whether, wounded, he crept into some underbrush and died. We came across four or five clusters of corpses before reaching Tjentište, and concluded from this that the Germans didn't succeed in capturing many of our troops. So powerful and all-pervasive a stench could come only from the multitudes that were killed in the woods and ravines. We talked of how we ought to bury them, but lacked the manpower and tools to do it. When I passed that way five years later, in 1948, the bones still hadn't been gathered. I called this to the attention of Ranković, Tito, and the leading generals. They took steps to have the bones collected properly: with the Sutjeska legend there also arose the need for its material enshrinement.

In the afternoon we visited the German trenches on the slopes under Tjentište. They were expertly and solidly built. It was little wonder that the Sixth Majevica Brigade was cut in half before them, in the vain attempt to prove itself the equal of the Proletarian Brigade, to which Tito had hastily attached it.

At dusk we reached the village of Vrbnica. The peasants who had given quarters to the Supreme Staff in 1942 remembered my name, and knew that I was from the Staff. We found there the poet Goran Kovačić and the professor of medicine Simo Milošević. Goran and Simo hid out after the shattering of the Third Division, and later withdrew into the village. Vrbnica favored the Partisans and received them hospitably. The people of Vrbnica and the surrounding villagers knew Simo: he had given them medical aid in 1942, and he did this now again, though wounded in the hip.

We—mostly I—urged Simo and Goran to come with us. But they refused, worn out by the offensive and lulled by the good-natured people of Vrbnica. We were going into uncertainty. Besides, Simo was lame, though I insisted that we would get him a horse. I mentioned the possibility of withdrawing into Montenegro. "At present we are better off among these good people," Goran said, while Simo assured me that no danger threatened them since they were not soldiers, let alone party functionaries. Finally I said to the host, the most prominent peasant, that all his expenses on their account would be paid. I gave him a voucher to this effect, with my signature as a member of the Supreme Staff. But the peasant never came forward with this voucher after the war. The Chetniks from the neighboring village of Ljubinje fell upon them unexpectedly, dragged off Goran and Simo, and cut their throats—surely without any comprehension of the worth of their victims and the consequences of their act. Thus perished a Serbian scientist, and a Croatian poet who in his poem "Jama" (The Pit) immortalized the hell of the

Ustashi massacres of the Serbs. I heard that a large number of Chetniks abroad have boasted of that exploit; some would rather boast of evil than of nothing at all. What had brought Simo and Goran together, other than common suffering and deliverance? A gentle nature and wisdom, though different in each of them: Simo the rationalist, yet almost God-fearing, and Goran with his restrained irony and expressive wisdom. To this day I am sorry and feel guilty at not having saved them, especially Goran; no poet ever shared his fate. Yet the status and personality of both were such that I couldn't order them. Though I was suspicious of the toadying of some of Vrbnica's people, I wasn't convinced but that we would encounter trouble and even greater uncertainties.

We had resolved to find Sava Kovačević's body and, recognizing the symbolic significance of his remains, to give him a burial. However, in going over the battle ground for booty, the Vrbnica peasants had discovered Sava. They knew Sava well, for he had been in their village with his staff in 1942, and had made a powerful impression on them. The people of Vrbnica buried him—him alone. When I passed through that way in 1948, I visited his grave—a few paces from the spot where he was cut down, on a patch by the side of the road.

Out of curiosity I asked the peasants if they had also found the two Germans whom Raja Nedeljković and I had killed, not far from the place of Sava's death. The peasants had indeed found them, their throats cut, next to the spring, which they described exactly. The peasants had plundered plenty of weapons and ammunitions. Also shoes and clothing, and even horses. They were smart enough to shut up about that, though we did not begrudge it to them.

Thus not even in death did Sava Kovačević hold us back, because he was already buried. The next day we continued on our way to the village of Ljubinje, which we knew was Chetnik. We found there only two or three women, and they just barely summoned enough kindness to give us some milk. That same day we reached the village of Rataj and spent the night in the nearby woods.

In Vrbnica we had been given the name of a peasant who was on our side and could act as our guide for the rest of the way. We didn't get in touch with him until the next day, lest someone in the village alert the Ustashi and the Germans, who were patrolling the Foča-Kalinovik highway. In the morning we ran into a shepherd girl and asked her about our peasant and the general situation. At first the girl was frightened, thinking that we were Ustashi, but then she relaxed. She alerted the peasant, who sneaked out of the village and came to us as soon as it grew dark.

All bone and muscle, with a noiseless gait and keen senses, that peasant was the incarnation of the *hajduks* and *jataks,* the heroic highwaymen and rebels out of the folk ballads and my childhood daydreams. Here was the offspring of the irrepressible centuries-old rebellion against the authority, power, and restriction. Quick-witted and intrepid, he could not only sense danger, but was also able to meet it. He made us supper—to be sure, with the help of his friends, since there were too many of us even for an unravaged peacetime household. He also knew reliable men in the villages along the way. A network of trustworthy persons had been established everywhere, especially in Serbian villages, and one could move through that network securely, perhaps because it was largely the creation of the people's misfortunes and courage. The people are like underbrush: the more it is burned and slashed, the mightier the shoots that rise from it.

The peasant led us over pathless ground and good roads, over the highway and through the village of Miljevina, which was occupied by the Ustashi. He bedded us down before dawn in the woods, next to another village, to catch some sleep, while he looked for his contact and prepared some food. We trusted him, but—just in case—we moved to another spot and set up a guard. That didn't surprise or offend him: that was how he himself would have acted. That same peasant looked me up in Belgrade in 1970 and 1971, to get from me a letter confirming that he had helped the Partisans during the war—in support of his claim to a government pension. He acted with an amiability that was restrained and measured: apparently the message had got to him that the authorities have certain devices by which they see and hear everything. I promised him the letter, with the warning that I was in disfavor and that it would do him little good. He replied, "I know you're out of favor, but your signature is good for recognition of wartime service."

Cautiously but dauntlessly, amid destruction and disorder, we continued our search for the Supreme Staff—along a trail marked by the tales of peasants, by shell holes, and the wake of troops tramping through grain fields and across meadows. Though there was no fighting, it was a tortuous and strenuous journey as we picked our way between Ustashi and Chetnik patrols, with meager rations and sleeping on the bare earth. Suddenly the murky night covered us in the thick forest. It also began to rain, but we wouldn't have dared to light a fire anyway, since an enemy might be near. We slept propped against tree trunks, on dry leaves. The cold morning mist awakened us. Water was trickling under me, but I was still so tired that I hadn't even noticed it. My left side had become numb, but it revived as I walked.

We passed through the village of Zakmur in broad daylight. In 1942 the men of Zakmur had been volunteers in our army, and they received us cordially now. As we were leaving, a barefooted and half-naked little girl ran out of the yard crying, "Comrade Djido, take me with you!" It was the daughter of Dr. Gutman, a woman who had perished in the encirclement. The child had slipped through only because she was not with her mother at the time. We took her along and somehow clothed her. In 1952, as I was passing through Užice, I was approached by a shapely brunette—the little girl from Zakmur. She was married to Djurdjić, secretary of the Party Committee, and had a child. She invited me to a party in her home. I accepted. She was overjoyed, and her husband was honored by my presence.

But the most interesting incident happened in the village of Jabuka, below the peak of Jahorina Mountain. When we entered the village at dusk, we found not a living soul, though the fires still glowed in the hearths, the cattle were bedded down, and here and there milk had been set to boil. It was a Serbian village, and it occurred to us that they might have thought us Ustashi. Finally, in one of the old houses, we found a mute and lame old man who didn't seem so dumb that we couldn't get through to him. He understood—probably by our five-pointed stars—who we were, and that we intended no harm, and off he limped into the darkness. We settled down for the night in two or three houses—those suitable for defense—and set a heavy guard by the fireplace of the house from which, so they told us in the morning, a woman with typhus had been taken away.

The old mute returned at daybreak, but wandered off again. The village had to live, so women came back to do the milking, and children to take the cattle out to pasture. And before long, on a rocky knoll above the village, about fifty men armed with rifles showed up. They were afraid of reprisals by the Ustashi because of our stopover, and threatened to shoot unless we left immediately or if, like the units before us, we took any booty. With the village full of women and children, we concluded that they couldn't shoot without endangering their own people. We replied that we would leave as soon as we ate. We also fooled them into thinking that we were the advance party of a brigade. We denied any looting by our predecessors, though we suspected that there might be some truth in their complaints. Perhaps the pressure had been great, as in the encirclement of Durmitor, and the Supreme Staff units hadn't had time for any fair or planned requisitioning. But the women swore that the Partisans—that is, the Supreme Staff units—took away whatever livestock and food they came upon. Still, the women came to our aid by confirming that we ourselves hadn't touched a thing, and said that they would feed us. They did, even though the village had become so destitute that they didn't have enough milk.

These peasants of Jabuka were Chetniks, and we didn't trust them: they might shoot at us the minute we left the village. So we sent a small guard out with a machine gun, which they set up on the side of the road. Then we began to evacuate our men—one by one, and spread far apart—so that the first man reached the machine gun before the last had said good-by to the village women.

Once reassembled, we invited the Chetniks to a parley. We made an exchange: for their bad tobacco we gave them ammunition, with which we were loaded down. Mollified, they even gave us a guide, with a rifle slung on a rope.

"What kind of army is it if you aren't fighting against anyone?" I asked him.

"By God," he replied, "not much! We're Chetniks now. And if the Partisans take over tomorrow, we'll be Partisans." He admired the Partisans: "You can hold out even against the Germans!"

I tried to persuade him to join us, but he hesitated: he would have to tell his folks first, and take care of certain pressing matters. But he brought us to Sjetlina and there turned us over to an honest man, who took us across the Sarajevo-Višegrad railroad.

In the morning we arrived on the heights of Glasinac. We had just dozed off when out of nowhere a royal sergeant appeared who wore nothing of his uniform except a threadbare blouse. He was sprightly, cordial, and chatty. Though a Chetnik, he seemed well disposed toward the Partisans and lamented over the Chetnik-Partisan fighting, which he saw as a Serbian "feud." He hustled the peasant women into preparing us a meal as quickly as possible. Finally he insisted that we meet with Lieutenant Djokić, the commander of the Rogatica Chetnik brigade, who didn't want to fight Partisans.

Taking all necessary precautions, Vukanović and I went to meet Djokić in a lonely house. Swarthy, bearded, and bedecked with cartridge belts, Djokić was neither cordial nor offensive. He was confused. He didn't reproach the Communists for anything except that we were against the King. My response was, "We aren't raising the issue of King now; our only condition is the armed struggle against the invader." To my observation that now that the Germans were disarming them, they ought to see finally that the invaders were everybody's chief enemies, he had no reply. To my reproaches for Chetnik collaboration with the invader, he replied, "Those are tactics!"

"But Serbs are paying with their heads for such tactics," I insisted. When I invited him to join us, he replied that he was under oath to the King and wouldn't trample on that oath. It was a vacuous conversation in which Djokić at times argued bitterly. Finally he asked, "When are you leaving?"

"Tomorrow," I replied. "We're tired and have to rest tonight."

Soon thereafter, when I met Tito and Ranković and told them of that encounter, Tito gestured almost reproachfully, as if to say, "A bad business!" As for Ranković, he pointed out to me after the war, as if in passing, that an item had been found in the Chetnik archives regarding my conversation with Djokić.

Anyway, we didn't trust Djokić. Keeping an eye on the village in which he was staying, we noticed that, one by one, the Chetniks were leaving in various directions. We suspected that, whether on his own or on orders, Djokić was sending couriers to the Chetniks in the neighboring villages so as to organize an attack against us. We could have attacked him, and successfully at that, because our men were more numerous and better armed. But I wasn't about to do this since I didn't want to expose our leading men to danger, and also because the Chetnik peasants were restless. We didn't spend the night there, to be sure, but at nightfall slipped away into the deserted slopes of Glasinac.

Early the next morning I bagged two rabbits with my German rifle. It was better to eat game and drink plain water than to have an empty stomach in a lean and desolate land.

Among the peasants it was rumored that the Partisans had brought destruction on the Ustashi of eastern Bosnia. Indeed, units of the Supreme Staff had taken a string of little towns from them there: as we descended into Vlasenica, we ourselves saw a heap of bare corpses on a field below the road.

In Vlasenica the Partisan command received us with the consideration reserved for rare guests and martyrs, and the Montenegrins in some units went on a binge when they heard that I had arrived. We learned there that Tito had been wounded in the arm on Mount Ozren—a light wound, but already legendary. Vlasenica had changed hands many times: there was nothing left to confiscate and no one to kill. But the Partisans did take some booty and were in the process of collecting grain and livestock in the neighboring villages. We could have taken a good rest there, but were in a hurry to get to Tito and the Supreme Staff in Kladanj. Along the way, Vukanović and I took comfort on learning that the units assigned to Vukanović, though broken up, had not in fact suffered much greater losses than the units under the direct command of the Supreme Staff.

We came upon Tito and Ranković on July 3, in a little cave overlooking Kladanj. The two of them had lost a lot of weight: Ranković looked like someone about to die of tuberculosis, and Tito's fingers were so thin that the ring he had acquired in Moscow to serve him in need had slipped off and was lost.

I first asked Tito about his wound. "It's nothing—they just grazed me," he replied nonchalantly. He no longer had a bandage around his arm and used it as if he had never been wounded. The tale of how the dog perished while protecting his master with his own body is pure myth: if the dog did press against his master, it was out of fear and not because he knew anything about ballistics.

Tito listened to our reports in despair, but without a single word of reproach or criticism. Nor did he reproach me later, at least not while I was in power. However, I was loath to speak of the Sutjeska, especially concerning details.

When I said that Mandić's failure to secure the passage across the Sutjeska caused our own failure, Tito observed with bitterness, "What else could he do? As if those who got through ahead of us waited up for us!" With this remark our discussion of the "Mandić affair" was finished. Later I asked Mandić what actually happened, and while telling me the story already related in this book, he commented sadly, "I was powerless. The soldiers wouldn't stop once they had broken out of the encirclement."

Tito's remark that they didn't wait for his group either was a reference to the failure of the First Proletarian Brigade, under Danilo Lekić's command, to wait for Tito and the Supreme Staff after the decisive breakthrough at Balinovac. The First Proletarian Brigade formed part of the First Division of Koča Popović, who insisted that the Supreme Staff and the other units move as soon as the breakthrough was effected. According to Koča's guarded account, the Supreme Staff—meaning Tito—hesitated because their passage was not sufficiently secure. Koča and the even more highly placed Sreten Žujović, who had been assigned to the First Division as a member of the Supreme Staff, continued to advance for fear that the circle might close again. It all ended well, because the Germans were so scattered and demoralized that they didn't join forces before Tito and the Staff had passed through. When Sreten Žujović declared himself for Stalin in 1948, this incident was held against him—mostly on the initiative of Ranković and over my objection. What was the truth? It seems that both sides were right. With Koča Popović there could have been no ulterior motives—nothing save the need to act fast before mortal danger. And Žujović? In him, surely, there was repressed hostility toward Tito. But just as in 1948 I opposed the resurrection of long forgotten "sins," so to this day I don't believe that Žujović's animosity could have been so malicious.

The chief of the British mission, F. W. Deakin, asked me to give him information concerning my group. I did this on Tito's orders, without markedly minimizing the failure or our losses. But I was ashamed that, though fresh from battle, I had to look so ragged before the representative of a foreign state.

Tito was extremely angry over the fact that our units continued to requisition food—a practice which had been permitted them on Durmitor out of necessity—now that there was no longer need for it. All warnings had gone unheeded, and just a day or two before my arrival in Kladanj, Tito had finally called officers to a conference. Ranković presented a wealth of facts. Tito was so upset over this issue that he declared he would not command a plundering army. This marked a transformation in the units.

I was not able to contribute to that transformation: I suddenly came down with a temperature of over forty degrees centigrade which lasted several days. I kept sweating and shivering on the cold, damp forest floor, and once we got going, I could barely stay on my horse. We were hungry, too: once again, it was meat boiled in water. We had no doctor with us—not that a doctor could help. I thought it was an intestinal infection, whereas my comrades were sure it was typhus, probably from that woman in the village of Jabuka. I wasn't frightened of the enemy: I kept my pistol handy—unaware that Ranković had secretly emptied it, so I wouldn't kill myself in my delirium.

Finally, during the ceaseless night marches, I began to recover. My morale was strong, had in fact improved despite my all-too-slow physical recovery. Our spiritual superiority over the Germans, and over our own physical condition, seemed to me even then to be the most essential factor in our survival and victory. Our losses in the Fifth Offensive were enormous: around 7,000 picked men, or almost every other comrade. But regardless of losses, our fighting spirit and experience grew. Moreover, we had something more indomitable than any other army, including the German: a spirit, an ideal, perseverance.

The cat-and-mouse game between the Supreme Staff and the Germans and Ustashi in the mountains of eastern Bosnia became senseless, coming down to the mere preservation of one's life. This made us ill tempered and petty. Early one morning Tito wandered off with only two escorts. Meanwhile, as we were making ourselves at home in two rather lovely houses, I remarked that Tito would be angry when he got back, because we had provided for ourselves in such elegant fashion, apparently without a thought for him. Actually, we sent out patrols to look for him. Tito didn't show up till very late that night. Angry with himself, he turned on us: "So, the gentlemen have already made themselves comfortable!" Everyone was offended—just imagine, calling one's comrades "gentlemen"! No one said anything to him. After he had slept I went to see him on business and it all blew over: we were back to normal.

The quartermaster got it into his head to slaughter "Tito's cow," which had stopped giving milk and had to be replaced by another. The Central Committee members clung to the inherited folk attitude toward cows as half-sacred animals not to be slaughtered until they are completely done for. As for this particular cow, it was the pet of the entire Supreme Staff. It had made it through fierce offensives without complaint. When Tito heard that the cow had been slaughtered, he was furious and ordered that the major be stripped of his rank. Orders had to be carried out, but it made no sense to bust a major over a cow, so Ranković quietly spirited the major away.

When everything had quieted down a couple of days later, I said to Ranković in jest, "Tell the Old Man that the cow had broken a leg."

That evening, as we sat around the fire in a friendly mood, Ranković said to Tito in an off-hand way, "You know, that cow had a broken leg."

"Why didn't anyone tell me?" Tito asked indignantly.

"You blew up and they got scared," said Ranković.

So that cow, like so many others, got eaten, and the major kept his rank and was later promoted.

Everyone was jittery in those days. But not Ranković—he was controlled and calm. I was known for liking to exchange possessions. In one such glorious transaction I had exchanged my boots for a pair of sandals! But at this time I had nothing to trade with—we didn't have a thing.

One morning we found ourselves by a little stream, amid the sun-splashed ferns next to an abandoned mill. One of our units had got hold of some coffee and sent it to Tito. The poet Nazor happened to be with us, and Tito had some coffee made in his honor. Nazor liked to talk with the most important people, particularly Tito, in whom he saw a favorite of fate because he had made his way up from the masses. Over coffee, the conversation turned to the role and place of the artist in wartime. Nazor maintained that the artist had to be on the side of justice, that is, with the people, but that art, no matter how committed, had to remain independent. Tito agreed that the artist's bond with the people was essential, and stressed the importance of dealing with the artist delicately. I regarded participation on the side of the embattled people as the highest priority of the artist, something without which his work could not have integrity.

Such a conversation inevitably touched on Miroslav Krleža, the most renowned Yugoslav writer. Krleža was a leftist, and connected with the party until the beginning of the war, but he hadn't joined the Partisans. As a result, he was the frequent object of our sorrow and anger. This was particularly true after the arrival of Nazor, who had never been on the left, but whom the fascist wartime frenzies had moved to a patriotic and humanist revolt. That morning both Tito and I mourned over Krleža, perhaps more because the uprising was necessary to him as a humanist without equal in our land, than because he was necessary to the uprising. Nazor had no such regrets, though he esteemed Krleža. He told us that he had met Krleža during the Spanish Civil War and asked him, "How come you're not in Spain? You're a Communist."

And Krleža had replied, "I have a horror of death, corpses, and stench. I had enough of it in Galicia during World War I."

In reflecting today on Krleža's failure to join the Partisans, I wonder if the problem isn't deeper and more complex. Didn't it also involve Krleža's skepticism with regard to "historical changes," his consistency in feeling horror at all violence?

We soon realized that our wandering all over eastern Bosnia was useless. Tito came up with the idea that he, Ranković, and I ought to go to Croatia, while the rest of the Staff stayed in eastern Bosnia. Other leaders also recognized the futility of this military spinning in a circle. Peko Dapčević and Mitar Bakić proposed that their Second Division return to Montenegro. The Supreme Staff immediately accepted this proposal, as much to expand our base for the proposed breakthrough into Serbia, as to prevent a Chetnik revival in Montenegro. The return of the Second Division to Montenegro proved to be one of the best and most far-reaching decisions: a concentration was created there which played the most crucial role in the battles for Serbia.

We were on the march in the direction of western Bosnia and Croatia when on July 25 we heard of Mussolini's downfall. That evening, sitting by the fire, we agreed that this was the prelude to Italy's leaving the war, and that new, more favorable prospects were in the offing for us.

At that time the British had begun to send us aid in nighttime flights—mostly food, which we needed above everything else. To be sure, there was also complaining—as always among allied armies: the shipments were small, the clothing inadequate, not nearly what we deserved. But that aid, and the mysteriousness of those planes, made an astonishing impression on the peasants.

We continued on our march. The following morning we emerged on the mountain meadows above the Bosna River and the Brod-Sarajevo railroad. In a surprise attack on the village of Bijele Vode one of our units captured Golub Mitrović and thirty of his Chetniks. Though a Partisan commander, Golub Mitrović had organized a Chetnik mutiny in his unit, killed many Partisans, and concluded an agreement with the Germans to guard the railroad. After he was captured, Mitrović tried to escape. He might have gotten away—as is often the case, when shots are fired at random—had not a soldier from Serbia caught up with him and bashed his head in. When we heard this story, Žujović remarked, "The pigeon has sung his swan song!"—the name Golub means "pigeon"—and we all laughed inanely. The Supreme Staff didn't get involved in this incident, although we knew that the appropriate staff was interrogating the prisoners. Early that afternoon we heard some guns fire above us. We thought they might be executing the condemned, but didn't attach any special importance to it. Soon thereafter, we got going again. Not far off, in a clearing by the road, we came across some ten executed men—spilled brains, smashed faces, contorted bodies. The English mission was with us, and we all felt awkward: Tito even remarked, "Couldn't they have done it somewhere else?"

One Staff member said, "Well, let the English see how we deal with traitors! They know anyway."

As we were descending toward the railroad and the Bosna River, I asked the political commissar of the division, Mijalko Todorović, "What happened to the rest?"

"We took them into our unit—youngsters, politically naive," he said.

"That kind won't stay in the army," I observed.

"Well, some will."

When a civil war has permeated every pore of the nation, there can be no replenishing of units except with adherents of the opposite side. For a private, the transfer from one side to another is far simpler than the ideologized mind can comprehend.

We halted on Vlažić Mountain and made camp. There on the mild, sunny heights, at last we were able to relax.

The New Regime

1

Though we largely took refuge in backward regions, we still were astounded at the backwardness of the peasants, especially the women, on Vlašić. Many of them had never even been in the nearby towns. They wore hand-woven dresses open down to the navel, so that their breasts flopped out. They greased their hair with butterfat, parted it in the middle, then tucked it up over their foreheads. Their vocabulary was meager, except concerning livestock and the like. The quartermaster of the Supreme Staff, a rascal quite typical of the Serbian towns, exchanged all kinds of trinkets with the peasant women for excellent cheese. During the trading they would naively question him.

"Honestly, now, are these real gold?"

And he would say in mock seriousness: "They are, so help me God and all the saints!"

We laughed at this scoundrel, and probably not even the deceived peasant women were angry: even if they weren't gold, the trinkets glittered. Here was an opportunity made to order for our feminists: they pleaded with the brothers and husbands to permit their women to attend meetings, and the women did go to them, in groups and all dressed up as on feast days. The men were on a markedly higher level than the women, for they had seen something of the world in the army, on jobs, and through trade. Yet the relationship between the sexes was based more on a natural division of labor than on sexual servitude. Several armies had already passed through this village, but the peasants greeted ours as their own; we spoke their language and fought against the German evildoers.

Before our coming to Vlašić, and at my suggestion, pseudonyms were adopted by Tito and the more prominent leaders, to prevent enemy informers from tracing our movements and locations through the peasants by our names. But one day I found the guards arguing with four peasants who had two roast lambs on spits with them. "There's no Tito here," the guards shouted at the peasants. The peasants sensed by the guards' deferential attitude toward me that I was an official, and when I asked what it was all about, they replied joyfully, "We heard that your supreme commander Tito is here. We've brought him a treat, and we would like to see him with our own eyes since fortune has brought him through our godforsaken place."

I led the peasants to Tito, without announcing their visit. Tito interrupted his work and spoke at length and warmly with them about the suffering of the people, about the prospects of ending the war, about postwar reconstruction. And we leaders admired the unpretentiousness and eloquence of our supreme commander. But we also waited impatiently for the conversation to end, our hungry eyes fixed stealthily on the roast lambs leaning against the beech tree. After this encounter it never occurred to anyone to conceal the names of the leaders. The legend was mightier than prudence and reality: the news of Tito and his army spread in spite of all our precautions.

The Moslem villages now received us for the first time not only tolerantly but cordially. And our image of the Moslems suddenly became brighter: they were genial, wonderful people.

We were on free territory which stretched across the Sava to Slovenia—even across the Italian border and as far as Hungary and Austria. The German power was visibly declining and fear of the Germans was letting up. The Central Committee judged that Italy would soon surrender, which would greatly improve our position. Our plans changed accordingly: Tito and the Supreme Staff stayed put, which gave them a greater choice of routes in the event of either an offensive or Italy's fall. Žujović and I were scheduled to tour Croatia, in addition to which I was to start publishing *Borba,* so Mitra and Dedijer were assigned to me. But we never got to issue *Borba* in Croatia; the technical facilities were lacking, and after the fall of Italy the Central Committee convened again to take up more important assignments.

The Seventh Banija Division was sent with us, as well as Dr. Ivan Ribar, the president of AVNOJ; several members of the Executive Committee; and the poet Nazor, who had been elected president of the Council of Croatia. Velimir Terzić was also sent with us to assume the duties of chief of staff in Croatia, because Arso Jovanović had meanwhile joined us as chief of the Supreme Staff. Terzić deserves credit not only for his devotion and courage during the Fourth and Fifth Offensives, but also for strengthening the influence of the Supreme Staff over the main headquarters of Croatia. Tito's former wife Herta also joined us. So the Supreme Staff unburdened itself of idle bigwigs, as well as the occasional tears of an abandoned wife. Yet all of this was only temporary: success would bring new bigwigs, while Herta's departure would only render Zdenka's rages and fears all the more unrestrained.

On the night of August 10–11 we set out from Petrovo Polje for Croatia. The march lasted twenty-eight hours, with less than an hour's pause to eat. We were driven to such effort by necessity: we had to cross two enemy-controlled highways, a railroad, and the Vrbas River, before we could consider ourselves out of danger. Human strength and endurance are limitless when men are sustained by conviction and hope. The soldiers of the Seventh Banija Division, who had already grown weak and dispirited, discovered in themselves an incredible energy and zeal on their home ground. Only Nazor was so weak that he could barely cling to his horse.

The wind and the sun had dispersed the fog above the rocky plateau of Janja by the time we finally came to a stop, on the morning of August 12. Poverty, backwardness, and the devastation of war were everywhere. But the local peasants were proud to receive their own army, though it meant depriving their children of milk and corn meal, and their sheep of lambs. The old men and young girls in the local government and organizations—the young men were all in the army—worked hard and competently. Everyone obeyed them without coercion, which kept the community together and carried on the war. In addition to deprivation and hard work, we also brought some joy to the villages of Janja: we reassured them that the units from their region which were attached to the Supreme Staff were very much alive, and that the Chetniks were deceiving only themselves by spreading rumors that these units had been destroyed.

At Glamočko Polje we parted company with the Seventh Banija Division, which took a more direct route back to its home base. It was here, too, that Nazor collapsed. He held us up for a whole day, in the vain hope—more his than ours—that he would recover sufficiently to be able to stay on his horse. Dedijer wasn't feeling well either; wounded in the head during the breakthrough, he complained of dizziness and headaches.

The grain was ripening; brigades of girls and boys would steal out at night to within range of enemy bunkers to harvest and bring in the sheaves. The villages around Glamoč had been burned down repeatedly; the peasants improvised sheds whose frames they were able to dismantle and hide in the woods at the sight of the enemy, then quickly reassemble as soon as the enemy left. War became a way of life, unnatural and therefore only temporary: man cannot conceive terrors which man is also not capable of overcoming.

We decided to carry Nazor. Our escort was small, so village committees helped with the transfer from village to village. I tried to cheer Nazor up: "You've pulled through all the bombings and offensives, so you'll pull through this time, too!"

To which he answered, "Can there be a worse death among the Partisans than diarrhea?"

Just before the village of Boboljuska we were met by a group of twenty sturdy girls, white-skinned and dark-eyed, typical of the girls in those parts. There were no men left in the village, so the people's committee had sent the girls to help carry Nazor. Nazor had become very popular through radio reports and the Partisan press. The girls took charge of his emaciated body as if he were a saint from their village. Nazor lifted his head and declared with renewed strength, "Serbian girls are carrying a Croatian poet! What can be more beautiful?"

At dusk we began our descent to the Una River—into the territory of the Croatian Partisans. Behind us we left Krajina (western Bosnia) a little less turbulent and exciting than it had been in 1942, but more steadfast and self-assured. We were planning on retaking several little towns which had suffered a great deal during the winter offensive—much more than we had imagined: troops and people caught in a deadly game amid mountain snowdrifts and frozen wastes. We inquired about Drinići, the headquarters of *Borba* in 1942. (Who could ever forget one's own arduous exploits?) Apparently, the Chetniks had descended on Drinići and burned down the house in which the press had been located, as well as the houses of the Communist peasants. The Chetnik commander, Mane Rokvić, ordered that my old bed be taken out of the house and burned, because "Djilas slept in that bed!"

The Una River with its cascades appeared awesome and mighty in the night, but also comforting, in contrast to the bare mountain crags. Here we could anticipate a more methodical and rational warfare. We were met at the Una by the political commissar of the main headquarters for Croatia, Vladimir Bakarić, who was to become a prominent politician and leader of the Croatian Communists after the war. He arrived by car.

We immediately got into the car, so that the morning wouldn't find us on the road, which was checked daily and bombed by reconnaissance planes. I don't remember who else was in the car; probably Žujović, Nazor, and the rest were picked up later.

The conversation with Bakarić was disjointed; even then his opinions were reserved and ironic. He had been nicknamed "the Corpse" more because of his lukewarm temperament and circumspect manner than because of his sickly fatness. Yet thanks to his keen intelligence, even this choppy conversation gave me some picture of conditions in Croatia. Growing out of the revolt of the Serbs under Communist leadership, the Partisan movement had taken root in all parts of Croatia, while in the Littoral and Dalmatia a national uprising was in the making. There was also pressure within the Croatian Peasant party for collaboration with the Communists. Žujović and I knew of this. We were far more interested in the state of the army and the main headquarters of Croatia. More precisely, Tito and Central Committee members in the Supreme Staff were convinced that the units in Croatia lacked sufficient drive, largely because of the stubbornness of the headquarters commander, Ivo Rukavina, a volunteer in the Spanish Civil War. For one thing, Rukavina had failed to carry out Tito's order to apply greater pressure on the Germans during the Fourth Offensive (Operation Weiss). The main headquarters' explanation for this failure was that they were themselves exposed to an offensive by the Italians at the same time. Such an explanation might even have been accepted, had it not been polemical by implying that the Supreme Staff was not familiar with conditions, and that its orders were influenced by its own situation. Apart from this, Rukavina was reputed to be sarcastically critical of the Soviet Union, though no one ever denied his intelligence or devotion. The Central Committee was aware of the amorous relationships that flourished rather freely at the headquarters and in the Central Committee of Croatia, though that carried no great weight, at least not with Tito, in determining his orders to Žujović and me before our departure for Croatia: "See who might replace Rukavina."

From Bakarić's account, it was obvious that the army in the free territory of Lika had reached a kind of stalemate with the surrounding fortified towns. Without heavier weapons and a more thorough subversion of the enemy's rear, it was difficult to alter the existing state of affairs. Yet that was precisely what was being done, though all too slowly in relation to our aspirations and general needs: the Ustashi had to be struck at other points as well throughout most of Croatia. The army was good and gaining strength in every way. Bakarić's judgments seemed realistic to me. As for Rukavina, we were cautious, I more than Žujović, since he was more involved with military affairs. Bakarić, however, was most cautious of all, though he sensed our intentions with regard to Rukavina. Bakarić indicated that they in Croatia had had greater difficulties during the Fourth Offensive than we on the Supreme Staff could gather from their dispatches—worse, in fact, than we had.

Žujović and I agreed that Žujović would concern himself mostly with the army, and I with the party. We were staying together in the command post of the town of Otočac, where at the end of the day we exchanged impressions and planned policy.

This territory—Lika, Kordun, Banija—had managed to stay free up to now, and would remain so until the end of the war. The Ustashi didn't have the strength to fight there, while for the Germans it was a nonessential territory devoid of communications and raw materials. But the strength and durability of that freedom rested on the unity of the people and on their all-embracing, tightly knit organization. Roughly speaking, the Serbian people—especially from the villages—made up the majority, while the leadership consisted mostly of Croats from urban centers. Ustashi rule meant extermination for the Serbs, as well as for the Croatian Communists and antifascists. The link between the groups was vital, inevitable. The Chetniks cropped up here and there but, being confronted by a co-ordinated and functioning apparatus, they were nipped in the bud.

Nowhere was a power structure as conspicuous and as real as on this liberated territory. It was evident not only in the better dress and food of the staffs and agencies, but also in the official, bureaucratic mode of operation. ZAVNOH, the Regional Antifascist Council of National Liberation of Croatia, was headed by my former prison mate Pavle Gregorić, a long-time Communist; it had every appearance of an assembly and a government, though Gregorić was as accommodating and informed as one could wish. All kinds of schools were operating; agencies exchanged reports and circulars. But one could detect, even within the party, a certain anxiety regarding the far-reaching and tight intelligence service. The distinctiveness and effectiveness of a secret service lies in its being something above and beyond—all the more inexorable in that its ideas are "infallible" and its authority "irreplaceable." This was not yet the case with the Supreme Staff, because of constant battles and marches, and certainly because of a different cultural and political heritage: the Supreme Staff and Central Committee administered from necessity, and their power structure was based largely on personal standing and capability.

Our arrival caused some commotion. The Central Committee of Croatia was practically nonexistent. Some members had been killed, others were scattered throughout the country and bogged down in their own sectors. All the business of the Central Committee and all the more important decisions were handled by Andrija Hebrang—with the authority of a Tito, but without Tito's noninterference with the initiative of his associates. Hebrang didn't even stay in Otočac but farther off in the woods, in a house built for him and equipped with a printing press, where he had his own staff and security.

In the propaganda, one detected an inadequate stress on Yugoslavia, or rather too great an emphasis on Croatia. Hebrang explained this by pointing to the need for winning over the followers of Maček among the peasant masses. At the same time there was a noticeable tension among the Serbian cadres: the uprising and the long warfare had attracted many Serbs to the party, and the most devoted and intelligent among them gravitated to the top or made their way up. By placing the primary emphasis on attracting the Croatian masses, Hebrang gave the party a predominantly Croatian tone. This was out of line with the policy of the Central Committee of Yugoslavia: though the party always took account of national representation, the inviolable principle was that Communists assumed positions on the basis of party responsibility, ideological soundness, and personal ability.

I observed all this, but only gradually, without reaching any final conclusion. It bothered me greatly that while in prison before the war I hadn't been on friendly terms with Hebrang: he was a member of the right wing, and I of the left. I was afraid that he might think me capable, or even accuse me, of carrying on our prison feuds with my criticism. Moreover, I was not sure how Tito would react, since he was an old friend of Hebrang's and believed in his experience and steadfastness. Nor did I see eye to eye on this matter with Žujović, either because he noticed less than I did, or because Hebrang's arguments convinced him. We didn't argue about it, but didn't lend each other support either.

Nevertheless I didn't escape friction with Hebrang—over a rather incidental question, but one not devoid of significance at the time. Specifically, Hebrang's authoritative style of operation was linked—perhaps among leaders this is always so—to a taste for a luxurious way of life. During the taking of Otočac, the sequestered shops had to stay closed until Hebrang's wife Olga, who was considerably younger than he and attractive, could select for herself and her entourage whatever she liked. Also, Andrija and Olga both dressed in city clothes which were frequently changed and always elegant, with hardly any Partisan insignia. Several prominent women who knew me from my underground activities in Zagreb, welcomed me all the more because they could now complain against Olga and Andrija. Occasionally an old prison mate or underground comrade who was dissatisfied with the state and style of affairs, would also come to complain. These disclosures forced me to come out, at a meeting of the activists, against the aloofness and aristocratic ways of those at the top.

I understood that my criticism of Hebrang for his lavish tastes would not strike a responsive chord in Tito, for whom such things were incidental as long as the party line was being promoted. But I also knew that I could persuade Tito that lavish living had become such a problem in the party that it hindered the attainment of goals. And that was precisely the case, though I hadn't yet fully understood that this luxury and exclusiveness not only intensified the oppressiveness and lack of criticism within the party, but emanated from them. Such relations in the party pitted "Serbs" against Croatian "newcomers," and caused the Serbs, who were proud of the uprising and embittered by their sufferings, to be suspected of Chetnik sympathies.

My declaration aroused everyone. It encouraged the discontented, surprised Žujović, and embittered Hebrang. To be sure, Hebrang tried to restrain himself when we met a day or two later, but he had merely suppressed his anger. I felt that I should be frank and make clear what I was aiming at. I conveyed the bitterness of the women over the closing of the shops until Olga could make her selection.

Hebrang interrupted me caustically: "You don't expect my comrade to go around bare-assed, do you?"

And I replied, probably offended, "Who said anything about that? We're talking about moderation and the way things are done! While other comrades walk around in tatters, and the wounded lack bandages . . ."

Hebrang swallowed all of it and even became more obliging, assisting me in visiting various institutions and giving reasonable consideration to my opinions.

However, Kardelj soon returned from Slovenia. He was not burdened by prison feuds, and had a greater and more direct authority than I in the Central Committee of Yugoslavia, and over the Croatian Central Committee and Hebrang. I was glad to see Kardelj. We hadn't seen each other for a long time, and I could express to him all my feelings, suspicions, fears, and ideas. Kardelj already had a rather definite negative impression, so that my drastic details only contributed to his forming conclusions regarding Hebrang and the situation in Croatia. This was the beginning of Hebrang's end. Later on, Tito and the Central Committee were bitter over ZAVNOH's declaration annexing Istria and the Dalmatian islands–hitherto Italian–to Croatia, thereby assuming a sovereignty which belonged to Yugoslavia alone.

At that time another change took place in the émigré government in London: a "government of clerks" under Božidar Purić, but in which Draža Mihailović continued as war minister. I presented an analysis in the Croatian weekly *Naprijed* (Forward), asserting that "general disintegration" had ensued among the émigrés, and that "Greater Serbian politicians" had brought officials into the government in order to continue their previous policy covertly.

Tension developed between the Croatian leadership and Nazor as well. Nazor interpreted his function as president of ZAVNOH to be executive rather than merely symbolical. He began to summon officials to report, and issued orders. I was chosen to have a talk with him for two reasons: I was a Central Committee member close to Tito, and I had taken care of him during the Fifth Offensive. He received me, sensing that it concerned something important. I inquired into his health and accommodations, then made a start, choosing my words carefully. Yet the awkwardness wouldn't have been dispelled, if he hadn't understood immediately. He agreed, without discussion, not to mix into minor matters, but to let Gregorić take care of them while he himself consulted with Hebrang over more important affairs. Nevertheless Nazor became cool to me, believing that in this matter I was the initiator and not just the executor.

At that time we received a radiogram from Slavonia containing the offer by several leaders of the Croatian Peasant party (HSS) to collaborate with the Partisans, on condition (1) that a "Radić Brothers Brigade" be organized; (2) that the brigade have not just Communist political commissars, but also commissars from the Croatian Peasant party; and (3) that the brigade be used to fight not the Home Guards, but only the Ustashi and the invaders. After midnight of August 27 Hebrang, Kardelj, and I hurried to the main headquarters, which was located in a nearby village. The headquarters moved every day because of air raids, while the party and other agencies functioned unmolested in Otočac. The discussion was lively, but without discord. The proposals by the Croatian Peasant party were considered acceptable, on condition that the "Radić Brothers Brigade" be subject to the orders of our own high command and that it not be exempted from fighting against the Home Guards. It was also decided that I and Anka Berus, a member of the Central Committee of Croatia, go to Slavonia for more detailed conversations with the Croatian Peasant party.

2

Negotiations with the Croatian Peasant party turned out not to be the only reason for my going to Slavonia. It was decided that a Central Committee member should visit our Slavonian comrades, who were fighting in isolation, though in a region rich with food and vital communications. I wanted to do this myself, and Kardelj agreed I should go.

We set out two or three days after the meeting at the main headquarters, probably on August 30. We were joined by the chief of operations Bogdan Oreščanin, more recently ambassador to China and Great Britain. Though Oreščanin was regarded as a conscientious and capable soldier, he acted more like a nonconformist intellectual. Lively and witty, he gently teased Comrade Anka, telling her her mouth was so big that her loquacity could encounter no obstacle. Thus Oreščanin made our interesting but tiring trip all the more interesting and much less tiring.

On getting into the forests, we came upon the Plitvice Lakes, which looked like lost emeralds. They cascaded into one another as enchantingly as before the war, but now not a single living soul was in sight, only the charred walls of the tiny café and hostel. The blackened ruins of the villages and small shops along the way seemed all the more horrifying in these little plains in the rock: here, apparently, man had tried to wipe out every trace of himself.

Our car didn't stop until we got to Kordun—to a small house in a grove by the road, in which the staff of the First Croatian Corps was stationed. I wanted to meet the commander, Ivan Gošnjak, and the commissar, Veco Holjevac—but especially Gošnjak, whom Žujović and I had proposed to Tito as Rukavina's replacement as commander of the headquarters.

Later I came to know Gošnjak and Holjevac well, but my first impression was in no way altered. Gošnjak had been a student in the KUNMZ* and a Spanish veteran, and became known to the Central Committee for his resolute struggle against defeatists in a camp for Spanish volunteers in France. His firmness and self-control reminded me of Ranković, though his horizons were broader. Holjevac had sprung out of domestic conditions, out of the industrially and culturally developed town of Karlovac. He became known for his bold, well-planned operations against the Ustashi in his native Karlovac and for his organization of the uprising in Kordun. Handsome, moderate in speech, and practical in his judgments, he was commissar under the peasant Partisan leader Ćanica Opačić. The outcome of the war was already certain, but the fate of the revolutionaries was not. After the war Gošnjak was war minister for a long time, while Holjevac infused Croatian nationalism into official socialist establishments.

I said nothing to Gošnjak about the possibility that he take over the command of the main headquarters. We shared views and dinner—a modest dinner, but nicely served.

We went on to the spa of Topusko. The homes and boarding houses had been burned or looted, but the pools were restored to working order for the wounded. The next day we continued by wagon through lush fields of corn which concealed the desolation and wretchedness of Banija.

Before noon we arrived at a school in which the local authorities were stationed. A group of elderly peasants were seated in front of the school, somber and tight-lipped, summoned to have their quotas determined. We couldn't continue on our way until dusk, because not far from there began the semiliberated territory through which we could move securely only by night. Struck by their discontent, I approached the peasants. But they were not complaining against the requisitioning, though it hit them hard: they understood that this was wartime and there was no other way. What upset them was the return of their Seventh Banija Division: only every third man had survived. There was hardly a home that didn't have someone to mourn. Many had perished without a trace. Moreover, each village in that region had been subjected to at least one massacre and one burning. In Lika there is a gorge called Jadovno—

* Party school in Moscow. The initials stand for *Kommunisticheskii Universitet Natsional'nykh Men'shistv Zapada:* Communist University of National Minorities of the West.

Grief—as if some secret providence had chosen this abyss to be remembered by that name world without end. We were told that the Ustashi didn't manage to fill the gorge of Jadovno, not even with the hundreds of unarmed and innocent Serbian victims. In Kordun, in the little town of Veljun, they killed around 450 Serbs early in May 1941: the extermination of the Serbs was not provoked by Serbian resistance but was an integral part of Ustashi policy. And not far from that village, in the little town of Glina, the Ustashi herded the more prominent Serbs into the Orthodox church—the Ustashi prayed to the same God, but in another church—then slaughtered them and dynamited the church. Each household had its own tragedy, its own bloody history.

I too was involved in their misfortune—as a participant—so I addressed these peasants in the language and values of their heritage, knowing that this could best console them: "And how do you think that the Serbs can ever free themselves, except by death and sacrifice? When did a Serb ever free himself any other way? Remember Kosovo! The heroes of Kosovo also chose death in order to preserve the Serbian name! Do you remember the saying, 'He curses like a Serb on a stake'? And would Karadjordje have freed Serbia, had he not hurled it into blood and fire? And as for Montenegro—who would ever have heard of it had it not shed its blood for centuries? And would there have been a Yugoslavia without Serbia's victories and sufferings in the last war, without the Golgotha of the retreat across Albania?"—I don't claim that I said these very words, or that I didn't go on at greater length.

Then one old man lamented, "Why don't *they* speak to us like this, instead of using words we don't understand: *afeža,* anti, anti . . ."*

His fellow villagers comforted him, though they had tears in their eyes.

At dusk we set out in two-wheeled carts, with about ten soldiers who were experienced at getting across the Sava River and various roads. However, we soon had to send the carts back for fear that their racket might alarm enemy patrols. We descended from the war-torn highways to the gentle byways in the fields, emerging from shady groves into clearings. Occasionally our guides stopped to consult with reliable informants, calmly listened to their reports, and then resumed our march. And so it went all through the night.

It was daylight when we arrived through the mists of the fields and meadows in Crkveni Bok, on the banks of the Sava. The village lay in a

* The peasant was alluding to the AFŽ—the Antifascist Women's Front.

cluster, with spacious two-storied houses. The region was obviously fertile and rich. The village committee received us with a practiced routine. However, on learning that we were leaders, they honored us by putting us up in the best house, with good, clean beds. The village was near the highway and they advised us to sleep in our clothes.

"And how do we escape?" we asked.

"If the trucks are coming, we'll see the dust on the road. There are shepherds around who signal if it's the enemy. You then go by way of the cornfields, down by the Sava and into the marsh."

We ate breakfast and slept. In the afternoon the committee members were eager to tell their story. Just recently the Partisans of Slavonia had retreated across the Sava River in the face of the offensive, and turned up in the marshy forest below the village. The Germans got news of this—probably inaccurate, because the Partisans outnumbered them—and surrounded the forest with a motorized battalion. However, this didn't dismay the Partisan commander, Petar Drapšin. He waited for the Germans in the forest, then broke through the encirclement at three or four places simultaneously. Split and disorganized, the Germans scattered. Several groups fleeing across the Sava were peppered with shots by the Ustashi, who thought they were Partisans. Several small groups fled into the village, and the villagers hid them in their stables and storage bins until the Partisans arrived, whereupon they killed them off together. That evening the Partisans retreated toward Banija, while the villagers buried about twenty Germans—in a common grave, so they wouldn't stink up their fields and yards. Three or four days later, a German motorized battalion showed up in the village. The villagers—our committee members included—assured them that they had nothing to do with the Partisans and the fighting. At the same time they loaded a truck with pigs as gifts for the Germans.

"Where are the dead Germans?" the German commander asked.

"We buried them, as is proper," the peasants declared.

The commander then ordered the grave dug up, convinced himself that what they were saying was true, and put four or five bodies in coffins for a token funeral. Fêted, the Germans left the village, well pleased at the villagers' attitude. This sort of thing was possible in villages where there were no spies or hostile elements.

The peasants of Crkveni Bok, as well as of other villages in Slavonia within reach of the Ustashi, had been converted to Catholicism. When we jokingly reproached them for their "apostasy," they smiled abashedly: it was to save their necks, but they continued to observe their Orthodox feast days and rituals, and sent their youngsters to join the Slavonian Partisans. Crkveni Bok remains unforgettable for its resourcefulness, and its awareness of its own and the national destiny. The people managed to survive, even though the Ustashi later burned down their village.

When it grew dark, boatmen from Crkveni Bok pulled out their rowboats, which they kept submerged during the day. Silently we descended to the muddy bank, and silently crossed the motionless river. Silently, too, we were received on the other side by ten or so local Partisans who had expert knowledge of the land, the railroad crossings, and enemy patrols. Through a clear moonless night, we moved swiftly over lanes through primeval oak forests, or along wide lowland roads.

We crossed the Zagreb-Belgrade railroad with extreme caution, for the Germans and Ustashi had lined it with bunkers. Their patrols were on constant guard because the Partisans attacked it frequently. It was the most important communication line between the Reich and the Balkans. While we were crossing, the squad leader whispered to me, pointing to a red light some two hundred meters away, "That's an armored train!" After we had crossed, he added, "This is the safest crossing. They're sitting there in the armored train right now, and can't imagine that we're so close."

We found ourselves in a Croatian village. The squad leader pointed to a building on the opposite side of a rather spacious square.

"That's an Ustashi gendarme station."

"And where are the gendarmes?" I asked, suppressing my astonishment.

The squad leader explained to me, without the slightest sarcasm, "During the day they're there in the building, in their offices. Evenings, they're afraid we'll attack them, so they hide in the woods or in the cornfields—we haven't found their hiding place yet."

We passed down the long street of still another village. There was no gendarme station there, or at least no one pointed it out to me.

"Are there spies and pro-Ustashi in this village?"

"Sure there are," the squad leader replied, "but there isn't a peep out of them. They don't even dare report that they saw us, or they'll be blamed for not having done anything about it or for not reporting it in time."

It was all an unheard-of, inconceivable jumble, the kind that only real life could concoct: by day the Ustashi ruled, and by night the Partisans. The Croatian people in ethnically solid Croatia—in Slavonia, Podravina, and Zagorje—were slow in joining the uprising, yet they didn't accept the Ustashi regime either. Perhaps this was one of the reasons why the massacre of the Serbs in Slavonia did not assume the disastrous and frenzied proportions that it did in Hercegovina, Lika, and Kordun, regions with mixed populations that were at odds over basic values and bound by nothing vital.

We didn't reach Papuk Mountain that night. The sun caught us on a dew-covered path on the slopes. We were drowsy by this time, though we didn't dare to stop; it was free territory, but by day it was within reach of the enemy. The corps commander Petar Drapšin and the corps political commissar Duško Brkić rode out on horseback to meet us. That afternoon we arrived with them at staff headquarters, on Papuk Mountain. This welcome sprang more from pleasure at seeing us than from protocol: there was a lot they wanted to hear about, and they had much to tell as well.

Commander Drapšin also had personal reasons for wanting to hear my opinion and to gauge my reaction: in 1942 Drapšin had been expelled from the party, along with Vlado Šegrt, commander of the Hercegovina Brigade, for excessive and injudicious executions in Hercegovina. How much Drapšin was to blame and how much the committee, I don't know, and I don't believe this was ever thoroughly investigated. While the punishment was very severe, it was not strictly administered: both Drapšin and Šegrt retained their command duties, and when I arrived in Slavonia, Drapšin had already been readmitted to the party. However, Drapšin was not only a gifted soldier but a man of sensitive and inquiring intelligence. He was anxious to find out what they thought of him now in the Central Committee. I could sense this, and recognized his extraordinary qualities. As with few others, I was on close terms with him from the moment we met.

I met political commissar Duško Brkić, of whose heroic conduct under torture by the Belgrade police I had already heard. I became friends with him, too, for the first time, though not on cordial terms: Brkić was withdrawn, with deep, inexpressible feelings. Without boasting or complaining, he described for me the incredible difficulties which he and Pavle Gregorić had had to overcome in the beginning with a small group of Communists and peasants, in the midst of the Pannonian Plain deluged by an Ustashi-Nazi reign of terror. At that time no one denied that Brkić was the most energetic and most courageous figure in Slavonia, though one would never have guessed it from his manner.

An inconspicuous but important person at the headquarters was Jevto Šašić, chief of the intelligence service. That service was so well organized that the staff was able to learn within a few hours not only of any development among the enemy, but also what the enemy was finding out about the Partisans. The German and Ustashi had recently massed troops for a thorough clearing of the mountains, mountains so gentle and low that motorized units could go up to their peaks. Accurate reports of this led the staff to evacuate its men across the Sava, so that the enemy's efforts were futile. As I have already recounted, Drapšin crushed one German battalion at Crkveni Bok. Accurate reports also made possible sudden powerful attacks on neighboring towns and the destruction of their garrisons, largely Ustashi. The information came mostly from Home Guard officers. Everyone was mobile, continuously on the alert. Underground shelters for the wounded and for food were built with great ingenuity; enemy troops had passed through the area without learning a thing. Such skill and conspiratorial methods rubbed off on the villages as well. I was shown shelters that had been dug into roadsides and camouflaged with weeds and vines. They even found ways of making dogs lose their sense of smell. After the war, Šašić was for many years the chief of counter-intelligence (KOS) in the Yugoslav army. It was his successes in Slavonia that initiated his rise, and I certainly contributed to it.

The committee secretary for Slavonia was Dušan Čalić, a young man, but wise and experienced. The committee and the staff co-operated so closely that there was no visible difference between them except in the division of labor. We held long meetings, even though there were no great problems or essential divergences. The Slavonian leaders were interested in the Central Committee's assessment of developments in the country and abroad, while we were interested in their experiences and views. The Chetniks never got started in Slavonia: Chetnik emissaries would get caught in the web of the party and its intelligence before they could ever find followers, and would simply be swallowed up by the darkness. But the passivity of the Croatian Peasant party leaders, who had been the reason for our visit to the Slavonian Partisans, persisted: they didn't even show up at the meeting, but having again changed their minds, went back to sitting and waiting. What else could we do but seek to divide the Croatian Peasant party as persistently and cleverly as possible, and win over those who were for the armed struggle?

I also visited a brigade. In this brigade I met a certain Demonja, a rather short young man, in big German boots, with an embarrassed smile at being the object of our attention. Nowhere did our army look as fit and as ready.

In Slavonia there was no problem with food supplies, not only because of the region's productivity and richness, but because the war had stopped any commerce between town and country. The problem was to store the gathered foodstuffs and make it impossible for the enemy to get them. Instead of destroying the threshing machines, the Slavonian Partisans had requisitioned them and kept the millers' fees for themselves. Columns of trucks brought in sheaves, and columns of trucks took away loads of sacks into the villages or to forest and underground storage dumps. There was food in such abundance that the whole brigade sometimes got a goose stew for dinner.

But the enthusiasm stemmed from the determination to fight, within that tight, almost impenetrable encirclement. Those people, those soldiers, had crossed the threshold of death inside themselves, and were glad to breathe and to fight. The fighting force continued to be mostly Serbs. When I inquired whether at least a third might be Croats, Drapšin replied, as if ashamed of his failure, "Well, we're getting there."

Enthused by the simple, hearty bravery of the Slavonian Partisans, I asked the staff for a squad of picked soldiers for the Escort Battalion of the Supreme Staff. The staff was glad to oblige us. The selected soldiers took leave of their native region with sadness and pride, and set out with me into "foreign" parts and "high" responsibilities.

In connection with the diversions on the railroad, Ilija Haris—"Gromovnik" (the Thunderer)—enjoyed the greatest reputation. He had learned his trade well in Spain, but was able to practice it on a large scale only in Yugoslavia. When we were there, the Thunderer was on one of his mysterious missions, so we never met him. Most of the time the staff didn't know where he was or what kind of diversion this self-willed, shadowy hero was engaged in with his little band. Among his great exploits was the demolition of a trainload of German officers returning from a furlough in Greece, and the killing of a Bulgarian minister. In all this destruction, innocent passengers suffered as well. This was inevitable in war, as was the reproach: why must they travel while the country lies in slavery and blood?

At that time the corps staff was carrying out its own diversions on the railroad, which it would attack along a stretch of fifteen kilometers or so, taking away everything to be found there.

A prewar Communist—a Croat who had become an outlaw on his own and engaged in his own war with the Ustashi—recounted to me the following incident: "I never believed that the folk tales about heroes who cut off heads were true. One day we captured some Ustashi and I sharpened an old cavalry sword, just to try. The Ustashi had to die anyway. I swung at the neck of one of them, and the head came off like a pumpkin." That comrade was so strong that it wouldn't have surprised me if he had cut off that head with the edge of his hand. He was a naive, simple-minded man who told his story as if the world about him held no horrors, as if he had not done anything terrible. I note this merely as a commentary on the mixing of Communist and Marxist perceptions with the legacy of folklore.

The staff of the corps—and particularly Drapšin, who was extremely sensitive of criticism—were stung by the reproach of the main headquarters that they had been premature in organizing a corps. The Slavonians sensed a certain jealousy in this reproach: the main headquarters had organized the First Corps, and they, the staff of the corps, organized the Second, without consulting the main headquarters. I supported them in their decision: recruits were arriving all the time, and the brigades were growing in strength.

The capitulation of Italy on September 8, 1943, found me in Slavonia. We judged this to be a turning point in the war, at least for us Yugoslavs. But we had already had enough experience with the war and German power not to anticipate the imminent fall of Germany, even though the Germans had already lost battles in Russia and Africa. We believed that the disintegration of the Chetniks and the émigré leaders, as well as the Croatian Peasant party, would continue. We suspected that the Germans would no longer have the strength for extended encirclements, and hoped that there would be no further need for those deadly breakthroughs of the kind just recently effected by the Supreme Staff units. These hopes were shown to be illusory, and in the cruelest fashion, precisely in Slavonia. When at the end of 1944, with the incursion of the Red Army, a front was established, the Slavonians found themselves in its immediate rear; this tripled their importance, but also made them a target for the terrible strength of the Wehrmacht. Constantly surrounded in a small hilly area, amid snow and frost, the Slavonians fought battles which were among the fiercest ever fought by the Yugoslav Partisans.

The British mission had already reached this corps headquarters, as it had other large units. The mission was staying in one of the little wooden houses erected in the forest for the staff. Relations with the mission were correct. There was also some grumbling, and with reason. The British were dropping food into Slavonia, though the Partisans there had more food than they knew what to do with, and were crying for dynamite; in Lika, on the other hand, they were dropping dynamite, even though everything there had been destroyed as early as 1941, and even the wounded were starving. Comrades at headquarters told me with embarrassed smiles about one event in connection with the mission. An earlier site of the staff and mission had been discovered and heavily bombed. During the lull between bombs, while everyone had fled into ditches, a courier got into the mission building, took out the mission's radio, and hid it. Nothing was done about it. Our people kept the radio, and the British obtained a new one. I warned the comrades, "Don't do that in the future. They are our allies. Just make sure they don't recruit our men." I never heard later of any transmitters being stolen, or that the British military mission had undermined our troops.

I stayed in Slavonia a week—perhaps a day or two longer. Before we left they even took us on a hunt in the prewar game preserve of Count Pejačević, with the help of the count's gamekeepers. We were more beguiled by the Count's game preserve than by the hunt itself. It was wartime hunting: young, unlawful deer were killed. We didn't even suspect that this moment of relaxation between battles was the beginning of the postwar prerogative of high functionaries to restricted hunting grounds with rare game.

Staff comrades suggested that we visit neighboring Moslavina as well. We agreed, all the more readily since the comrades in Moslavina would have been offended had we not found time for them. We started out in the afternoon, on good mounts. We were supposed to reach the Ilova River at night, since the Germans maintained a checkpoint at the bridge by day. Drapšin and Brkić came with us. We reached the bridge on foot, by way of a long embankment between fish ponds that reflected the moonlight like mirrors. In the shadow of some buildings an automobile awaited us, and we arrived at the headquarters of the Moslavina unit by ten o'clock.

We could hardly avoid a comparison: the Slavonian free territory seemed stable and vast, compared to Moslavina. Consisting of low, gentle uplands near Zagreb and other towns, this free territory extended as far as the enemy's vigilance or our own vanguards permitted. That territory could change at any time, and it expanded considerably at night. We also saw young peasants in a Croatian village join the Partisans—without song and music, but with an anxious sobriety, just as they always went into the army and war. Yet the commander, Vlado Matetić, radiated confidence and resoluteness, while the gushing of the political commissar, Marko Belinić, complicated the situation, though it made our job less difficult. Moslavina was like a flue which connected the fires of Slavonia with the Partisan centers around the capital of Croatia, Zagreb. In Zagreb at that time the joke went the rounds that Pavelić's state extended from M to M—that is, from Malta, the point where the municipal customs duties were collected, to Maksimir Zoo, at the opposite end of Zagreb.

On the day we arrived the Moslavina headquarters received a report that the Germans had set up a machine-gun post on the bridge over the Lonja, about an hour after we passed. The next day Drapšin and Brkić received a report from their own headquarters in Slavonia that the Germans had been informed that some high Partisan officials would be coming through. It was obvious that someone had reported our journey to the Germans, and that we had been spared either by our enemy's inefficiency or by the speed of our good horses. After the war Brkić told me that we had been betrayed by the radio operator at headquarters, a deserter from the Home Guards. He was a pale, overly obliging young man who found all too frequent reasons for being among the leaders. Our case led to his being exposed.

On the second day after our arrival in Moslavina, as soon as night fell, we took cordial leave of our comrades from Slavonia and Moslavina, and turned back toward Lika.

With us came a pale young woman, the district secretary of the Communist Youth. She was from that region, and since our first stop was to be her village on the Sava River, she came along to be of help, and to hear what she could from these important comrades. Her husband had been killed recently, and she was expending her bereaved strength in a feverish zeal.

We proceeded briskly, on foot. We crossed the railroad below Sisak and arrived in the dead of night in a Croatian village which was to furnish us with a wagon. The village was on a drunken spree, and looked just like a Brueghel painting. A buxom, barefoot girl was dispensing brandy with a ladle, while a hefty young man with a Colt revolver at his waist blocked our way: "Who are you?" he asked boldly. "What do you want? Do you know that I'm the Communist Youth secretary here?"

Our young guide was embarrassed, though she was used to such scenes. She tried to reason with the young man. Anka Berus got into the heated discussion as she demanded a wagon. The girl began to swagger: "Who do you think you are? I've been a Partisan for ten years!"

When the young fellow began to slap his Colt, I broke in angrily. Taken aback, our guide took the fellow by the arm and whispered to him, "Comrades from the Central Committee of Croatia and the Supreme Staff!" He calmed down in an instant and withdrew into the crowd, which was in the beer garden bawling songs against Pavelić.

We started to look for the committee members. But even the committee members were drunk and enjoying themselves—all except the duty officer. He immediately went off to get the wagons ready. I went with him to keep an eye on him, and my escort followed. Along the way he said to me, "During the day the committee hides in the brush. So do the guards, the secretary of the party cell, and any compromised villagers. When evening comes, they want to have a good time. But the committee doesn't neglect its duties by day in the brush, or even at night while drinking."

Indeed, as soon as the committee member told a peasant to hitch a wagon, he did so instantly and without a complaint.

The time passed faster than it should have: the dawn overtook us on the plain of the Lonja River. Even so, we drove without a care. The enemy were late risers, and there were no main roads around. Our guide, the district Youth secretary, made excuses for the young man's rude behavior in the village: "He's a good fellow—brave and devoted. Only he's still raw, undisciplined. We'll straighten him out yet."

When we arrived in her village, she put us up in her house. She told us that we could undress and go to sleep, because the children were keeping watch in the fields, and there were willow groves and river channels around the house. Still, we didn't undress, more from habit than caution. When it grew dark again, we set out across the Sava River and took the road to Banija and Lika.

In Lika, in Otočac, everything was turned around. Our units had disarmed two Italian armies and liberated practically the entire coast. The booty in weapons, food, and motorized vehicles exceeded all imagination, if not all hope. Croats from the Littoral, from the islands, from Istria, were joining the Partisan army. Old units were being filled with fresh manpower, inexperienced but vigorous. New units were formed on the basis of experience and around a core of experienced men.

Yet the fall of Italy and the spread of the uprising didn't seem to have any effect on the Ustashi. During the night of September 13–14 one of their battalions fell upon Otočac; disrupted our installations; killed over thirty soldiers, noncombatants, and wounded; and disappeared the same night. I was told that, in reply to our taunts to the Ustashi that they had lost the war, the Ustashi shouted back, "We know, but there's still time to rub out a lot of you!" They also sang: "Oh Russia, all will belong to you, / But of Serbs there will be few . . ." This was war with no quarter, no surrender, no letting bygones be bygones.

It was then, when I had absorbed the reality and experience of Slavonia, that I first felt the uprising, the revolution to be something alive, something that permeated every pore, the very being of the people. This was no longer an ideological analysis and heroic trial, but the very course of life, murky and irrepressible.

3

On Kardelj's invitation, Žujović and I set out from Otočac, probably on September 25, to attend the Slovenian Antifascist Assembly. Dr. Ribar also left, separately.

We traveled by car and wouldn't have had to spend the night anywhere along the way, had not unexpected events confronted us in Crkvenica. The Germans were advancing from Rijeka and had already retaken Sušak. Our rapidly growing units were not sufficiently trained to hold them back. The most urgent tasks were to get supplies out of the threatened towns, and to save the Jews whom the Italians had placed in a camp on the island of Rab. If I remember correctly, Žujović was more engaged with the evacuation of supplies. However, I remember for sure that I took charge of the Jews and their removal.

The Jews had in fact already been evacuated from Rab, and were housed in two or three abandoned boarding schools. There were several hundred of them, if not more, mostly from Croatia. Not all could fit inside, so they huddled around the buildings—mostly old people in city clothes, with leather suitcases. They were organized and heeded their committee's instructions. The committee consisted of an old, gray-haired gentleman, an elderly lady, and one or two younger people. None of them was a Communist, though their young people flocked to Communism. Yet one could work with such a committee—all the more easily once they were frankly told how matters stood and given reasons. Even less threatened and less perceptive people could have sensed the danger; yet there was no panic among them, or submission to fate.

I met with the Jewish committee late at night. The resolution of their situation was dragging out. True, the Germans couldn't surprise us, but we couldn't hold them back much longer. We agreed immediately to evacuate everyone from the coast into Lika and to distribute them among the villages, unless they were fit for the army or auxiliary services. But it had to be done quickly; the military authorities helped with vehicles captured from the Italians. And so it was done. Some ten days later, when we returned from Slovenia by way of Lika, many Jews had already been absorbed into the auxiliary services or the army. But many also engaged in business in the villages and thus interfered with the natural, "Partisan" economy. All complaints were to no avail, but the amount of trade was so insignificant that it was hardly a threat to the established hierarchical relations; it was simply accepted as a part of Jewish tradition.

The only disagreement that arose was over the young Jewish people who had already organized themselves into a battalion, though practically without weapons. Not only the Communists but society at large knew about the extermination of the Jews. The Ustashi, the Bulgars, and especially the Germans carried it out in our country openly, sometimes even with pride. Though we condemned this treatment and expressed our horror over it, these reactions were different from those over the massacres of our own people. The extermination of the Jews exposed the horrible nature of fascism, but this affected us intellectually more than it aroused us to vengeance and struggle. To be sure, the helpless, fatalistic attitude of the Jews contributed to this, as did the alienation between them and the peoples among whom they lived.

I opposed the existence of a Jewish battalion: not only would the Germans and the Ustashi lust after the blood of a separate Jewish unit, but such a unit would take on itself excessive tasks for the sake of vengeance. It pained me, it seemed irrational to me, that the Jews who happened to survive should be exposed to even darker perils than our other units. Though I knew that the Jews were a separate people, I accepted them as a part of our own peoples—the peoples with whom they lived. And that was, I insisted, why the Jews should not be in a special unit like, for example, the Czechs or the Magyars. The young people reluctantly accepted my position. Not so the committee. It had succumbed to pressure by the youth, and to the wish to have the Jews fulfill their obligations like all the rest. But eventually the committee did accept my position, with relief and without protest.

Žujović and I arrived in Slovenia on September 27. At first we stayed in the forest base at Rog, in comfortable wooden cabins which were skillfully camouflaged. Here were the leaders of the party and of the Liberation Front of Slovenia. Tomšič, Zidanšek, Šlander, and many of my other acquaintances from prison and underground activities had either been killed or were in prison. But the rest were happy to see me—Boris Kidrič, Lidija Šentjurc, Luka Leskovšek, and the others. They were all busy with preparations for the assembly—with reports, resolutions, and organizational tasks.

There I also met the most important non-Communists among the leadership of the Liberation Front—Josip Vidmar, Josip Rus, and Edvard Kocbek. Vidmar didn't represent any political group, but was all the more prominent for it: a renowned literary critic, influential among the liberal non-Catholic intelligentsia. At that time he wasn't a party member or even a Marxist. He accepted the revolution as a historic opportunity for the Slovenian people, while he looked upon Marxism as the most significant modern social teaching—albeit a defective one, especially in its aesthetics. He was a person with a vast European cultural heritage, with a lively and refined manner and expression: a positivist and a patriot, and as such, a fierce opponent of clericalism and fascism; also, an opportunist in his persistent efforts to inject aesthetic tolerance into Communism. Kardelj and Kidrič consulted him on unimportant matters, though he had already espoused the historic role of the party out of a sense of friendship and democracy. In view of the dominant role of the party within the Liberation Front, this was obviously a formality. Kardelj told me at the time that he and Kidrič had only recently been engaged in long and painful discussions with Vidmar and Kocbek, in an effort to convince them to accept the leading role of the party in the Liberation Front.

I didn't understand why our allies in the antifascist movement had to renounce their independence even formally, and asked Kardelj, "Why did you have to press them so much?"

Kardelj explained that the situation within the Liberation Front had changed: it was the party that sustained the army and the local authorities; therefore the Liberation Front, though it needed to be a broad, united organization, had to lose the elements of a coalition.

My opinion was that the elements of a coalition would disappear in time, simply by virtue of the fact that no organizations parallel with the party would be created. But in Slovenia the situation evolved in a manner and at a pace different from the rest of Yugoslavia, as did the uprising itself. Elsewhere there was no organization like the Liberation Front, or any common program like the one there agreed upon by the party and the other groups in the Front. At the time of its formation, in April 1941, eighteen groups had entered the Liberation Front, but they had gradually merged in a common movement where the Communist leadership grew in strength simply because it was the most ideological—the most resolute, the best organized, the most realistic. The Central Committee of Yugoslavia didn't hold the past against the adherents of the bourgeois parties, but accepted them with open arms—to be sure, only if they supported the armed struggle. The formation, or rather the restoration, of the dissolved parties was not even considered; this would have disrupted the unity of the uprising, and might even have presented a lever to its enemies. The participation of individuals or groups from these parties was for show and symbolic; it broadened the base of the uprising and facilitated transitions. But in essence it never went further than that—except in Slovenia, in the beginning. Yet not even there did these groups ever have the influence, much less the organizational strength or militancy, of the party. This is how matters stood in the Liberation Front, in my recollection, though I must admit that to this day I have not substantially increased my understanding of the matter. It seems to me that Tito was right, though he said it thirty years later: "Something similar took place in Slovenia. It was, after all, a civil war. But we didn't wish to speak of it during the war, for it would have done us no good."*

Josip Rus would have had an even slimmer prospect of a political role than Vidmar, had it not been for the kind of war it was. Yet unlike Vidmar, Rus didn't show any inclination to play a real political role. He was already along in years, but robust. He had a slow, set mind, but he was also very steadfast. He was a leader of the Sokols, a gymnastic society that originated in Bohemia under Austria-Hungary and spread throughout the Slavic lands. The Yugoslav idea was an essential and fixed principle of that society. As a result, the society played a varied role among the Yugoslav peoples, depending on the orientation of the national leaders: in Serbia it was predominantly Greater Serbian; in Croatia, predominantly centralistic and "anti-Croatian"; and in Slovenia, pro-Yugoslav. It was the Yugoslav idea—the bond between Slovenian destiny and the survival and restoration of Yugoslavia—that brought Rus and the Sokols to co-operate with the Communists, and later to be their comrades in arms as well. The Sokols evinced a militant perseverance and national consciousness. It was precisely this that led these ideological innocents to flock to the Communists, especially after the uprising

* *Vjesnik*, Zagreb, May 24, 1972.

faltered in the autumn of 1942, and the party remained the only unshaken, cohesive force. At the time of my stay in Slovenia, Rus was a prominent and uniquely ornamental figure, though younger ideologized Sokols had already made their way into high positions in military units, the administration, and even the party.

Kocbek, on the other hand, preserved his own spiritual core, probably because he was intellectually and metaphysically religious. His withdrawal into himself was intensified by his spiritualized, ascetic appearance. He never cared for any recognition or outer display. He possessed a thorough education and vast culture, though this became evident only in lengthy, exhaustive discussions in which he engaged reluctantly. Both Kardelj and Kidrič esteemed his literary talent and acclaimed his language and style. His integrity and sincerity were not only incontestable, but lent him an esteem and significance beyond that of the forces which he represented, since with the appearance of the collaborationist White Guards, organized from among the Catholic masses, the prospects of a leftist, socialist Catholicism were dim. The Catholics who joined the Partisans were assimilated, engulfed, and absorbed. The dissolution was in progress of those social and national visions of a Catholicism which Kocbek had evidently seen as something popular and Slovenian, and this was his personal tragedy. It was perhaps that tragic link with nation and faith that endowed him with a quiet steadfastness which we Communists could comprehend from our experience, but which, being unswerving in our own aims, we couldn't share. Kocbek was to have come to Bosnia in late 1942, to become vice-president of AVNOJ. I never heard, and I don't believe now, that the chief purpose in this was to get him out of Slovenia. But even if there were such intentions, our efforts at the time to attract a broader following, particularly in our prominent central bodies, are beyond dispute. Then as always, there was manipulation in politics. But Kocbek—like others who accepted the armed resistance and socialism, though not also the total spiritual monopoly of the party—was not in favor of certain developments: notably, radicalization in the direction of atheistic socialism. Christian socialism belonged to a religion which had lost its chance in a time that was decisive for the Slovenes, and not for the Slovenes alone.

Though it was damp and cold at Rog we would have stayed there, had we not had access to incomparably more comfortable quarters in the Soteska Manor, which had belonged to Prince Karl Auersperg. Not just the manor itself, but all the furnishings and the paintings on the walls, were preserved. Never before had I seen such a tasteful edifice, let alone stayed in one. The manor contained some two or three centuries of Central European art, and had housed perhaps an even longer line of owners, from whose midst had sprung a renegade poet given to liberal ideas and to the folklore of the Slovenian peasantry. On top of everything, the manor had been equipped for modern living with electricity, running water, and bathrooms. The war had brought its own adaptations: the count had invited the Italians in, had ringed the palace with bunkers, and himself took part in the fighting. Our wounded were quartered in some of the manor chambers, and we functionaries in others. A few of the old servants remained, as accommodating as if new counts had arrived. This obliging service led us to the most superficial and dogmatic discovery of an awakened class instinct in the servants. The smell of mothballs was stifling, but the freshness of the Kupa River and of the woods soon breezed through the chambers.

A day later Kardelj moved into Prince Auersperg's study, along with all his drafts of reports and his secretary, while I picked for myself a smaller and even more tasteful room, probably the Princess's study. I copied down the charter of the palace, and noted the names and descriptions under the portraits. There was Empress Maria Theresa with her ancestors and her son, Joseph II. That night the idea for a story came to me—that the aristocratic lords and ladies in those pictures should come to life and mix with the Partisans, thus providing a confrontation between epochs, classes, and personal destinies. I lost those notes, but the idea came back to me later in prison, filling me with notions such as the similarity between ingrained nobles and ingrained party bureaucrats, or the sympathy which the neglected, unhappy counts and countesses might have felt for the nurses and wounded. But the plot based on this motif was never elaborated; like other ideas beyond one's personal experience, it went unwritten and continues to remain so.

The Auersperg manor also is no more; it was burned down with all its forty or so chambers filled with unique portraits and antique furniture. I was sorry for the manor. And I remained unconvinced—perhaps because of the idea that I carried around in my head—when Kidrič explained to me after the war, "We burned it down to keep the Germans from using it as a stronghold." Among the Slovenian Partisans there was a special animosity toward manors and castles—an intellectual heritage from the peasant wars, as well as the result of their battles with the Italians and the White Guards, who used castles as strongholds against the more poorly armed Partisans. "The castle burns—the count has fled," is an old saying which in those days I frequently heard from Kardelj and Kidrič.

It was lively in the manor: couriers, functionaries, officers, meetings. The better part of Slovenia was free, and Slovenian units were pushing into the Slovenian parts of Italy, but the Germans were advancing as well. The White Guard was falling apart. We had taken the castle of Turjak, in which some seven hundred hardened White Guards had taken refuge. Our side brought up some artillery captured from the Italians, battered down the castle walls, and took most of the White Guards prisoner. Angered over their losses and by the incorrigibility of the White Guards at a time of national rejoicing over the fall of Italy, the Partisans wiped out all the Turjak prisoners. I once inquired of Kardelj, when this was brought up in the Auersperg manor.

"But why did you have to kill them all?"

"That ought to demoralize them!" Kardelj replied with a knowing smile.

Massacres are the fruit of bitterness, and of the calculations of leaders. I accepted this judgment myself, not so much because Kardelj stood above me in the unofficial hierarchy, but because of ideological implacability. The part of me that rebelled against retribution regardless of personal guilt, was an age-old humanistic legacy unmolested and unblinded by practicality and purpose.

Kardelj was under the strong influence of the resistance in Ljubljana. It was as if Ljubljana had become his native city. "I feel more secure in Ljubljana than I do here in the forest," he observed. Kardelj had rickety legs from childhood, and all that marching over the mountains was harder on him than on the rest of us. But that was not the important reason for his prolonged stay in Ljubljana. Until the spring of 1942, Ljubljana had been the seat of the main headquarters of Slovenia. This may have retarded the growth of our units, but it strengthened the resistance within the city. That resistance assumed proportions and forms as nowhere else in Yugoslavia, or even in Europe, except perhaps in Warsaw during the 1944 uprising. Apart from the leaders of the clerical party and hired spies among the riffraff, the entire population was not only on the side of the Liberation Front, but participated in the resistance movement. The Liberation Front published newspapers, periodicals, and cartoons. Occasionally even the Liberation Front radio station could be heard. The Italians were in effect surrounded inside a city which was to be the center of their newly annexed province. Armed activity and espionage penetrated the police, the church hierarchy, and the Italian administration. Once Kardelj received a report that in a house on a certain street there was a person who should be watched, because he was not a part of the Liberation Front. It was Kardelj himself.

The nerve center of the entire movement was the Security Intelligence Service, or VOS (Varnostno-obveščavatelna služba), which was managed by Kidrič's wife Zelenka, who seemed reserved and conceited, but was in fact just right for such a job, unerring and enterprising. VOS not only carried out a series of "justifications" or executions—even of the former viceroy Natlačen, his Italian guard notwithstanding—but also made arrests and conducted interrogations, all within an occupied city with a population of no more than 100,000. By the time I got to Slovenia—that is, when Kardelj was telling me all this in the manor of an Austrian noble—the Liberation Front in Ljubljana had been weakened, while VOS, though still reliable and effective, was hardly the equal of a good intelligence service. By taking extraordinary measures, the Italians had succeeded in suppressing the resistance and in disrupting many links through mass deportations and court-martials.

On the road along the Kupa River, just below the manor, small units frequently passed by. The soldiers here were better dressed and better fed than in Bosnia and Lika. There were also more workers in the units, which gave the appearance of greater homogeneity and solidarity. It was my impression, from what I heard and saw, that the Slovenian brigades were not at all inferior to the very best units of the Supreme Staff which had fought in the recent battles. On the other hand, Tito and the comrades around him had said that there were too few brigades in Slovenia, and that they lacked sufficient manpower. During my stay there the units grew in size, and new ones were formed. Yet the quiet dissatisfaction of those on top over the small number of fighting men in Slovenia persisted until the war's end. The reasons for this relatively small number were many: the proximity of Austria and Italy, a well-developed communications system, the clericalism of the peasantry, and the distaste of the city people and workers for guerrilla warfare (which even Kardelj pointed out).

On one occasion after the war, when at some celebration of the Liberation Front the Slovenian leaders asserted that the Slovenian people had created their state through their own efforts, I remarked that such a declaration was unsuitable, first because of the comradeship in arms of the Yugoslav peoples, and also because of the non-Slovenian units which were harassing the Germans, the Ustashi, and the Chetniks in Slovenia at the end of the war. This took place at an informal gathering at Tito's; Ranković was also present. Tito commented with ironic anger: "Like hell they could have liberated themselves, if the Serbs and Croats hadn't come to help!" Although this judgment was a partial one, like any expressed in the heat of emotion, it confirmed my impression that the spread of antifascism in Slovenia was not reflected in the size of its military organization. Why was this? I am not competent to give an answer.

The Slovenes, and their struggle against the invader, were something special. Yet there would have been no struggle if the leaders hadn't been convinced that they were bringing about a turning point in the national destiny such as leaders before them had only dreamed of. In no other Yugoslav land, among no other Yugoslav people, was there such keen awareness, such enthusiasm over the creation of one's own state.

I myself first became aware of this during the meeting of Slovenian representatives which began on October 1 in Kočevje. This gathering was more impressive than all the previous ones. The setting, the food, the decorations of the hall were all as if one conqueror hadn't ruled there till yesterday, and another still more formidable one weren't on the way. Among the 562 representatives from all parts of Slovenia, the number who were prominent in their field or occupation lent the session an extraordinary historical significance. Kardelj and Kidrič had the principal role, which they acquired by virtue of their sacrifice and political talent. Yet no one adulated them; there was no personality cult.

The cult was Slovenia itself, a unanimous surge toward statehood as the crowning fulfillment of nationalism and the beginning of socialism. When Kardelj, as chief speaker, remarked that foreign rulers referred to the Slovenes as a nation of servants, the hall murmured in anger, only to explode with rapture when he praised the Partisans or spoke of a free Slovenia. Perhaps even more moving was the delirious unanimity of the committee members from the villages, the underground workers from the cities, and the soldiers with wounds still fresh—all with their own little affairs, their own fears, yet fearless, surging inevitably toward a national and social ideal. The speakers and all present were caught up in a moment of immortality.

Above the podium was emblazoned a quotation from Cankar, the Slovenian man of letters: "The people shall write their own destiny." Cankar has written that before the October Revolution, when socialism was regarded as the self-assertion of a benign people following the downfall of bourgeois rule. There were very few, if any, at that gathering who didn't know from their own experience that the people had to be led by an avant-garde ideologically capable of conducting a people's government and constructing a new society. All of them regarded that historically predetermined role of the avant-garde to be a popular one, since the people themselves were not conscious of their destiny, the ideal society. The slogan "The people shall write their own destiny" dazzled and enthralled minds because it joined, indeed identified, the destiny of the people with the role of the party: all that we Communists were doing was in fact the destiny which the people were writing for themselves. And the slogan was all the more enthralling and prophetic in that it sprang from their own Slovenian socialist writer.

The session was held at night, because of the danger of an attack from the air. The night, and the isolation from the world outside, contributed to the self-containment of the gathering and its surge toward a single aim and an unquestioned unity. No one before us Communists was ever so scientifically convinced that they were not only transforming a given state of affairs, but giving men and nations an ultimate and unalterable direction. All development and movement were seen as the self-fulfillment of the ideology and the party. To be sure, the course of life was not denied, but inasmuch as it was teleologically understood, it had to be directed, constructed. What was left for spontaneous, blind existence, but to submit to an omniscient and vital consciousness? Our assemblies were even then unanimous, zealously obedient to the leaders, with a sense of historic self-awareness. Yet the assembly at Kočevje was the first to attain a total, conscious, and wanton fascination with itself, with the ideas, battles, and leaders from which it sprang.

Every word resounded and cut into the tense silence, and with each mention of the party, the Soviet Union, the Partisan army, or Slovenia, the assembly responded with unrestrained, frenzied ardor. After the war there were even more clamorous, more enthusiastic gatherings, but none so spontaneous, so unequivocally unanimous as this one. Totalitarianism at the outset is enthusiasm and conviction; only later does it become organization, authority, careerism.

At Kardelj's suggestion, I appeared in the name of the Central Committee of Yugoslavia at the first-night ceremonial opening. The announcement of my appearance and my arrival on the lighted stage were greeted by a torrent of shouting and applause which subsided only after my fifth or sixth attempt to speak. But the speech was a complete failure. I had written it out, and Kardelj had liked it when he read it in Prince Auersperg's manor. However, that speech was not suited to the moment, to the mood and excitement of the delegates. It was, if I may say so, an overly intelligent and literate speech. Besides, though a member of the Central Committee, for the Slovenes I was another brother from the south whom they liked in a romantic way, but who didn't grasp all the special features of Slovenia. I was upset. I felt as if I had forgotten a poem of mine that I would never again be able to recall. Obviously, I had composed the speech in such a way that I couldn't reach the audience and draw it to me through our common Yugoslav and revolutionary feelings. Oddly enough, I regret it to this day. Is not the total intoxication with an idea easier and more delectable than critical and individualistic reasoning? The other speakers from the south were even worse. Especially Dr. Ribar, who spoke on behalf of AVNOJ. He didn't even write out his speech but depended on his oratorical talent, though all he had was a leonine appearance and a big voice.

But the assembly in Kočevje was not all applause and clamor. It lasted three nights, and though the shouting continued to the very end—indeed, that was when it reached its culmination—this didn't hinder but rather inspired and illuminated the analyses and conclusions. The assembly at Kočevje—like all our ideological and constitutional gatherings—had its practical, realistic side: an analysis of the situation, and a determination of the main guidelines for the administration. At the time I didn't grasp the incongruity, the irreconcilability between the utopianism of the ideology and what could be accomplished—the realities of our politics. Ideology and realities seemed to me to be all one, and to evolve each from the other. Moreover, the more total our ideological fervor, the more successful and substantial were our performance, our organizational structure, and our resources.

4

The Germans advanced from many sides. They were especially active in Croatia, where they were protecting their lines of communication toward the sea and along the coast. This put our speedy return to Bosnia, to the Supreme Staff, in jeopardy; yet in his dispatches Tito urgently requested a meeting of the leadership, including Kardelj and Ribar. We had to move fast in any case. We were detained by a party council held on October 8, 1943, in a schoolhouse not far from the Soteska manor. Kardelj presented a political report in which he easily combined new realities and current assignments with ancient teachings about imperialism, the leading role of the U.S.S.R., and the class struggle. I envied him, because I myself was having a hard time deriving a new theory—or at least freshening up the accepted one—in the light of reality. Kidrič expounded in great detail, and passionately, on the significance of the new regime and on the duties of Communists under it. The council was terminated quickly, but it familiarized the leading functionaries with new possibilities, and brought the council delegates more in line with the ideology.

On the next day—October 9—Kardelj, Žujović, Vidmar, Kocbek, and I set out by car for Croatia. We drove on side roads. In some places not even the commanding officers knew how far the Germans had advanced. Kocbek and Vidmar came along as the guests of the second session of ZAVNOH, the Regional Antifascist Council of National Liberation of Croatia. But they turned back to await further reports, when it looked as if the German advance might jeopardize that session. However, we members of the Central Committee had to go on. The changes in Yugoslavia, and in the attitude of the Allies toward Yugoslavia, were so significant that it became imperative for AVNOJ to make crucial decisions regarding the royal government and the country's future regime.

Kardelj is not completely open, though not secretive either, unless he feels threatened in his position and ambitions. On this occasion he was extremely quick and nimble in his thinking, as well as extremely receptive and tolerant in discussions. This was what most attracted me to him. Whenever we were faced with important decisions, it was with Kardelj that I reached agreement most easily. Within the Politburo he was perhaps the only one with whom I could discuss theory—which life constantly stirs up, as it does my mind. Through theorizing, we grew close to each other. We also confided to each other our critical view of the First Session of AVNOJ, in Bihać in 1942; we both felt that it hadn't been prepared properly, and that it hadn't resolved basic questions: the future form of the state, the future government, and the foundations of the social system. Kardelj maintained that the next meeting of AVNOJ should establish a government and possibly declare a republic. I agreed with him. But these were no categorical stands, simply matters for discussion with Tito and the other Central Committee members.

On the morning of October 10 we arrived in Otočac, and went on by car the next day toward Bosnia. Two details, however insignificant, are still fresh in my memory.

We had to terminate our journey by car before we got to the Bosnian border, because there was no road: the Germans had cut off the side roads. However, we had horses waiting for us. Crossing sunny dales between rocky crags overgrown with yellowing thickets, by around noon—probably on October 11—we arrived in a clearing near Donji Lapac. We had to wait there until nightfall, because the Germans controlled the road ahead by day. Feeling good after riding on horseback through the fresh gentle breezes, and even more so after a lively conversation about our imminent victory and future happy life, we made ourselves comfortable in the clearing. The nearby village had been burned down early in 1941, and the people lived in sheds or lean-tos against the charred walls. In the clearing there was also a lean-to constructed of saplings, propped up around a beech tree and covered with branches and ferns. A young girl lived there with her four brothers and sisters, whom she had rescued from her destroyed home when her parents had been murdered.

This girl had been knocking about for over two years, but had managed to keep the family together. She had never learned to read and write, but knew the names of all our organizations as well as their functions. The names of these organizations, which contained terms she didn't understand, seemed to her unschooled mind to possess something magical, perhaps because their effect on her life was immediate and comprehensible. She was half-naked, with only a tattered dress of coarse cloth to cover her. She squatted there near the road, so she could dig up potatoes at night in a field by the side of the road, and load them on a little hack pony that she called *Parip*—most likely a word from her dialect, but one that sounded funny to me, since the term is used for powerful riding horses. The girl was running about eager to serve us, to treat us to some baked potatoes and to milk from her one goat. She knew practically everything about the enemy and their movements, and was more fearful of the Ustashi, a more familiar and maniacal danger than the alien and transient Germans. The memory of this girl from Donji Lapac long obsessed me as an example of inexhaustible human effort and sacrifice. Let it be preserved from oblivion here, in this fashion.

The second episode took place in a village in Lika, not far from the Bosnian border, after a twenty-four-hour march beyond the girl from Donji Lapac. We arrived in the early afternoon. The entire village—the older men, the women, and the children—gathered to greet us on a street of razed, gutted houses. Only the house of the local committee was rebuilt, and there they put us up. They were dressed nicely and neatly, the men in a combination of military and peasant attire, the women in threadbare but clean smocks and skirts, and the girls in white kerchiefs. They knew that we were members of the Supreme Staff and the Central Committee, and greeted us with an easy deference. We were their people, their own government, but already high and mighty. As we approached the peasants, our comrades hailed them with the accepted militant greeting "Death to fascism!"; the peasants responded zealously, "Freedom to the people!" I arrived a few minutes later and shouted—partly in jest, but partly out of awareness of the sufferings of the Serbs—"Good day, brother Serbs!" The peasants cried with enthusiasm, "Good luck to you! Welcome! Welcome!" Kardelj laughed heartily and quipped to Žujović, "I didn't know Djido could be such a demagogue!" To which Žujović, replied, "And how!" We sat late into the night with the peasants, who were more curious about world events and the prospects of the British withdrawing their support from the royal government, than about when they would rebuild their homes.

The next day we reached the road, where a truck awaited us. We were in Jajce in the afternoon. As we passed by Kupres, I felt sorry that we had taken that Ustashi Alcazar and Partisan battlefield so ingloriously: after the capitulation of Italy one of our armored units simply entered the little town, which the Ustashi had wisely abandoned.

I expected that in Jajce we would find the Supreme Staff engaged in wider and more significant activities than when I left it in the woods on Mount Vlašić. However, what I found far exceeded all my expectations. A triumphant animation penetrated the wartime gloom over this little town situated beside the cataract of the Pliva River and around the medieval fortress. The streets swarmed with Partisans of all ranks, and members of the now sizable British mission. Military festivities blared far into the night, drawing into their vortex Moslem girls in baggy pantaloons, while the Theater of National Liberation had made a jump from agitprop sketches and recitations to productions of Nušić's play *The People's Representative,* Gogol's *Inspector General,* and a ballet—just like some renowned company in an established country. Tito had also changed: he had suddenly become heavy, never again to regain that look of bone and sinew which made him so distinguished-looking and attractive during the war.

The Supreme Staff was housed below the fortress, next to a little underground church which the Bogumils had dug out of the sandy soil in the Middle Ages. Liberal and Marxist theoreticians assessed the Bogumils as a progressive sect for their time, but we chose the spot next to their place of worship because it was considered safe from bombardment. Tito was staying in a tiny barracks right next to the entrance, while Ranković was in a cottage nearby. Žujović and I, and Kardelj as well, moved in with him. The sculptor Antun Avgustinčić also joined us, and on a ledge in the front of the church immediately began a bust of Tito in clay. Radovan Zogović, however, lived on his own, in a cottage surrounded by a garden and an orchard, not far from the Pliva cataract. Ranković whispered portentously to me that Zogović was writing a poem about Tito. And indeed Zogović, who had been purified by his sufferings in the war and the party, later read me certain parts of it.

We didn't stay there long. Enemy planes flew over Jajce frequently, and one afternoon showered it with bombs. Žujović and I rushed to the church shelter as the bombs burst by the walls below us, and machine-gun fire splattered all around us. Tito and the rest were already inside, their flashlights lighting up its dark, angular recesses. The roaring and the din lasted quite a while. I warned Tito that the ceiling was thin and made of loose stone shingles, and that the bombs would have brought it crashing down on us had they not missed us. Tito listened to me in silent confusion, even as I told him how in Belgrade medium-sized bombs had blown five-storied concrete buildings to bits. Anxiety crossed his face only after the staff officers supported me. Soon thereafter we began the construction of two barracks behind headquarters, one for Tito and one for the rest of us, in front of the entrance to a shelter which had been dug into the stone cliff before the war.

Two or three weeks after we had moved to our new quarters, the planes suddenly roared from behind the mountain and began to dump their bombs. Jovanović and I played the hero and stayed outside, until a bomb burst around the corner and struck us with its blast. A bomb severed the leg of a young soldier who was accompanying Koča Popović to the Supreme Staff. A Moslem soldier from the Escort Battalion of the Supreme Staff, on guard duty at the time, was wounded in the stomach. The wounded were brought into the shelter. The doctors immediately attended to the young man's stump; as the victim moaned, one of them gave him his own blood. Meanwhile the soldier from the Escort Battalion sat down and leaned against the wall. I asked him,

"Does it hurt?"

"So-so," he replied with a forlorn look.

He was a veteran warrior, wounded before, who considered complaining shameful. He was operated on later, right there in the cave, along with the soldier who was accompanying Koča Popović. The former died at dusk, the latter late that night.

Though we suspected that the enemy knew exactly where we were, new problems and new vicissitudes soon crowded out our suspicions. After the war these suspicions were confirmed: Ranković told me that the Germans had received reports from Jajce by way of carrier pigeons.

When Kardelj and I arrived in Jajce, Tito had already issued a declaration through the Soviet government, in connection with the conference of the foreign ministers of the U.S.S.R., Britain, and the United States, announcing nonrecognition of the royal government because of its support of Mihailović, and the decision that the King would not be permitted to return to the country. In the same dispatch Tito announced, "The English general has informed us that the English government would not insist very much on the King and the government-in-exile."* But nothing came of all this: the Soviet government never

* Vladimir Dedijer, *Tito* (Belgrade: Kultura, 1953), p. 379. [*Note:* See also Vladimir Dedijer, *Tito* (New York: Simon and Schuster, 1953), p. 205, for another version in English.—Trans.]

brought Tito's declaration before the conference. We were not even told what the ministers had said about Yugoslavia; only after the war did we learn from Western documents that when Eden insisted on a reconciliation between the Partisans and the Chetniks, Molotov merely expressed the wish to send a Soviet mission to the Partisans.

Even though the Soviet government kept us in the dark concerning the extent and form of its support, we received a good deal of information by radio and by ever growing contacts with represesentatives of the Western powers. In this effort, as well as in our relations with the Western Allies, a significant if not decisive role was played by Brigadier Fitzroy Maclean, head of the British military mission.

Prior to our arrival from Slovenia, Tito had held a series of talks with Brigadier Maclean. It was as a result of these talks that we were able to provide the previously mentioned information to Moscow that Britain would not "insist very much on the King," which was merely a realistic interpretation based on Maclean's statement that Britain would not mix into Yugoslav internal affairs, and on the ratio of forces in Yugoslavia, primarily between the Partisans and the Chetniks. The Russians were still far away; we could count only on their diplomatic assistance. And in the event of a British landing in support of Mihailović, not even open Soviet diplomatic assistance was assured us. The British were concerned about pressure on their front in Italy, while we were concerned about their support of Draža Mihailović, that is, their intervention against us. In other words, the British wanted us to tie down as many German and satellite divisions as possible, which suited us, on condition that we be given aid and recognition as the only Yugoslav force. Such a relationship would not have been possible had Mihailović possessed an effective anti-occupation army. The British had no choice but either to carry out a landing in order to fight the Partisans, or else to come to an agreement with them on a rational, mutually profitable basis. They chose the latter, cautiously and without enthusiasm, while our own dogmatic ideological distrust kept us from understanding them, though it also preserved us from any hasty enthusiasm. With Maclean's arrival the aid increased, particularly to our base in Italy, and later also to the refugees and units on the Dalmatian islands.

Above all, Brigadier Maclean impressed both the leadership and the Partisans generally: a member of Parliament with his own "partisan" exploits in Africa, reserved but accommodating, and with a political gift for words. We found it strange that, on ceremonial occasions, he wore the Scottish kilt. Yet he was not eccentric or amusing. Rather, he was extraordinarily brave. I myself was amazed, during an air raid on Drvar in February 1944, at the sight of his tall, bony frame moving about on the road as if nothing were happening.

Tito didn't conceal from the Central Committee members the content of his talks with the British. The hypothesis that he concocted with the British his later separation from the Soviet Union, at the price of their renouncing the King and Mihailović, doesn't jibe with the actual state of affairs. True, we in the Central Committee didn't know all the details, all the complexities and turns of his negotiations with the British, especially later on. Occupied with our own duties and having confidence in Tito, we didn't even take a particular interest in these negotiations, except insofar as they affected our own assignments. We lived and worked together harmoniously. Our knowledge and information came from that collective activity. It is correct that Tito, like the other Central Committee members close to him, stressed the indigenous nature, the independence of our uprising, without denying an ideological link with the U.S.S.R.: it would have been strange to deny this independence, since our entire propaganda stressed it, and it was one of the essential features of our party. I don't believe that anyone in Britain or America—not even Churchill—was so farseeing as to anticipate the Soviet-Yugoslav clash of 1948; all that the West could anticipate was the formation in Eastern Europe of states which would be more or less dependent on the Soviet Union. Even we, whose ideology enjoined upon us unification with the U.S.S.R., did not view our relations with it as anything but voluntary and gradual.

Tito's hopes for understanding and assistance from the United States, aroused by talks with Major Louis Huot, reflected our desire not to be left to the British alone, and not to lend our relations with them greater weight than those with the States. Huot came up with the idea of sending to the Partisans on the islands a ship loaded with war material, eight times more than we had received from the British. This effective generosity, as well as Huot's belittling of the British, contributed to Tito's rapture: the United States were not involved in the Balkans through economic or political interests, and therefore would approach our struggle with greater objectivity and selflessness. But these hopes, these feelers for still another support in the West, were quickly cut off by the Americans: under various pretenses they sent a mission to Mihailović, while on December 31 President Roosevelt solemnly presented four Liberators to royal aviators, and announced that for the duration of the war he would recognize only the royal government. A fatal, pragmatic shortsightedness even then deluded Washington into thinking that it would be the heir to spent and ruined empires, and would achieve this by supporting discredited reactionaries and outmoded structures.

The letter I wrote to Vukmanović-Tempo in Macedonia on behalf of the Central Committee just after my arrival, called for caution and independence toward the British. The occasion for this letter was the creation of a "Balkan headquarters" announced by the BBC. Vukmanović-Tempo claims in his memoirs that it was his intention—and the sincerity of that intention and of his account is indubitable—to get the Greek Partisans away from British control by means of a joint "Balkan" headquarters. The Central Committee, on the contrary, saw in this a ruse by British agents to submerge our Supreme Staff in an abstract and ineffectual Balkan headquarters, and thus diminish its significance and role. Perhaps for the Central Committee members in Jajce, the perfidy of the sons of Albion grew in proportion to their distance from us. At any rate, one thing is certain: it never occurred to the Central Committee, that is, the Supreme Staff—to submit to any command, even a Balkan one. As for Vukmanović-Tempo, as soon as he received the Central Committee's letter, his revolutionary zeal in gathering all Balkan movements under one hat left him. The Central Committee insisted on co-operation, but also that the movement of every people remain independent, within the framework of established state boundaries.

We no longer bothered to take guesses as to how long the war would last, probably because its end was, so to speak, physically palpable. This was perhaps even more obvious on our domestic front than on foreign fronts. Of course the Chetniks continued to be of the utmost importance to us, because of the support which they still enjoyed in the West, which could make them into a bastion of anti-Communism. But after the fall of Italy their confusion and disintegration took on fatal and cataclysmic proportions. The death sentence which they pronounced on themselves, and on Serbian bourgeois nationalism, by their collaboration with the enemy, was being methodically executed by the Partisans. And now, when their Western allies were within close range—when finally their time had come to beat back the invader—the panic-stricken and scattered Chetniks sought refuge with the Germans! And the Germans, plunging into the pit that they had dug for others, grasped at the Chetniks as auxiliary units under their control.

In Montenegro the fall of Italy left exposed to the Partisans the main Chetnik strongholds in Vasojevići and Kolašin: the Partisans were now holding rallies there and organizing new units. The same was true in

Hercegovina and Sandžak. Moreover, a group of Italian prisoners of war in Montenegro was organized into a small but compact Garibaldi Division which, according to Dapčević, demonstrated high qualities: it is ideas, not national traits, that lead men into either mortal combat or a miserable existence.

Cries of joy greeted a dispatch from Montenegro that the leaders of the Montenegrin Chetniks had been executed, including Commander Bajo Stanišić—the same Stanišić who in July 1941 rejected my offer to command the uprising. All this took place at Ostrog, the Montenegrin and Orthodox holy place which all faiths and ideologies revered, each in its own way. During one of our postwar journeys across Montenegro, Blažo Jovanović told me that the surrounded Chetnik leaders gave themselves up following negotiations and promises that they would be tried after the war. The negotiations on behalf of the Chetniks were conducted by the lawyer Bojović, who had personal confidence in Blažo Jovanović because Blažo had once worked for him as a law clerk. Twenty-three Chetniks surrendered, and the Partisans shot them all, including Bojović, who, according to the accounts, had joined the condemned. The only one not to give himself up was Stanišić, along with three of his kinsmen. But a Partisan machine gunner spotted him at a window and mowed him down. His kinsmen committed suicide. The Chetniks might have held out: Ostrog monastery, which is built into the cliff side, can be reached only by a narrow flight of stairs, and the Germans were advancing at the time; in fact, they arrived there a day or two later. Having satisfied my curiosity with that bloody account, Blažo Jovanović asked me, "Well, what do you think? Did we do the right thing in not keeping our promise?"

I couldn't answer him immediately, not because the Chetniks kept promises they made to the Partisans—they didn't—but more out of an inherited respect for one's word of honor. At last I replied, "What else could you have done? You had no prison. You could hardly have dragged them along with you for months in the hills. You would have received an order from the Supreme Staff to do what you did anyway."

Scattered on all sides, the Chetniks pulled themselves together and made for Serbia. This was confirmed by a report from Blagoje Nešković, party secretary for Serbia, which we read over and over again, particularly Ranković and I, as though it were some ancient tale of a massacre taking place then and there in our homes. The Chetniks knew they weren't lost as long as they didn't lose Serbia, just as we knew that we wouldn't win unless we won in Serbia. In Serbia itself a methodical

German and Nedić reign of terror in the towns had combined with an unrestrained Chetnik reign of terror in the villages. Partisan units were reconstructed with difficulty, though in southwestern Serbia, occupied by the Bulgars, our units maintained themselves the whole time. The tyranny of the Chetnik commanders and leaders, of the black troikas and the wanton riffraff, far exceeded the Communist danger and Partisan strength, and took on insane forms and proportions: whole families were slaughtered, and in some districts tens and hundreds of "Partisan" women. The Chetnik leadership was obsessed from the start with "Kerenskyism"—that is, moderation; in 1942 Draža Mihailović had reproached Bajo Stanišić for it in Montenegro, as if the revolution could be averted by moderation or ruthlessness—as if the Russian Communists hadn't triumphed because of their slogans against the war, while the Yugoslav Communists grew ever stronger because of their summons to war.

That bloody nightmare was quickly rationalized in Western evaluations: Serbia was Chetnik and western Yugoslavia was Partisan. This aroused our anger, but also the realization that such an evaluation threatened us with new and far-reaching dangers. Tito indicated to the staff in Montenegro that the penetration into Serbia was their most important task.

The Central Committee members lived in fairly close quarters, and they ate together. But decisions were made at meetings and not haphazardly, as had frequently been the case previously. Top assignments were made as the need arose, but in accord with personal inclinations and previous experience. Thus Ranković, who had managed cadres and illegal connections, took on intelligence and investigatory matters as well. Kardelj and Pijade occupied themselves with preparing the AVNOJ session, and I with the press and culture. Among other things, I had to straighten out the situation in the Theater of National Liberation, which became poisoned with intrigue and rivalries as soon as its members resumed their normal city life and their professional activities. Also, the Central Committee assigned me to deliver the address at the anniversary observance of the October Revolution.

This was the first time that the Russian Revolution was commemorated in our country in this fashion—at a gathering of the highest officials and distinguished guests, with a political report and a cultural program. Celebrations of special events were customary with us as well, but an address which analyzed the past and outlined future tasks was on the Soviet model.

In any case, I needed a good deal of time and thought to put that speech together, because I had to combine standard Soviet Leninism with our own blood-drenched paths and hopes, and to expound on this. Tito and Kardelj looked the speech over and made minor comments. The basic line had been set before I began writing: an enthusiastic identification with the October Revolution, but also an analysis of and emphasis on our own conditions and problems. And no criticism of the Western Allies, except that certain circles were supporting the émigré government and proposing a landing in the Balkans as a second front.

Everything was ready for our ceremonious gathering on the evening of November 6. The news from the Eastern front accidentally provided the occasion for the pathos of my speech and an inducement to applause and hurrahs: after the final collapse of the German offensive in the Kharkov salient, the Red Army was advancing in the Ukraine; the fall of Kiev was imminent. Tito's radio received Moscow poorly, but that evening I pressed my ear against the set and made out from the well-modulated sonority of the announcer that Kiev had been liberated. I rushed out to the rampart—we were still stationed next to the little underground church below the fortress—and fired three shots from my pistol. The signal was understood: the firing spread through the little town, soldiers embraced one another and danced, and some distance away an artillery piece began to thunder. Though overjoyed, Tito remarked, "A lot of ammunition was just used up!" As for me, it never occurred to me that my rejoicing had roots in the same consciousness that had prompted that peasant in Montenegro to be truly dumbfounded on hearing that the Germans had occupied "holy Kiev."

My speech contained all the important positions of the forthcoming session of AVNOJ. This was the first time that these positions were publicly stated. All of them proceeded from conversations and discussions in the Central Committee; I only formulated them.

Yet for me today, the significance of that speech lay perhaps even more in the fact that it combined analytical and categorical conclusions with an unrestrained pathos. I was probably influenced by the session in Kočevje. At any rate, there was a decisive finality in our positions concerning the changing of the political system in Yugoslavia. Despite suffering and losses beyond measure, the attainment of that aim seemed much easier now that it was our will that determined the course of events. More than any of my previous speeches, this one, which was more intellectual and profound than my presentation in Kočevje, evoked a euphoric applause which caught up even Tito and Kardelj. The faces were hazy, and in the dimly lit little hall in which the Second Session of AVNOJ was about to take place, the walls receded from view as if nothing existed in the universe except this determined and alluring intoxication with supremacy and triumph. Everyone felt relieved and a little tired, as after some exceptionally exhausting pleasure. I took special pride in those stormy outbursts which swelled into wave after wave of applause, even though a little corner of my consciousness or some dim perception cautioned me that I was creating a spiritual uniformity: the manipulation of fervor is the germ of bondage.

That dualism, that reality which shaped our beliefs and our undertakings, but which didn't promise freedom even for ourselves, did not disappear; it was only covered over by the war and by illusions. The glitter of success and the endless measure of daily tragedies turned doubts and uncertainties into deceptions of love and strained hopes. How was it possible that the world which would arise out of such boundless tragedies and unassailable truths could be one of domination and betrayal?

It was around this time that Dedijer had to fly to Cairo, to have English doctors extract a piece of shrapnel from his head. Before his departure he left his *Diary* with me for safekeeping. And he wrote in his *Diary* that I told him I could live in no country except Yugoslavia after the war. A revolutionary war makes one's love for the people and for their happiness absolute.

The Central Committee held several meetings in preparation for the Second Session of AVNOJ; the most important of these took place in late October. It decided on the federal organization of the country, the creation of a provisional government, and a prohibition against the return of the royal government and the King. That meeting was held one sunny afternoon in front of Tito's little barracks, next to the entrance into the underground church. There were some differences over whether it should be a regular or a provisional government, whether to overthrow the King immediately or just to prohibit his return. However, all of us agreed with Tito to adopt the more moderate transitional course. Kardelj and I had fully supported Tito's view. It was the Serbs of Serbia—Žujović, Pijade, and Ranković through his silence—who were the most radically disposed.

The most important reasons cited for this more moderate course were relations with the Allies, primarily the British and the Soviets. The thinking behind this was as follows: we should make it easier for the British to break with the King and the government-in-exile; at the same time, we shouldn't complicate the Soviet government's relations with Britain. The latter was the more important for us, in view of our ideological dependence on Moscow. Yet there were also internal reasons for such a course: the broad masses, particularly the peasantry in Serbia, were not yet ready to renounce monarchist traditions. Radical decisions had already been carried out with the creation of a new authority; any lag in forms corresponded to the lag in consciousness of the masses, and to our need for the legal mobilization of soldiers and for international recognition.

The same reasons prompted the decision that a final determination concerning the King and the monarchy be made in postwar elections. The Western Allies had promulgated the Atlantic Charter, providing for the free choice of the form of government, and we were satisfied with a prohibition against the return of the King. Actually we weren't in any doubt about the final fate of the monarchy, or about the results of an election which would be held under our supervision and on the ruins of the old regime. The moderate course with its transitional forms was inspired by our desire to make it easier for our people, but also to establish and legitimize our own new regime.

The name "National Committee for the People's Liberation of Yugoslavia" was also decided at that meeting, with due regard for the reaction of the Allies. In pondering what name would be the most suitable, Kardelj recalled that the French of de Gaulle had adopted the name "National Committee." Tito liked that at once, so it was adopted—along with our addition of "people's," since with us everything had to be of the people.

At this important meeting there were observations that the Russians wouldn't understand, that they wouldn't approve all our decisions, particularly those concerning the King and the royal government. A day or two after the close of the meeting, Tito remarked—Kardelj, Ranković, and I were there—that we should not keep the Russians informed of all our decisions, because they would be opposed and would undermine the entire session. So Tito informed Dimitrov, which is to say, the Soviet government, that a provisional government would be formed. He did this relatively late, on November 26, only three days before the session of AVNOJ, while passing over in silence the divestiture of the royal government's legality and the prohibition against the King's return. The new regime started out by breaking with the old regime—and by being "unfaithful" to its spiritual fathers.

Tito hurried the session along, partly because of conjectures that the royal government and the King might take advantage of the meeting of foreign ministers in Moscow and return to the country, though it wasn't clear where they could establish themselves without becoming allies of the Germans or prisoners of the Partisans. The session would doubtlessly have been held earlier, had the delegates from Montenegro and Slovenia been able to come.

At one of these meetings Pijade put forward a proposal for the territorial autonomy of the Serbs in Croatia. He had just returned from Croatia with a bagful of statistics concerning the number and distribution of Serbs in the Croatian lands. He even had borders marked on a map for this Serbian autonomy, comprising districts with a Serbian majority in Lika, Banija, and Kordun. On Pijade's map this territory looked as twisted as an intestine, and failed to include the Serbs in Slavonia as this would have meant incorporating numerous Croatian districts, too. The idea was a new one, and the Serbs in Croatia had served valiantly in the uprising. Everyone was silent, perplexed. I think I saw dejection even on Tito's face: perhaps, as a Croat, he found it awkward to oppose the idea, perhaps conflicting thoughts clashed within him. I was the first to come out against Pijade's proposal: the segregated territory was unnatural, lacking a center or viability, and moreover provided fuel for Croatian nationalism. Kardelj immediately agreed. Pijade had the reputation of being the most zealous Serb among us, but Ranković squelched him by remarking that the Serbs and Croats were not so different that every district had to be divided. Tito calmly accepted our stand, while injecting the class motivation: "With us this will be more of an administrative division, instead of fixed borders, as with the bourgeoisie."

The party had heretofore held the view that Bosnia and Hercegovina should have autonomous status, rather than become republics. This plan implied autonomy under the Republic of Serbia. However, the war had turned Bosnia into a battleground between feuding Ustashi and Chetniks, as well as a base and haven for the Partisans. Autonomy under either Serbia or Croatia would have encouraged further strife and deprived the Moslems of their individuality. The Bosnian leadership, too, like every authority that grows out of an uprising, insisted on their own state, and later even on their own historical outlet to the sea. But the republican status of Bosnia and Hercegovina was not decided at that time, or during the session of AVNOJ. It was decided in early January 1944, at a meeting during a march, after the retreat from Jajce. Ranković reported that the Bosnian leadership proposed a republic, Tito agreed, and so did all the rest of us, as if this were something acceptable on the face of it.

Delegates to the Second Session of AVNOJ were chosen by republican assemblies and local authorities. But all of them were verified in one way or another. Party agencies made an effort to include well-known non-party people, patriots who didn't question our aims or methods. The majority of these patriots eventually became party members, and some were members secretly already, but the Central Committee chose the delegates for Serbia from military units. As for Macedonia, since contacts with it were poor and information concerning cadres lacking, two delegates—Poptomov and Vlahov—were designated on the suggestion of the Bulgarian Communist Shteriu Atanasov, known as Viktor, who had landed in our midst on the way from Moscow to Bulgaria.

Except for the suggestion that Poptomov and Vlahov be invited, which we accepted out of ignorance—and Poptomov never did come to AVNOJ, preferring to stay in the Bulgarian party in Bulgaria—neither Viktor nor anyone else had any influence on the work of AVNOJ. Atanasov, as it turned out, was a combination of a romantic Bulgarian and an NKVD man; the more zealously he served the NKVD, the more fervently he imagined that he was carrying out some exalted assignment. Tito knew him slightly from Moscow, and didn't have a high regard for his intellectual capacities. I became convinced of this after I had read one of Viktor's reports in which all his simplicity was lost, and only semiliterate phrases learned by rote in party schools were left. Perhaps it is worth recording a detail, related to us by Viktor, concerning the prisoner exchange that involved the Rumanian Communist Ana Pauker. The Soviets didn't have a suitable prisoner for the proposed exchange, so they arrested a renowned Romanian professor, an old man who not only wasn't guilty of anything but had no desire to leave his work and comfort. Nevertheless they brought him to the border with the promise that he might later return, and in exchange for him they got Ana Pauker.

The greater part of the work in formulating the resolutions of the impending session of AVNOJ was done by Kardelj. He was helped by Pijade. The future parliament found its man in Pijade, and Pijade found himself in it. Pijade's lively and imaginative intelligence seemed to come to practical expression with AVNOJ and later with the Assembly. My part in the Second Session of AVNOJ hardly went beyond stylistic and editorial revisions of Kardelj's texts.

The rapid strengthening of our position prompted the British to accept a military mission from us. In view of British tendencies to get rid of compromised royal governments, that mission had a greater political than military significance. The Partisan army was thereby formally recognized as an adversary of Germany. Paving the way for the recognition of the insurgent regime was more important than gaining direct military aid.

At first it was expected that Vlatko Velebit, who had meanwhile mastered English, would be chief of the mission. However, at that time Lola Ribar felt that he was too mature to continue as secretary of the Communist Youth, and was ready for a new assignment. He first raised the matter in a conversation with Ranković and myself, and later the two of us persuaded Tito and Kardelj. Lola also knew English. Thus he was made chief of the mission. Velebit took it well because he wasn't vain; besides, Lola was a Central Committee member.

The mission was to take off on November 27 from Glamočko Polje in a Dornier 17, two days before the Second Session of AVNOJ. This plane had been brought to our territory by several Home Guard aviators who had deserted to our side. Just as the mission was piling in, a Storch reconnaissance plane swooped down and attacked with bombs and machine-gun fire. Lola jumped out of the plane, but at that instant a bomb exploded and killed him. Two British officers and a Partisan were also killed. So was the pilot, inside the machine-gun turret. Velebit just barely made it out of the tail of the burning plane. Lola's body was scorched by the heat. It was never established whether the enemy knew beforehand of the mission's departure. Had they known, the attack would probably have been carried out by faster and better armed planes. In all probability the Storch was looking for the abducted plane and happened to fly over just as the mission was boarding. Five minutes later and the mission plane would have been airborne and the Storch powerless, or indeed its victim.

Lola died in the morning, and we in Jajce received the news the same day, around noon. Silence and sorrow gripped all the Central Committee members. The news spread in no time, though horror and disbelief stilled any outcry or commotion. Lola was the youngest of Tito's close entourage and the first to be taken from it, and on an occasion for which all of us had worked and which we felt to be historic.

Just as in some ancient tragedy, a historic event became intertwined with a personal tragedy. On the day Lola lost his life, his father Dr. Ivan Ribar, president of AVNOJ, arrived from Slovenia with fresh impressions, eager to talk about the forthcoming session. A month earlier Dr. Ribar's younger son Jurica, a painter, had lost his life in Montenegro. In consultation with Lola, we had decided not to tell old Ribar about

Jurica's death; it had been left to Lola to do so at some suitable time. Now Lola also was dead. There was no reason or possibility to conceal from the father the death of his only children.* Strangely enough, the fact that the Ribars came from a well-to-do bourgeois milieu only evoked in us Communist leaders an even greater sense of tragedy. Life and history had really played a game with them. In 1920 old Ribar had been the president of the Constituent Assembly that had approved the ban on the Communist party. Now his sons had fallen for the Communist cause, and he was president of the assembly that would legalize the power of the Communists.

On the same day, at dusk, at a meeting in Tito's little barracks, the matter of informing Dr. Ribar of the death of his two sons came up. Kardelj said it would be the most appropriate for Tito to do this; the rest agreed. And Tito said, as if summoning up courage for a major decision, "All right, I'll tell him."

As soon as the meeting was over, Tito invited Ribar and told him that Lola had perished that same day. At this the old man said, "Does Jurica know? Have you told him? It will be hard on him."

Tito took Ribar by the hand and said, "Jurica too lost his life—a month ago in Montenegro."

Dr. Ribar stayed with Tito for quite a while. Supper wasn't served. Everyone grabbed a bite as best he could. No one expressed sympathy to Ribar; our muteness and dejection were sufficiently eloquent for us and for him. Though during the war we weren't as close as we had been earlier, perhaps of all who were in Tito's circle, Lola's death hit me the hardest. Lola often came to mind later, especially in prison, sometimes joining me or sometimes standing to one side, but never condemning my break with the party—in accord with my moods and with the Lola who endured inside me. His father too felt the special closeness between Lola and myself; after the war it was to me that he turned most frequently regarding unofficial matters, his memoirs, little things concerning his adopted daughters.

Again at dusk—I think it was the next day, November 28—Kardelj informed us at a meeting with Tito that the Slovenian delegation would propose that Tito be proclaimed a marshal. Tito turned red in the face, and stood up as if to think the matter over while pacing: wasn't that too much? And might not the Russians be offended?

We immediately discounted Russian displeasure: after all, we too

* Dr. Ribar's wife had three more children from her first marriage who were reared in her second home. She was executed by the Germans.

could have a marshal. As for the magnitude of the honor for Tito, this we didn't even discuss: if anyone deserved it, it was Tito, and the army and the movement would thereby gain a new "Soviet" head, as a counterbalance to the traditional monarch. I believe it was Josip Vidmar who conceived the idea of proclaiming Tito a marshal.

Most of the delegates were assembled by November 28; the Montenegrins had arrived a few days before. The briefing of the delegates on the draft resolutions had been completed in a shelter near headquarters by the morning of November 29. Every so often the Central Committee members flocked around Tito to sort out the delegates' comments with him and with one another. In the discussions with the delegations, Kardelj was the most active. Criticisms were few, but some of them were important. The majority of delegates were disposed to have the republic proclaimed at once, and had to be dissuaded—particularly those from the former bourgeois parties. At one point Kardelj, who was enthusiastic over the course of these councils, whispered to me, "Proclamation of the revolution!"

I said nothing in reply; I shared with him a belief in the historic significance of the session, but looked upon the session's resolutions as a way of legitimizing an existing state of affairs, and not as a turning point. Koča Popović characterized the discussions with the groups of delegates in his own way: "Of course, politics are conducted in corridors."

And indeed, at the session itself on the evening of November 29, there was no disagreement, no real discussion. There was a rapt attention which kept straining to explode into applause and the shouting of slogans, tumultuous rhythmic outbursts, which subsided in pleasurable relaxation. The delirious clamor was intensified, and sometimes started, by the young people, staff officers, and officials who stood along the edges of the crowded hall.

Only Tito's speech introduced a businesslike, rational note, though it too was interrupted with applause. Tito read more fluently than usual, with a natural, easy confidence. In his manner and particularly his voice, one felt a certain dignity, even tragedy. His speech was moving; it had seemed a bit dry to me when I first read it, but that turned out to be its virtue; the summation of a two-and-a-half-year war and civil war whose horrors nobody could have imagined, but whose end was now in sight.

I was also a delegate. I sat to one side, on the right of the stage. Though I followed the proceedings of the session, and particularly Tito's speech, I could not get Lola out of my mind, sealed in a tin coffin and lying in the chapel of the Jajce graveyard, nor did my eyes ever leave his father in the president's chair, whose deliberate movements and calm I interpreted as petrification and numbness. I was sure that all the delegates felt this, at least those who were close to the leadership and the Ribars, though nobody mentioned the latter.

However, even I forgot about the Ribars in that part of the session that proclaimed Tito a marshal. It was this moment that aroused the most intoxicating unanimity, even though everyone knew about the proclamation and we were all inwardly keyed up for that unanimity. Though I had myself stirred up such collective raptures and had felt enraptured myself, I was never carried away by them; my consciousness always remained intact. As a matter of fact, I think I became more self-controlled and cool as the expressions of unanimity became more frenzied and stormy. I applauded zealously and enthusiastically. Only later—but frequently then—would my prolonged applause, like that of many others, be prompted by fears that someone might think I was against the party and Tito. But I wondered at the time what things would be like within the Central Committee and the future government, now that Tito had been made a marshal. Tito had always treated the Central Committee members as his collaborators; comradeship and personal dignity never suffered, while Tito's short temper in discussions seemed unimportant, since there were no essential differences among us. Yes, the army and the government, and the creation of the new Yugoslavia—all benefited from the higher standing of an individual who was already playing the principal role, and whose capability and independence were proven. But what did this mean in terms of relations, particularly since the bearer of the highest title was inclined to personal power, and the title was being introduced just so that he alone would get it?

Following Tito's election, the presidium withdrew behind the curtain of the stage. Only the new national coat of arms remained above the stage. Actually it was not yet the coat of arms, but everyone knew that it would be. Djordje Kun had designed it, partly on the Soviet model, partly out of his own head, and partly on the suggestions of Central Committee members. That coat of arms, with Pijade's later additions, became the national emblem, while the date of the session—November 29, 1943—became the national holiday of Yugoslavia.

I too was on stage behind the curtain, though I was not in the presidium: Ranković invited me, or else I just went along to congratulate Tito on his election as marshal. There behind the curtain, congratulations were being eagerly extended. With the arrival of Ranković, myself, and other functionaries, the eagerness turned to an elation which was all the more unrestrained in that we were all crowded together. We embraced and kissed Tito—first we Communist leaders, then the nonparty patriots, who could hardly avoid it. The enthusiasm gripped Tito as well. The ecstasy became intoxication. We Communists continued to hug and kiss Tito without any measure of order, while the nonparty members watched us with an incomprehension that verged on bewilderment. Caught up in the joyous, frenzied vortex, Tito returned every greeting. Eyes glistened, chests heaved, sweat streamed over flushed faces, hearts were full, minds feverish. I too was absorbed by the delirium, conscious that my own future was being decided then and there, that by my surrender to this frenzy I was willingly accepting Tito as my ruler, my master, despite my ideal and desire for a world without masters, despite my own integrity and my own vanity.

It is strange today to hear anyone say that Communists—some among them, at any rate—didn't want Tito for their leader. An idea, a party, or a government finds a leader who is appropriate to it, and in them the leader finds himself and his own creativity. Without doubt Tito was not only glad to accept the leadership, but insisted on it. Yet his popularity and his popularization began before he was chosen marshal, and before his popularization became the regular responsibility of party functionaries and party apparatus. This process first started within the party—and among the people and lower party ranks, before the party began to promote it. Imitation of the Soviet Union played a certain role, though not a decisive one. An ideological movement and an insurgent people felt the need for an infallible leader and benign protector.

We in the leadership were aware that the popularization of Tito gave strength to the movement, and even to those closest to him, so this became a charge on us, or at least an obligatory ritual. We were all devoted to our leader, each in his own way, and for the sake of his own position in the party. Kardelj had become bound to Tito when they were training themselves and others in Moscow in 1934–1935. True, Kardelj considered himself the superior theoretician. If theory is understood as fitting new realities into old truths, then he was undoubtedly superior. But Tito was so overpowering as a personality and as a leader that in the 1950s, when differences arose among us, one had the impression that Kardelj was physically afraid of Tito. As for Ranković, he was unconditionally devoted to Tito, sentimentally and idolatrously. And I? I have already described my feelings about him: from time to time, however, a betrayed idea rebelled within me: the sense of my own unworthy role. If a heretical notion crossed the mind of any one of the three of us—and this happened most often to me—the other two would correct and dissuade him. The other leaders were in a far less promising and enviable position: they could only envy Kardelj, Ranković, or myself for our special closeness to Tito.

The delirium of congratulations to Tito behind the curtain lasted for quite a while—maybe twenty or even thirty minutes—until someone remarked that the session should continue. We returned to our seats tired, and perhaps even inwardly ashamed; at least I felt that way. All the resolutions were accepted with enthusiasm and tumultuous accord. The new government and the new informal chief of state were established with forethought and method, and proclaimed with hopes for final happiness and freedom.

The next day, on the afternoon of November 30, we gathered in the cemetery to escort Lola's coffin. It was gloomy and chill. We had agreed that Pijade should speak: the oldest Communist leader would eulogize the youngest. Pijade was distraught. His voice was hoarse, and his thoughts scattered as never before. But from his erudition he pulled out the unforgettable sentence of some Frenchman: revolutionaries are dead men on leave. Old Ribar was given to oratory: he couldn't resist even this occasion. But he only uttered platitudes in a somber voice—our struggle is a hard one, our struggle demands sacrifices, nothing can stop our struggle—as if straining hard to attain each one. Finally, like any father he blurted out, "Farewell Lola, farewell my sons!" Lola was buried in a secret spot so the enemy couldn't desecrate his grave.

Along with many others I burst into tears, while life and the war went on even more intensely, though now on an official course.

Facing the Outside World

1

The Central Committee and the provisional government—the National Committee—vigilantly awaited reactions to the Second Session of AVNOJ. The first positive response came from the West, from the Associated Press and the *New York Times,* even though Moscow was the first to have been informed, directly and in detail. Tito had also informed Moscow of the introduction of the rank of marshal, though he passed over in silence—so as not to arouse Kremlin jealousies—the fact that it was he who had received the rank.

It was more awkward than strange that Free Yugoslavia, the Yugoslav radio station in Moscow, did not report the decision which forbade the King's return. Veljko Vlahović, the Yugoslav party representative in Moscow, later told us that all the news which the Yugoslavs prepared for Free Yugoslavia and Radio Moscow after the Second Session of AVNOJ was placed under strict censorship. Moreover, Dmitri Manuilsky let Vlahović know that the "boss" (*khoziain*)—Stalin's name in the inner circle—was extremely angry, and that he viewed that decision as a stab in the back of the U.S.S.R. and the Teheran Conference, which happened to be meeting at the same time as the Second Session of AVNOJ.

We Communists had grown used to Moscow's caution and slow response, but it was difficult to explain this to the patriots in the government and in AVNOJ, many of whom expected a more active and unselfish support from Moscow than did we Communists. But we Communists also sensed and knew that Moscow's support would come, even if with delay; over there they could be angry with us only as they might be with their own people. Indeed, the awkward situation soon ended: in the middle of December 1943, the Soviet government issued a communiqué expressing support for the Second Session of AVNOJ, and announcing that it would send its military mission to Yugoslavia.

As for the Teheran Conference, we regarded it—and to this day I maintain that we were right—as confirming our independent Communist decision-making and the decisions of the Second Session. The Teheran decisions spoke of armed assistance to the Partisans in Yugoslavia "to the greatest possible extent." For us, that signified the end of Western support to Mihailović, and also the gradual abandonment of the King and his government, since what were they without an army—particularly in a civil war? True, the Teheran decision concerning Yugoslavia also spoke of support to "commando operations." But that didn't worry us too much: commandos are small, specialized units inadequate for anti-Communist intervention, even if they should land here and there on special assignments.

In early January, having managed to save Banja Luka, the Germans advanced on Jajce. The force of the German offensive against the Supreme Staff was not nearly that of the Fourth and Fifth Offensives (*Weiss* and *Schwarz*). According to the calculations of the Supreme Staff, the Germans no longer had the strength or the time for extensive encirclements; rather they were striving to block the growth of our units and ward off our pressure on important regions. However, this doesn't mean the offensive was not "Teutonically" stubborn and severe in some areas—for example, in eastern Bosnia, judging by the accounts of participants.

On the evening of January 7, we evacuated Jajce in an organized and calm fashion, though with some trepidation. These marches hindered the orderly work of the newly established higher authorities. Moreover, the National Committee and the presidium of AVNOJ were partly scattered, so that the decision-making was again left, even formally, to the Central Committee and the Supreme Staff. Nothing much happened in the course of our withdrawal through the forests and across the mountains on the way to Drvar. Then one day the snow-covered hills and the blue infinity above them resounded dreadfully and monotonously: hundreds of Allied planes, gilded by the sun, were flying serenely northward to Germany. Our column stopped, amazed, enthralled. I knew that the tense drone heralded destruction, and the butchering of women and children. But this was a realization devoid of any pangs and sorrow, as if the Germans were no longer human beings. I too rejoiced at these death-dealing flocks that were so unerringly and inexorably bent on violence and hatred.

One night, on a plateau covered with snow and mists, we ran into "Malčik," the chief of security for Croatia. The Central Committee members knew that Malčik had been recalled because his men had begun to spy on our own committees and staffs, thus giving rise to suspicions and ill will. The internal language of these agents used a special contemptuous terminology for our generals and committee members. Those who knew Malčik in the U.S.S.R. recounted that he had worked in the Soviet security, and so was made-to-order for the same service with us. But he had brought with him the Soviet methods, which didn't suit our innocent comradely relations and our view of the party as an immune, sacrosanct institution. Out of nowhere in the middle of a winter night, that power-loving, satanic potentate appeared before us, in the form of a scrawny, frozen manikin that could barely hang on to his horse. Only Malčik's uniform was new—and of course, strictly regulation. And as Malčik, dead tired, slid off his mount, his submachine gun got caught in the stirrup; as the horse jerked, the gun fired several short bursts. People scattered across the field and behind a house, while I grabbed the machine gun with one hand and helped Malčik dismount with the other. A fellow from Serbia, a member of the Escort Battalion—one of the few survivors of the raids of the winter of 1942, and therefore all the more unforgettable—dropped the submachine gun in the snow; the butt struck against a rail, the unsecured bolt slipped, a bullet ripped through the young man's heart, and he dropped dead right next to the members of the Central Committee. Even so, Malčik was kept for some time to work on the Supreme Staff; the party leadership had placed the security service in Croatia under its direct control.

It was in those days that Churchill's son Randolph joined the British mission. We of course felt honored, though it did occur to us that Randolph might be the gray eminence of the mission. But he himself convinced us by his behavior that he was a secondary figure, and that his renowned father had decided on this gesture out of his aristocratic sense of sacrifice and to lend his son stature. Randolph soon enchanted our commanders and commissars with his wit and unconventional manner, but he revealed through his drinking and lack of interest that he had inherited neither political imagination nor dynamism with his surname.

After a fifteen-day march we settled down in Drvar; all kinds of ever more numerous auxiliary units came tagging after. During the day Tito would withdraw into a cave on the right bank of the Unac River. It was in that cave that a German raid found Tito on his birthday—May 25, 1944*—so it now bears his name. Soon after, they built him a little wooden hut at the entrance, and he spent his nights there as well. The little house below the cave belonged to Ranković and myself, and occasionally to other Central Committee members and commanders as they came to report.

Time went by for us uneventfully, in peaceful work and easy mutual warmth such as we had experienced, even in warfare, only on rare occasions. Ranković reacted to some insult of Zdenka's by a complete break: he never spoke another word to her. This was unpleasant for Tito, though he didn't let it stand in his way or ours: every evening Ranković and I would climb up to Tito in the cave, and if we were late for any reason, Tito would send a courier for us. There we most often played chess and chatted, while the sheet metal stove popped, and an aide turned the dial in search of Free Yugoslavia or the BBC; even though they trusted the Soviet government more, the Supreme Staff preferred to listen to the BBC rather than to Radio Moscow, because of its fresher and more accurate news.

Tito, Ranković, and I were about equally bad chess players. Sometimes Tito beat me and sometimes I beat him, while Ranković and I most often ended in a draw. For a while I was beating Tito steadily. Tito endured defeats without resentment. Nevertheless, Ranković remarked to me that it made no sense, since the Old Man got upset and slept badly. I refused to throw games—I didn't even know how to do it! I agreed to have Ranković play instead of me, for he lost or played draw games so convincingly that I myself wasn't sure whether he was faking or not. The worst player of all was Arso Jovanović. He was absent-minded and hasty, and lost dozens of games one after another. Once Koča Popović also lost against both me and Ranković, and commented ironically, "Rather awkward . . . the top commanders!"

During one of those happy evenings Tito informed us that he would confer the rank of lieutenant general on us Central Committee members. I asked him, "Why so high?"

He replied, "Well, no one should outrank Central Committee members."

* Tito was not in fact born on May 25, so Dedijer, who researched this, has told me. Tito held that he was born on that day, and since it began to be celebrated on May 25, he considered that it ought to remain that way. Any difference in dates must be insignificant—a matter of a few days.

Only Kardelj and Pijade were not given ranks, not because they were insufficiently military, but so the National Committee and AVNOJ wouldn't be overloaded with generals: Kardelj was on the National Committee, and Pijade was vice-president of AVNOJ. After the war, during the clashes with the U.S.S.R., they too were given rank, as were the other Politburo members.

I was in charge of the political periodical *Nova Jugoslavija* (New Yugoslavia), along with Zogović and Mitra. This was a substitute for *Borba,* conceived to reflect the new government; it was literate and firm in its stands, and almost devoid of dogmatic and party verbiage. In one article I portrayed Maček's passivity as comfort for the enemy and the Ustashi. In another I defended Zogović from attacks that his poem about Tito copied Mayakovsky. This literary critique of mine—like others—was in fact political: a defense of one "truth" or another, beyond the pale of literature.

Hierarchical relations began to take on a hardened external form at that time. Until then, at public occasions and meetings care was taken only where Tito and Ribar would sit. Also, everyone could see Tito or the Central Committee members if they weren't busy. Then appointments began. There were dinners with missions, and the seating arrangements brought us unprecedented headaches. It all fell on Ranković because of his gift for detail, and his involvement in organizational matters. He in turn dragged me into it, more because he felt unsure in matters of protocol than because the job had a propaganda aspect, too. Late into the night we would move all sorts of functionaries around diagrams of tables and halls—tables made of boards slapped together and covered with bed sheets, and halls with floors reeking of kerosene! When we had put it all together we went to Kardelj, and sometimes even to Tito. One suggestion from them would throw everything out of whack, and impose more hours of stultifying rearrangement. Although there was no formal state hierarchy, an unofficial political hierarchy existed. Worse yet, functionaries began to get sensitive about rank and about the places assigned them. The most sensitive were the women comrades in the lower echelons, who would be separated from their friends or husbands and relegated to the impersonal middle of the hall. They were particularly unhappy if some "meritorious" woman were given a place of honor in the front two rows, or invited to a formal dinner. Yet none of us had a single change of underclothing, and we all ate tasteless stews, were bombed every day, and lived in constant fear that we would be driven into the woods. But power and authority impose their forms in misery as well as in luxury. Ranković and I developed a protocol more out of concern lest we offend some patriot than that we might do an injustice to some party functionary, since we could always appeal to the latter. I also once found Tito in the cave practicing his signature—as befitted his new role as a ruler, and the atmosphere of titles, honors, and idolatry.

Still, there was some vague and not very determined resistance, especially among the comrades who worked with me in the propaganda section. Zogović was incurably contemptuous of social climbing and pomposity, while Mitra was inwardly resentful of my neglect, though we lived only five hundred yards apart. Yet even those two suppressed their discontent and "understood" after cordial ideological explanations.

Disputes with the Central Committee of Croatia, on the other hand, were an extension of deeper disagreement. I pointed out to Tito and Kardelj what the Croatian press was writing about Yugoslavia as an artificial creation of Versailles—what state is not artificial?—at the very time when blood was being shed to preserve it. A critical message was sent to the Central Committee of Croatia, then Kardelj went there himself—but without any real results. Men's minds are the most incorrigible when they serve "missions" and when they are carried away with power.

The dispute with the Central Committee of Croatia was an important reason, though not the only one, why the Central Committee of Yugoslavia on January 30, 1944, sent a directive in the form of a letter to all party organizations, regarding relations with the Western Allies. That letter was signed by Tito, but it was drafted I believe, by Kardelj and myself: I wrote it and he edited it. That letter pointed so firmly and in such detail to "various blunders" which might create "serious difficulties for the new state which we are creating," that it signaled a turning point: ideology was accommodated to reality, to the West; though capitalist, the West was not inevitably the enemy. An end was put to the poorly supervised newspapers of local bodies. It was ordered that all our allies, including those of the West, be mentioned at public meetings, and that their social orders not be criticized. It was also ordered that a change be made in the tone of "the press and agitation, in the sense that all questions be approached from the standpoint of a newly created independent state which is nobody's affiliate, but the work of the struggle of our peoples." Regarding the King and the monarchy, the directive called for adherence to the resolutions of AVNOJ and "not lending the King a greater importance than he had."* This was, without doubt, political

* *Hronologija*, p. 647.

state centralism. But it was also an act of national political independence from Moscow as an ideological center.

That letter, with its sharp delineation of positions, was welcome because of the impending arrival of the Soviet mission, though it hadn't been prompted by that arrival but rather by a change of views at the top. On February 23, 1944, the Soviet mission was greeted with a happy curiosity, though the manner of its descent didn't impress the Partisans: the mission had delayed its arrival by several days so it could land more safely in gliders, instead of dropping by parachute like the British. However, that was explained away by the "wounded" foot of the chief of mission, General Korneyev.

Nor did the initial reception of the mission at Tito's go as cordially as we had hoped: the chief of mission and his subordinates behaved with an almost cold formality which we attributed to apprehension that someone—though only we Politburo members were present—might tell the British that the Soviet mission was acting in collusion with the Yugoslav Communists. Later this relationship changed: the mission's tasks and our ideological fervor became intertwined. Two or three weeks after the mission's arrival, Tito recounted to us with mock bashfulness that one night, when they were left by themselves, General Korneyev drunkenly kissed him and cooed, "Oska, Oska," the Russian pet name for Joseph.

In general, the mission was well organized and competent. Its acting chief, General Gorshkov, was strong and steady, and experienced in guerrilla warfare. Former staff members of the embassy in Belgrade, Majors Grigoryev and Sakharov—if those were their real surnames—also knew Serbo-Croatian. The chief of mission, General Korneyev, was the most intellectual and dignified of them, but given to drink and, when drunk, to brutishness; he was unsuited to our still ascetic and generally idealistic mood. Some of the British also liked their liquor, but this didn't bother us because we didn't identify ourselves with them or see in them loftier socialist beings. When General Korneyev's orderly, a muzhik with the rank of sergeant major and a good-natured fellow straight out of a Russian novel, found sympathy among our guards because of the general's abuse, the idealization of these new Soviet men began to wane. We comforted ourselves with the thought that the Russians hadn't yet lived down their despotic past! And we began to accept the Russians as they were, sentimental and brutal—toward themselves perhaps more so than toward others.

The mission quickly found it bearings. Maintaining an official exterior, it secretly became thick with us and penetrated into our affairs and our commands and organizations. The mission was aided in this by Yugoslavs as well: from Moscow with the Russians came the daughter of the well-known Zagreb physician Šilović and the former wife of the Communist journalist Otokar Keršovani. The mission was assigned local cadres for intelligence purposes.

The British also had officers of Yugoslav origin. However, they used them mainly as interpreters: the British didn't have the opportunities of the Soviet mission, and were themselves conscious of the limitations imposed on them by ideological differences and ideological distrust. To be sure, there might have been some Communists among the Allies—perhaps more in the imagination of us Communists than in reality—who favored Western and British forms. But any such persons shrank from being branded as spies by their own fiery patriotic milieu. The British were cautious not to compromise such persons or offend the Partisans. Everything was accessible to the Soviet mission, and all of us were predisposed toward it in spirit. And all of us would have remained devoted to it, but for its own Great Power standards of loyalty—standards which we in our leadership always found a way to justify: "They're afraid that the West will trick us, that the West will undermine us."

The arrival of the Soviet mission didn't change our relations with the British mission, at least not as far as any official stance was concerned. We saw to it that the Soviets had no difficulties with the British over us, and since the international recognition of the National Committee depended in large measure on the British, we took care that they not get the notion that we believed all our problems to be solved with the arrival of the Russians. Yet we also realized that our enthusiasm over the Soviet mission's presence couldn't be hidden. Indeed, we didn't try to hide it, aware that now we were no longer alone with the British, and that the scales in the world at large were tipped in our favor.

Two days after its arrival, Tito held a gala banquet for the Soviet mission. In a toast Tito thanked all the Allies, but he also underscored the extraordinary role of the Red Army, while our functionaries, including the very highest, visibly melted with joy at being finally side by side with their powerful protectors. The chiefs of the British and Soviet missions also offered toasts, with strained cordiality and banality whenever they touched on mutual relations.

At that time Tito wrote an article for the American press—for some news agency—entitled, if I remember correctly, "The Struggle of the

Peoples of Enslaved Yugoslavia."* To my comment that the article was too long for an agency, Tito replied curtly, "By God, they have to be told about our deeds and sacrifices! They've been exposed too long to Chetnik lies!" That article was later published in an abridged form in the American press, and as a pamphlet for Yugoslav leftist emigrants. To my comment that 300,000 for the size of our army seemed exaggerated, Tito replied, "Actually, not at all. If we include all units, the various small underground groups, there are about three hundred thousand." I think he was close to the truth, figuring like this. The article itself, though long and monotonous, was soundly balanced, being slanted equally toward the Yugoslavs and all the Allies.

Ranković and I were on some business in Tito's hut in front of the cave, when suddenly, and with a meaningful expression, Tito showed us messages from Churchill about which we had hitherto not known a thing. The messages had been translated, probably by Olga Humo, the daughter of the émigré minister Ninčić. Olga had just recently become Tito's translator, and accompanied him on his meeting with Churchill in Italy in the spring of 1944. Photographs of her with Tito were published, which started the rumor—as so often happens to famous men—that she was Tito's mistress.

In these messages Churchill marveled at the Partisans and Tito, but he pleaded that he had personal obligations toward King Peter, who was young and inexperienced, and had sought refuge with the British. In one message he felt Tito out on the question of whether King Peter, if he dismissed Mihailović as his war minister, could clear the way for friendly relations with the Partisans, so that he might later even join them in warfare.

Tito and the two of us all reacted ambivalently to these messages. We recognized the extraordinary benefit of dismissing Mihailović, but also a trap in the attempt to bring about direct collaboration with King Peter. But while agreeing with Tito, Ranković's reaction and mine, though not fully expressed, differed from Tito's by a shade: we were more fearful of the trap than happy over the benefit. For Ranković and me, collaboration with the King was incomprehensible and unacceptable, largely because it would create ideological confusion and weaken the party. Yes, it was important finally to settle our accounts with the royal government and Draža, but to the two of us this didn't seem feasible except through

* Josip Broz Tito, "Borba naroda porobljene Jugoslavije," in *Borba za oslobodjenje Jugoslavije* (Belgrade: Kultura, 1947), pp. 202–32.

strengthening the party, that is, the army led by the party. Tito saw in Churchill's messages an easier road to power, while Ranković and I—and I probably more obstinately than Ranković—insisted on ideological and revolutionary consistency. Tito discerned the essence of our reaction, and immediately stressed that Churchill was a great fighter and skilled statesman. And he added impishly, "He's a big devil. He wants to foist King Peter on us!"

The naiveté of Churchill's position was conspicuous: what possible significance could the King have in a Partisan and Communist army? We regarded this naiveté as intentional—the hidden trap of the old imperialist wolf. This was all the more so because from Churchill's messages there emanated the thesis that a majority of the Serbs were against us, that is, for the King. That thesis was also publicly kept alive in the West, in the form of a proposal that the two movements, the Chetnik and the Partisan, be divided territorially. We looked upon such ideas as academic but nonetheless dangerous schemes to save Serbia for the Chetniks and, on a broader scale, to prevent a linking up of our movement with the Red Army. Serbia continued to be of key importance, if only because we were still weak there; however, with the winning of Serbia all such schemes would have the ground pulled out from under them.

It was not for this reason alone that we were shaken by the destruction of almost the entire Third Serbian Brigade, which the Germans surprised at Prijepolje. That brigade had been created out of misfortune, and only after protracted combat was it able to get away to Montenegro. Its men had been snatched from the gallows, they were underground fighters and Partisans who had survived countless blockades and raids. It was an irreparable loss for us. Everyone experienced and expressed the same sorrow and anger in his own way—Tito with violent bitterness, Ranković with pressed lips, and I with pathetic generalizations. In its sorrow and fury the Supreme Staff ordered an investigation into this tragedy. In time the investigation petered out, having found as the "guilty" parties officers who were even more grief-stricken than we in the Central Committee and the Supreme Staff. To make plans involving Serbia became a bad business, ill-omened.

Yet the concern over Serbia, always lively, was becoming vital. The free territories in Montenegro and eastern Bosnia had expanded as far as Serbia. In March 1944 necessity and opportunity prompted the Supreme Staff to send two divisions into Serbia under the command of Milutin Morača. These units relied on Montenegro and were to receive support from Bosnia, though they didn't get it because of the German defenses on the Drina River. In the Supreme Staff there were no excessive hopes that those two divisions might turn the situation around in Serbia; chiefly they were assigned a scouting role. Though these divisions had to retreat after heavy day and night battles, the experience gained by their incursion was valuable, even decisive, in making plans for breaking out of Montenegro and Bosnia in the spring of 1944. From this bloody and relentless campaign we learned about the condition and capabilities of our opponents in Serbia. The divisions were strong enough to break out of the encirclements by Nedić's and Mihailović's forces. However, in order to arouse the broad masses to action, we needed to occupy a larger territory and to establish and hold dominion. Besides, the Germans held control over almost all of Serbia, and the Bulgars were not at all demoralized or spent. A turnabout could be effected only by considerable forces, and not from one direction alone. All operations in Serbia were planned and decided by Tito with expert officers. The remaining members of the Central Committee knew about them and agreed—as something dictated by the course of the war and the state of political affairs.

The penetration by those two divisions filled in our perceptions of the political situation in Serbia. The Communist organizations there had been almost wiped out, but Communist sympathizers, guerrillas, and small groups of party members increased their strength tenfold by their stubbornness and conviction. They constantly replenished themselves, being rooted in the society and in the suffering of the nation. On the other hand, the anti-Communist political groups were gripped by hopelessness and dissolution, or else clung to Draža Mihailović and the Chetniks with blind desperation. Anti-Communism, which served as the only cohesive force and platform of disparate and feuding groups, increased their numbers, but at the same time weakened the cohesion and striking force of them all, both collectively and individually.

For this reason the Central Committee didn't attach much importance to the Chetnik congress in the village of Ba. It was held, and not by chance, on the day of Saint Sava, the medieval Serbian statesman and educator whose greatness not even the Communists denied. The Central Committee did not even bother to take a stand regarding that congress. We spoke derisively of it as an imitation of AVNOJ, but not even as an imitation was it consequential or well thought out. The Ba congress agreed to a federal state structure, but only for the Croats, Slovenes, and Serbs; the Macedonians were left to Serbia, while Montenegro and Bosnia were not even mentioned. Draža Mihailović declared that the army would turn over power to the civilian authority after the war, though the January 6 spirit* and cliquishness of his movement could evolve only into a military dictatorship of a primitive Balkan type. The congress was attended by persons of secondary importance, most of them unknown. The most prominent was Živko Topalović, once a notable socialist but long since without a socialist movement. The most notable Croat, and almost the only one, Predavec, was better known as an engineer and as the son of a promiment leader in the Croatian Peasant party. Draža's congress didn't produce much of a stir abroad either. Despite our worries over foreign schemes regarding Serbia and the amalgamation of our opponents in Serbia, we could make fun of it all. Serbia was already a project and no longer just a hope. With strength, our hatred turned into contempt, and gradually into arrogance.

* The reference is the King Alexander's proclamation of the dictatorship on January 6, 1929.

2

At the beginning of March we decided to send our military mission to the Soviet Union. I was to be in the mission; Ranković told me that was the Old Man's idea. The news threw me into ecstasy and overwhelmed me with pride: I was going to the land of my Slavic roots and universal faith, as the representative of an army created in the struggle with the Nazi monstrosity, and that put its hopes in Russia, in the Soviet Union. The formal decision regarding the mission came on March 16. Tito chose me because I had been in the leadership ever since Tito became head of the party, and also because I knew Russian well and could explain to the Soviet leadership our special circumstances and needs.

The mission had, in fact, two chiefs, myself and General Velimir Terzić—the same Terzić who had been acting chief of the Supreme Staff during the Fourth and Fifth Offensives. I was the real chief only in that I was a member of the Politburo and the Supreme Staff. Terzić and I worked together, all the more closely since his was a personality suited to collaboration. Also on the mission were persons who, strictly speaking, had no place there, but who lent it a broader, more representative character: the physicist Pavle Savić, thus rewarded for his tedious labors as code clerk and guardian of our wartime radio station, and Antun Avgustinčić, a famous sculptor and a vice-president of AVNOJ.

It occurred to someone that we ought to take a gift to Stalin. But we had nothing worthy of this most renowned and adored personage. In our distress we decided that Stalin would not attach importance to the value but to the expression of love. And so gifts began to arrive from devastated villages, largely from girls' dowries, whereupon functionaries from the AFZ—the Women's Antifascist Front—chose the most beautifully embroidered towels, bags, and stockings—and the plainest sandals. However, this popular adoration was not complete without a military gift; we were barely able to find a single one of the thousands of rifles we had manufactured in Užice in 1941.

Tito assigned the mission the task of getting Soviet recognition of the National Committee as the legitimate government of Yugoslavia, and also Soviet military aid and help from the United Nations Relief and Rehabilitation Administration (UNRRA). We were also charged with seeking a loan from the Soviet government in the amount of $200,000. We didn't forget about the manufacture of medals either, and I took along the measurements and the sketched designs of Tito's marshal's uniform.

Tito set aside a special meeting with me, even though we saw each other several times a day. He ordered me to find out from Georgi Dimitrov, or from Stalin if I could get to him, if there were any criticisms of our activities. That order of Tito's was a formal one—the underscoring of our bond with Moscow—because everyone in the leadership, and Tito with most reason, believed that the Yugoslav party alone had passed the test of war on behalf of both the Soviet Union and its own people. Regarding Yugoslav émigrés—Communists who had gone to Russia before the war—Tito cautioned me not to get involved in their disagreements among themselves, or above all in their complaints about Soviet officials. He cautioned me against female secretaries, because there were all kinds. He said this obliquely, so that the stress was less on morals than on the special character of the Yugoslav party and the integrity of Yugoslav Communists. Later on, women in the Soviet service were to play a considerable role in demoralizing Yugoslav Communists and enticing them away from their leadership.

Tito never instructed me what reply to give to the Soviets if they brought up our negotiations with the Germans. Tito must have known that I would stick to our official version, which was that we negotiated only for the exchange of the wounded. Still, the question did come up with Ranković, perhaps with Tito's prompting. I replied with a smile that all I knew about was the exchange of the wounded, at which Ranković laughed roguishly.

The mission took along the Supreme Staff archives; we guarded them from British informers with fanatical vigilance, which alone might have attracted attention. I also took along Dedijer's war diary, which merited special care.

One night in early April a British plane bore the mission aloft from Petrovačko Polje. I was seized by an uncontrollable joy at the thought of seeing the Soviet Union, and yet an unrestrained sorrow on leaving my comrades and homeland in suffering and death, amid the blackness below from which only the snow-covered peaks emerged.* We arrived first in Bari, Italy, at the Yugoslav military base. From there we went across North Africa to Cairo, Habbaniya, and Teheran. From Teheran, on April 12, after a brief stopover in Baku, we arrived in Moscow. In Cairo, with the help of the Soviet minister to Yugoslavia, I had already visited the chief of UNRRA, Herbert Lehman, who received me with reserve; but UNRRA promised to consider our request for aid at the next meeting.

I had pictured the Soviet landscape as all pastels and boundless, but in that time of year it was stark and gray. Moscow seemed rusty and its buildings low. At the airport the reception with honors was conspicuously subdued, presumably so that the Western Allies wouldn't hold against the Soviet government the overly cordial treatment of a foreign state, albeit a Communist one. We were then put up in the Red Army Center, amid comforts which I could only dream about even in peaceful times.

But we waited and waited—much too long—to be received by anyone at all at the Soviet summit, let alone Molotov or Stalin. Everyone with whom we came into contact—among them prominent Communists such as Dimitrov, Gottwald, and Manuilsky—accepted the Kremlin's isolation as a given fact, virtually a higher law. Though they had declared themselves for Tito, the personnel of the Yugoslav Royal Embassy could do nothing for us except to make sympathetic, half ironic comments: "Patience, that's how it goes in Moscow." The Yugoslav émigrés couldn't help us, even though the most prominent among them—Veljko Vlahović, the manager of the Free Yugoslavia radio station—had connections and influence. No eminence could help us to the top, to Stalin's circle.

The structure of the Soviet system was no mystery to me. Furthermore the Communist party of Yugoslavia, in whose development I had long participated, had in the course of the war become transformed from an

* I refer the curious reader to my book *Conversations with Stalin* (New York: Harcourt, Brace & World, 1962), in which I have described in greater detail both this and two of my later journeys to the Soviet Union.

embryonic authority into a similar regime. It even crossed my mind that they were eavesdropping and checking up on me. But this didn't trouble me: our feeling for the Soviet Union and Stalin was one of fervor and sincerity. I was disturbed only at the slow pace, if not the lack of consideration, with which the Soviet top echelon approached our problems—which were their problems as well, since Communists were involved.

This did not mean that we were isolated or idle—on the contrary. Soviet organizations arranged receptions and meetings for all members of the mission, regardless of rank. We could get tickets for any theater, even the most famous. Our meeting ground was the Panslavic Committee, which had been created during the war and which was housed in an elegant mansion, probably once the home of some rich merchant. However, by virtue of either its leadership or its scope, the Committee itself enjoyed no particular status. Everyone knew that its importance and initiative stemmed from the Soviet government. The idea of Panslavism had become outdated and was reminiscent of tsarist imperialism. But under Stalin the meaning of events and things was changed, let alone of concepts. It was taken for granted that the Panslavic Committee was an anti-German façade for Soviet patronage over the Slavic peoples outside the U.S.S.R. For Communists, particularly for the Communists of the "Slavic sea," this was a welcome "discovery" of Stalin's if for no other reason than it offered the prospect of coming to power. The Soviet members of the Panslavic Committee—its real leaders—were for the most part unsuccessful high officers, while the administrators were NKVD* men. All of this was known and accepted as natural, given the system and the circumstances. Such was the Panslavic Committee that first received us publicly, at dinners and receptions.

Other meetings followed as well, but always outside the highest circle. Meetings were arranged between me and the Czechoslovak and Bulgarian party leaders. The Czechs and Slovaks listened attentively and soberly to my account of the uprising in Yugoslavia; Vlahović told me that the Soviet leadership was dissatisfied with Gottwald's contention that Czechoslovakia lacked the conditions for partisan warfare. At the meeting with the Bulgarian leaders, old Kolarov couldn't resist putting a question to me—not about the sufferings and destruction for which some Bulgarains were also to blame, but about Macedonia: was the Macedonian language closer to Serbian or to Bulgarian? Sensing covert and

* The abbreviation for the secret service organization—National Commissariat of Internal Affairs.

incurable pretensions to Macedonia, I replied, "I do not know whether the Macedonian language is closer to Bulgarian or Serbian, but the Macedonians are not Bulgars, nor is Macedonia Bulgarian." Dimitrov felt uneasy: "It is of no importance!" And he passed on to the next question.

I also met with two prominent Soviet party officials, Manuilsky and Alexandrov. Manuilsky was a little old man without any real power, but with the halo of a former secretary of the Comintern. He was well spoken, solicitous, and overawed by Stalin. Alexandrov was regarded as a leading philosopher, and—more significant for me—was in charge of agitation and propaganda in the Central Committee. In my accounts concerning Yugoslavia I indicated, both to Manuilsky and to Alexandrov, the character and the depth of the changes in Yugoslavia. I didn't use the word revolution—in Yugoslavia it was considered awkward and premature. But I emphasized that the army and the new government were in the hands of the Communists, and that no other force had any prospects in Yugoslavia; the bourgeois class and monarchist regime were dissolved and destroyed. Manuilsky listened to me with sentimental interest, and Alexandrov with bureaucratic reserve: as far as Soviet propaganda was concerned, we were waging a national—and only a national—war. The Soviet mind and propaganda saw our reality only as it served their prestige and interests. For them this reality could be changed only by a change in the stand of the leadership, that is, of Stalin and the Soviet government. To this day the peoples of Yugoslavia and the Yugoslav Communists have not earned Soviet recognition for having carried out a revolution.

Georgi Dimitrov was different in that he at least presented his own views and allowed for discussion. I believe that he really understood that changes were taking place in Yugoslavia which were of great importance for both Yugoslavia and the Balkans. He admitted that he had not expected the Yugoslav Communists to be so successful, and he rejoiced, without self-praise, that he had been so decisive in Tito's selection as secretary of the Yugoslav party. Though he was in failing health and gave the impression of a sick man in whom new and graver illnesses were ever being discovered, he still had a sharp mind and impressive energy. He celebrated our successes as if they were the realization of the dreams of his youth regarding the unification of all the South Slavs and Balkan peoples. At one point Dimitrov mentioned our negotiations with the Germans, in an unprovocative way and without casting suspicion: "We were afraid for you at the time, but luckily everything turned out well." I made no reply, but took this as a warning.

The Comintern had in effect been dissolved, but not the foreign affairs section of the Soviet Central Committee: in the greatest secrecy Dimitrov conducted the Central Committee's section for relations with foreign parties, which collected information and made recommendations to the Soviet leadership.

We also visited the Yugoslav Brigade in the U.S.S.R.—without enthusiasm, even though Soviet officers tried to instill some in us: the brigade was made up of converted members of Pavelić's unit beaten at Stalingrad, with Communist émigrés as political workers. Soviet officers served as instructors. The brigade passed through an extremely rigorous physical and political training program: it may not have changed their earlier views and aspirations, but it certainly wore them out. Moreover, the Soviet authorities kept on the old commander, Mesić. He was lost and demoralized, but did as he was told. Quartered in underground barracks in a forest, the soldiers appeared well drilled and toughened, but without esprit de corps or enthusiasm. Such a "re-education" of men was something we hadn't known until that time. Not even then did my consciousness accept it, though when something in the Soviet Union didn't please me, I always appealed to my "intellectualism" and "idealization." It's a big country, I would say to myself, a different culture; the aim is the important thing, the means are incidental.

On the other hand, it was with impatient eagerness that we went on our visit to the Red Army, and our eagerness was fully justified.

We visited the units of Marshal I. S. Konyev, which had established a bridgehead on the Romanian side of the Prut River. They certainly were out to impress us with their display of wreckage of German mechanized units and their bravest and most loyal soldiers. But on the way we also came across an abundance of unplanned German wrecks, and self-sacrificing officers and men who hadn't been specially selected. The Ukrainian region through which we rode was dull and somber, though less devastated than we had imagined. The Red Army had been advancing since the spring of 1943 and had an air of absolute confidence, but also of grimness and anger: they knew their failings, felt their defeats, and kept silent about it all. It was the officers that most impressed me, particularly those who had chosen the military as a career: they were broadly educated, full of initiative, and friendly.

It was in the Red Army, from the lips of General Korotayev, that I first heard the stupefying thought, not entirely alien to me: when Communism is victorious the world over, then wars will be fought with the ultimate bitterness. Hadn't I had similar thoughts that night after our disaster on the Sutjeska River? Hadn't I reflected that forces stronger than ideology and interests had thrown us and the Germans into a death struggle amid those wild ravines? And now a Russian who was also a Communist, Korotayev, was entertaining the thought that wars would be especially bitter under Communism—though under Communism, theoretically, there would be no classes and no wars. What horrors gave rise to these thoughts in Korotayev and myself? And how was it that he had the boldness so late at night, after supper and a cordial conversation, to express his thoughts, and I to listen in mute remembrance of horrors and reflections of my own?

Though we were not used to it, having been brought up in a teetotaling party, we had to drink Russian style with Marshal Konyev's staff, so as not to offend Russian hospitality and an army which was still shedding its blood for its homeland, our people, and humanity. In the morning they loaded us into a plane bound for Moscow, worn out from lack of sleep and from carousing, but with admiration and devotion for the Red Army filling the most obscure recesses of our consciousness.

I wrote two articles for the Soviet press: one for *Pravda* about the uprising in Yugoslavia, and one for *Novoye Vremia* about Tito. I had trouble with the editors over certain formulations, particularly those concerning Tito, partly because I really didn't understand their criticisms, and partly because I pretended not to. This went on until finally one of them, who felt more distressed than I, mumbled that it was their accepted practice to write in such terms only about Stalin. I replied, with as much naiveté as I could muster, "Well, it's all the same—both Stalin and Tito are Communists. But if you think it's so important . . ."—and I agreed to the corrections. In the U.S.S.R. they didn't permit anyone to be glorified except Stalin, particularly if a Communist was involved. I already knew this, but it was worth finding out that in this respect our sacrifices didn't count.

In the Soviet Union one achieved the state of blessedness gradually, from the bottom up. That was how Terzić and I got to see Stalin. It all happened in such a way that we didn't know that we would be received, let alone when. I had just completed a lecture at the Panslavic Committee and had begun to answer questions, when someone whispered to me that I had an important and urgent engagement. It was five o'clock in the afternoon. There was only slight confusion, as if everyone were accustomed to such emergencies.

A state security colonel informed us—after the car was on its way, of course—that Stalin would receive us. I thought of our paltry gifts: the infallible state security had already brought them along in the car from the villa we had been moved to in the outskirts. Suddenly I felt insignificant and chilled: Stalin was the incarnation of an idea—of a now suffering but imminently happy mankind. I realized, as never before, how much my meeting with Stalin was a part of chance, yet I also felt thrilled that all this was happening to me, and that I would be able to tell my friends about it. All misunderstandings and disagreements—everything unpleasant, insofar as we were capable of recognizing it—vanished before the stirring, inconceivable grandeur of what was about to take place, and what was already going on inside me.

Without complicated procedures or long waits, they led us into a study which Stalin entered from his room at the same time that we entered from the secretary's room. Standing there were Molotov and Zhukov, a general in the state security assigned to foreign military missions. As I shook hands with Stalin, I spoke my last name. He didn't reply. But Terzić clicked his heels, and along with his name and surname reeled off his entire title, to which Stalin replied, "Stalin." It all seemed a bit funny. Why would Stalin have to introduce himself?

Stalin was in a marshal's uniform. But there was nothing military about him, not even any of that majesty one saw in pictures and films. He was of small stature and disproportioned, his trunk too short, his arms too long. His face was pale and rough, ruddy around the cheekbones, his teeth black and irregular, his mustache and hair thin. An admirable head, though, like that of a mountain man, with lively and impish avid yellow eyes. His forehead was not as stark as in his pictures. One felt the intent, constant activity of the mind.

Stalin surprised me. Pleasantly and sadly. He had to be presented as strong and sturdy, yet he had grown feeble and worn in serving all us Communists, in advancing the cause. That exhaustion immediately vanished in conversation. Stalin constantly fidgeted and fussed with his pipe and his blue pencil, as he passed easily from one subject to the next. A bundle of nerves which never missed the slightest word or glance.

And he had a sense of humor—crude, sudden humor. This surprised me least of all, perhaps because I had heard certain anecdotes about Stalin and run across similar remarks in his writings. In fact, I heard a story that circulated among the top echelons in Moscow about Stalin's remark, when he heard that Konstantin Simonov's collection of love poems had been printed in an enormous edition: "Two copies would have been enough: one for her, the other for him."

When we sat down, I remarked that we were enthusiastic about everything we had seen in the Soviet Union. "And *we* are not enthusiastic," he replied, "though we are doing all we can to make things better in Russia." Several times thereafter he referred to Russia instead of the Soviet Union. I assumed that this was a current wartime emphasis on the role of the Russian people. Maybe that is how he actually felt. Today I would add: he had already learned, and learned well, that what he had acquired was more important than in what name he had acquired it.

Turning to Molotov, Stalin brought up recognition of the National Committee: "Couldn't we somehow trick the English into recognizing Tito, who alone is fighting the Germans?"

Molotov smiled smugly. "No, that is impossible; they are perfectly aware of developments in Yugoslavia."

However, when it came to material aid, Stalin's generosity went beyond what we asked. He approved the creation of a Soviet air base to supply our army. "Let us try," he said. "We shall see what attitude the West takes and how far they are prepared to go to help Tito." And he was very upset when I mentioned payment: "You insult me. You are shedding your blood, and you expect me to charge you for the weapons! I am not a merchant, we are not merchants. You are fighting for the same cause as we. We are duty bound to share with you whatever we have."

He examined our poor gifts cursorily. I think I detected a hidden sadness in his expression—at our poverty, or else because the gifts reminded him of his native province.

The unreality of that one hour seemed to linger as he led us into a dusk of northern lights. Reality appeared less important but more beautiful, and the world seemed crisper and better. Before us, Moscow lay darkened by war, yet all in expectation of higher rations and new victory salutes of fireworks—serene, uncomplaining.

Just before my departure for Yugoslavia I had still another conversation with Stalin. This one was more interesting and more important, though I had no idea that it would take place. On the night of June 5–6, 1944, on the eve of the Allied landing in Normandy, they simply put me in a car and took me to the Kremlin to see Molotov. It was nine o'clock at night. Molotov told me casually that we would have supper with Stalin, and we drove off to Stalin's villa near Moscow.

Along the way Molotov asked about the dangers that might result from the German raid on the Supreme Staff on May 25, 1944.* Our mission was in constant contact with the Soviet service that maintained

* In a surprise paratroop attack on the Partisan headquarters, the Germans had almost captured Tito before being beaten off.

communications with the Soviet mission in Yugoslavia. They sent us daily reports concerning the course of the fighting after the Drvar raid, and consulted us on what assistance might be given. Molotov was able to gain from me a clearer picture than the one described in the dispatches from Yugoslavia.

Stalin's villa near Moscow, a surprisingly small one, was situated in a grove of young fir trees. We entered the little vestibule, and Stalin appeared in a simple tunic buttoned to his chin, looking even smaller and less official. He took us into his study, which had paneled walls. It was cluttered. Immediately he asked about the fighting around the Staff headquarters. While one couldn't determine whether for Molotov this was more or less important than other problems, Stalin was wholly engrossed; he asked questions of us and himself, and replied without waiting for answers. He had a passionate nature, with many faces, and could just as easily show reserve as excitement. But he also knew how to restrain himself and keep silent with passion. And his passion quickly and unnoticeably spread to his entourage.

I somehow managed to reassure him that our army would not die of hunger. Molotov chimed in that Soviet pilots were no cowards, but that distance made effective help impossible.

I totally accepted Stalin's opinion that, in view of the growth and increasing complexity of our political tasks, Tito and the leadership should have a permanent and secure base. Indeed, the Soviet mission had already acted accordingly: at its insistence, Tito and a part of the leadership had been transferred on June 3–4 to Italy, and would proceed from there to the Yugoslav island of Vis. But Stalin had not yet received detailed reports when I was with him—only that Tito had temporarily gone to Italy.

Stalin stressed that we shouldn't frighten the English with red stars. But I was adamant that we couldn't give up the stars, since we had fought under them for so long. He stuck to his opinion, but without anger—as one deals with fretful children.

Stalin paced back and forth, while Molotov and I stood still. Then Stalin half sat on the desk and spoke, sometimes anxiously and sometimes sarcastically: "Perhaps you think that just because we are the allies of the English that we have forgotten who they are and who Churchill is. They find nothing sweeter than to trick their allies. During the First World War they constantly tricked the Russians and the French. And Churchill? Churchill is the kind who, if you don't watch him, will slip a kopeck out of your pocket. Yes, a kopeck out of your pocket! By God, a kopeck out of your pocket! And Roosevelt? Roosevelt is not like that. He dips in his hand only for bigger coins. But Churchill? Churchill—even for a kopeck. It was the English, they were the ones who killed General Sikorski in a plane and then neatly shot down the plane—no proof, no witnesses, and they wouldn't stop at Tito! What is it to them to sacrifice two or three men for Tito? They have no pity for their own! As for Sikorski, I don't say this, Beneš told me."

In the course of the evening Stalin repeated these warnings several times. Soon thereafter, I transmitted them to Tito and the leadership; later they played a role in Tito's conspiratorial flight from the island of Vis to Soviet-occupied territory, on the night of September 18, 1944.

Stalin then turned to our relations with the new royal emissary Ivan Šubašić, who, unlike his predecessors, promised an accord with Tito and recognition of the National Liberation Army. Stalin insisted, "Do not refuse to hold conversations with Šubašić—on no account must you do this. Do not attack him immediately. Let us see what he wants. Talk with him. You cannot be recognized right away. A transition to this must be found. You ought to talk with Šubašić and see if you can't reach a compromise somehow." I also transmitted this stand to Tito. Tito already held a similar position, so Stalin's recommendation simply confirmed him in his agreement with Šubašić.

I asked Stalin if he had any comments to make concerning our policies and our work, to which he replied, almost taken by surprise, "No, I don't. You yourselves know best what needs to be done there."

On the way to the dining room, Stalin stopped before a map of the world on which the Soviet Union was colored in red. Waving his hand over the Soviet Union, he harked back to his previous remarks about the British and the Americans, exclaiming, "They will never accept the idea that so great a space should be red, never, never!"

My glance fell on a circle drawn in blue pencil around a space west of Stalingrad. I sensed that this glance pleased him, though he didn't say a word. Then, probably associating such a deep penetration by the Germans with the fateful battle of Stalingrad, I said, "Without industrialization, the Soviet Union could not have preserved itself and waged such a war."

Stalin added, "It was precisely over this that we quarreled with Trotsky and Bukharin."

In the dining room two or three high functionaries stood waiting, but no one from the Politburo except Molotov. Everyone served himself from warmed silver dishes lined up along the upper half of a long table. Each person sat where he wished at the lower half; only Stalin's place was fixed, though he didn't sit at the head. There was plenty of good food, but only with respect to the drinking was there urging and prodding—in the form of frequent toasts, as is customary with the Russians.

No one served us. We ate during a conversation which lasted five or six hours, until daybreak. Stalin's associates were evidently accustomed to such dinners. It was here and in this way that Soviet policy was largely formed, usually in the presence of those who had some connection with the topics under discussion. Stalin ate with gusto, but not greedily; the quantities of food he consumed were huge even for a large man. He drank moderately, slowly and carefully, unlike Molotov and particularly Beria.

He dwelled mostly on the Slavic theme: Did the Albanians have any Slavic roots? Was Serbian similar to Russian? And what sins did czarism commit against the South Slavs? There were anecdotes; I told two or three. Stalin roared, while Molotov laughed with restraint, silently.

Concerning the dissolution of the Comintern, he said, "They, the Westerners, are so sly that they mentioned nothing about it. The situation with the Comintern was becoming more and more abnormal. Most important of all, there was something abnormal, something unnatural about the very existence of a general Communist forum at a time when the Communist parties should have been searching for a national language and fighting under the conditions prevailing in their own countries."

An officer brought in a dispatch. I had the impression that the dispatch had arrived some time ago, but was brought in for me to read in Stalin's presence. It contained Šubašić's conversation with the State Department: Šubašić stressed that the Yugoslavs could not turn against the Soviet Union because Slavic and pro-Russian traditions were very strong among them. Stalin remarked, "This is Šubašić scaring the Americans. But why is he scaring them? Yes, scaring them! But why, why?"

Then he remarked to me, "They steal our dispatches, we steal theirs."

Just then a second dispatch was delivered. It was from Churchill. He announced that the landing in France would begin the next day. Stalin began to make fun of the dispatch: "Yes, there'll be a landing, if there is no fog. Until now there was always something that interfered. I suspect tomorrow it will be something else. Maybe they'll meet up with some Germans! What if they meet up with some Germans! Maybe there won't be a landing then, but just promises as usual."

Stammering, Molotov began to explain that the landing would indeed take place. Stalin did not doubt it either; he simply wished to ridicule the Allies.

Stalin presented me with a most beautiful sword for Tito: the gift of the Supreme Soviet.

From the clump of firs around Stalin's villa there rose the mist and the dawn. Tired and solicitous, Stalin and Molotov escorted me to the entrance. I was filled with admiration for the ruthless, inexhaustible will of the Soviet leaders. And with horror before the endlessness of the cunning and evil that surrounded Russia and my country. And with the thought that it was the mighty and the wise who survived, and we little ones with them—in our own way.

3

I returned to Yugoslavia—to the island of Vis, where the leadership had moved—via Teheran, Cairo, Tunis, and Bari. That was on June 11 or 12, 1944.

It was dusk when I disembarked from the little ship. Our base in Bari had sent word of my departure, so a jeep awaited me at the dock. The leaders were all gathered together on the ledge in front of the cave which Tito occupied. The rocky heights, shimmering under the bluish glare of carbide lamps, resembled a gigantic backdrop in front of which some carefree but important action was unfolding. My head was bursting with impressions of Stalin and Russia. I was tremendously excited, for my account was anticipated with longing by comrades whom I had been close to for many years in our cause, our work, and our fervor.

They all stood up when they saw me mounting the slope, but nobody came toward me, in the expectation that Tito would be the first to greet me. It had occurred to me on the way that I would have to report to Tito as the supreme commander. Such was the influence of a bureaucratized and militarized Soviet Union. One adopts most easily and imitates most eagerly precisely what doesn't suit one. Since the occasion was an exciting and solemn one, I made my report to the Comrade Supreme Commander, aware that I was doing something which nobody expected of me and which didn't suit me in the least. Tito played his role: he listened to the report with a smile, then hugged me: "Come on, Djido, tell us what's new in the Soviet Union!" Everyone lined up to embrace me, and I could hardly wait to shed the formality I had imposed upon myself. No one held against me this artificial and awkward departure from comradely relations in the Central Committee: hierarchism had slowly permeated them all, and my gesture, however inappropriate, struck them as rather attractive.

The Central Committee members had gathered around Tito, and I gave them more of a narration than a report. I began by telling them that the Comintern really did not exist, that Dimitrov was only the chief of the Soviet Central Committee's section for foreign parties, and that we had to get along as best we could on our own—to be sure, relying on the Soviet Union and with devotion to it. This came as no surprise to them, but rather as a reflection of their own perceptions. What they were most curious to know was how Stalin looked, what he said, and what he thought of us. It was late and I was tired when dinner and the conversation at Tito's was over, but I continued to stoke the curiosity of Ranković and a few other comrades.

For the next two days I was busy telling Zogović, Mitra, Dedijer, and the actors of the People's Liberation Theater all about the Soviet Union, Stalin, and the Red Army. This curiosity was even more general and lively than concern over their own fate. This craving seemed to whet my own compulsion to tell: I wrote two articles—one about Stalin, and one about the Red Army. Meanwhile I had immediately reassumed my propaganda duties on the Supreme Staff—that is, in the Central Committee.

At first I roomed with Ranković, in a little house below Tito's cave. Later I moved to the little town of Vis, into a cottage above the road leading to the cemetery and the little beach beyond. By day there was no refuge from the heat that radiated from this bare rocky island. Only Ranković and Arso Jovanović were able to work in the noonday sun; the others, particularly the agitprop people and the actors, scattered to the beaches. I too went occasionally to the little beach behind the cemetery. In the distance loomed the gray mountains of the mainland. Fighting and dying were going on there in the dull silence, yet this was quickly forgotten in intellectual conversation or while swimming against the blue waves. Then I was stricken and drained by dysentery, which dragged on for almost three months despite expert help and British supplies of medicine and food and other comforts.

Ivan Milutinović also came bathing with me. Born far from any river, he didn't know how to swim, but splashed about in shallow water. We wanted to teach him how to swim, but he waved us off in shame: "It's too late for me." When, in October of that year, his little boat hit a mine in the Danube and he drowned, we regretted that we hadn't been more insistent, though in so short a time he wouldn't have learned enough to save himself. Besides, it happened at night, he was fully dressed, and the river was cold.

Throughout the day the English sailors, freckled and parboiled, lay around on the docks. In the evening they would sail out, with Dalmatian Partisans as guides, in pursuit of the slower and smaller German boats. The Partisans pointed with gruesome admiration at their commander, a rather heavy-set fellow with a reddish beard. He would catch a German boat in some strait or inlet, sink it, and ruthlessly wipe out with mines the Germans who jumped into the sea. He seemed to enjoy his lurid glory, and chuckled softly with self-satisfaction whenever he noticed that he had attracted attention.

A day or two after my arrival, I was deluged with stories of the raid on Drvar. Their principal theme was Tito's not leaving his cave. Žujović was the bitterest, or so he made out to be. He and Ranković had come to the cave as soon as the raid began. They then managed to get out, and had organized some sort of defense with other Staff members, when the situation grew ever more critical. The Germans were advancing into the vicinity of the cave. Žujović returned twice, and Ranković once, to talk Tito into getting out. To go out was risky; some Partisans had already lost their lives in front of the cave. Yet to remain in the cave was dangerous as well as embarrassing. Tito would not go. Žujović became incensed: the loss of face before the soldiers—the supreme commander! Ranković was milder: it was awkward. Kardelj stayed with Tito in the cave. He justified himself, saying, "I was for going out, but Zdenka fell into hysterics and kept pulling at the Old Man and bawling, 'They'll kill us! They'll kill us!'" Yet Zogović and Mitra got out of a still more exposed spot, the town; they both told me their story—Zogović with sarcastic overtones, and Mitra with humor. Tito and the rest lowered themselves by a rope along a crevice which channeled the water off during floods—but not until after the Germans had been beaten off. Tito was frequently interviewed about the raid. He blamed Arso Jovanović for having taken too lightly the reports of German preparations which came, so I heard, from the Soviet mission.

Tito never boasted of his behavior. Though he conspicuously avoided danger, I maintain that it wasn't a matter of fear—much less Zdenka's influence—but an inborn caution. Also, at that time, a conviction in the uniqueness of his role, and infatuation with the position of power which he had achieved at such trouble and risk. Be that as it may, the course of events and new tasks relegated the stories to the background, and past failings and weaknesses were forgotten—except by those who found them useful for opposition activity. Even then, Žujović showed signs of ambition. I heard that as Tito was boarding a plane for Italy, he asked Žujović to take care of his horse for him, and that Žujović said bitterly, "He's getting out, and he's leaving us to take care of his horse and to die!"

Two or three days after my return to Vis, the announcement came of the arrival of the King's representative, Dr. Ivan Šubašić, to hold talks with Tito. The National Committee designated a delegation from among its members for the negotiations. I attended the talks as a member of the Supreme Staff. If I remember correctly, Ranković also attended in the same capacity as myself. On the eve of Šubašić's arrival the Central Committee and the delegates of the National Committee adopted a draft of the agreement; I believe that Tito write it.

Šubašić arrived on June 16, before noon. He was received with military honors, but without the playing of anthems; to avoid the royalist anthem, we skipped our own. The English ambassador Ralph Stevenson accompanied him up to some ten yards below Tito's cave. We looked at one another: we knew that the British stood behind Šubašić, but such open display of patronage seemed tasteless to us. All the more so when Stevenson, who finally separated himself from Šubašić, smiled with a gracious duplicity that confirmed perfectly our notions of the diplomatic perfidy of the West.

The meeting with Šubašić began immediately after the introductions, though not in Tito's cave but the neighboring one, where meetings of the Central Committee took place. Tito assumed the presiding role. Šubašić began with superficial generalities: as a mandatory, he was willing to collaborate with us; all our forces must be united for the sake of the victory, and in order to obtain aid and recognition from all the Allies. I thought that Šubašić was talking this way deliberately to feel our pulse. In actual fact he was incapable of going straight to the heart of a problem; he was not unintelligent, but weak and irresolute—a quality that in politics is often worse than being dull-witted.

The reply was made by Josip Smodlaka, the National Committee's commissioner for foreign affairs. Smodlaka was already past seventy, but still robust and voluble. A Croat from Dalmatia, he had been a fighter for Yugoslav liberation since his youth. A man of broad culture and with a mind of his own, he had played a prominent role in the creation of Yugoslavia, and during the war joined the resistance movement in Split.

He had come to Jajce on the eve of the Second Session of AVNOJ, and pleased everyone—especially Tito—with his pro-Yugoslav zeal, which national rivalries and old age hadn't dimmed. He was constantly attended by his son Slaven, who was more flexible in his thinking than his father. We looked upon Slaven with a suspicion based on nothing but his unobtrusiveness and independence. Later he proved to be one of the most conscientious and intelligent diplomatic officers we had. No one in the Central Committee prevailed upon Smodlaka to speak as he did, nor could anyone call him a Communist or anti-Yugoslav. He could say things—and did—which we would have felt inhibited in saying: it was not for the émigrés and representatives of the crown to make offers, much less to set conditions; rather they should beg the National Committee and AVNOJ to accept them, so as to secure Allied help and international recognition for an existing government.

Smodlaka's speech was too strong for a Communist to have made. We could hardly conceal our elation, nor Šubašić his confusion. Nevertheless he stood his ground, with bowed head. I think he felt uncomfortable and humbled from the start, all alone in the midst of our units and guards, among uniformed and self-confident generals.

His lips quivering satisfaction, Tito declared that unless someone had something of importance to say, we should go on to concrete proposals. We Communists kept silent in a disciplined way—especially since, after Smodlaka, no one had anything to say. Tito then asked Šubašić to propose his draft of an agreement.

Šubašić began to read his draft, and we stared at each other in amazement. Šubašić's draft envisioned a coalition government in which Tito would receive the portfolio of minister of war. The King would be the commander-in-chief of all forces, that is, of our army as well. Royal insignia would be introduced. AVNOJ and the National Committee would in fact cease to exist. When we later asked Šubašić which forces, apart from our army, he was counting on, he mentioned the Croatian Home Guards—Pavelić's army. And when we asked whether he would command the Home Guards, he laughed quite sincerely.

In the oppressive silence which prevailed after Šubašić's reading, Tito announced that we too had a draft of an agreement, then asked Kardelj to read it. Our draft provided for collaboration with the royal government in the struggle against the enemy and in giving aid to the people, with the promise that the question of the monarchy would be taken up after the war. We did not agree to a coalition; later we did agree to participation in the government by patriots from the resistance movement, but not as the official representatives of the National Committee. The National Committee continued not merely as a parallel government but as the only one which we in effect recognized.

As soon as our draft was read, Šubašić picked it up, examined it, and said, "I accept this proposal as a basis." Thus there was no discussion of Šubašić's draft or our own. Šubašić proposed two or three minor changes—largely stylistic—of our draft, which Tito magnanimously accepted. Yet had it not been for Tito, not even these criticisms of Šubašić's would have been accepted, so embittered were we against his draft. The agreement with Šubašić was reached in an hour, or an hour and a half at most, but the meeting continued until certain details and the manner of execution were confirmed.

The nature of the agreement was such that we foresaw no problems with our soldiers and minority peoples, who had it in for the royalist Chetniks and distrusted the Greater Serbian monarchy. It was Šubašić who encountered difficulties—with his own party at home as well as with the émigrés abroad. This was a new and serious reason for our insisting thereafter on the agreement.

The British were also dissatisfied with the agreement, and didn't hide it in their contacts with us. They were displeased that a coalition royal government had not been set up which would have offered the King at least some sort of hope. Though incontestable, the British arguments that the King was young and innocent seemed strange to us: that he was a monarch was enough for us to depose him. Perhaps it was in part because of British displeasure that Šubašić didn't carry out the agreement, particularly in the matter of placing the royal navy under our command. It isn't clear to me to this day why Šubašić accepted our proposal so readily. If he was consulting with the British—and judging by everything, he was—how could he have agreed to all we proposed? The British surely knew that not even the Soviet government, however well disposed toward us, was sure how high a price we would have to pay for international recognition. Probably it was Šubašić himself who, sensing our rigidity and the absurdity of his own proposals, agreed to what we had to offer, in the hope of getting around it later.

Incidentally, the Soviet mission had been acquainted with our draft, yet even so they were annoyed not to have been consulted prior to the adoption of the agreement. But they accepted our explanation—all the more readily in that Moscow was very pleased with that agreement.

Friction with the British was expressed largely in the coldness of their conversations, their lack of initiative in collaboration, and petty mutual teasing. However, there were more serious disagreements as well. The British were dragging behind in supplying our units on the islands with heavy weapons. This aroused our suspicions that we were not looked upon as the only army on our soil, but that the British were planning action by their own troops. Inevitably, this would have led to a clash, in that they were supporting the monarchy and regarded it as legitimate. British propaganda still insisted that Serbia was not Partisan but royalist, in fact Chetnik. Without Serbia, we were threatened by all sorts of schemes: the partition of Yugoslavia into a western Communist and an eastern royalist part, with Allied intervention in relations between them and the evasion of our agreement with Šubašić. We considered such schemes unrealistic, yet also recognized that they could bring trouble and complications.

As a result of all this the Supreme Staff hurried Dapčević into entering Serbia. Meanwhile the Germans and Albanian nationalists—the Skenderbeg Division—undertook an offensive against Dapčević's units in Montenegro. Yet the Supreme Staff didn't give up the penetration, even at the cost of losing territory in Montenegro. The Supreme Staff was by now solidly organized, particularly its operational and security sections; Arso Jovanović directed the former, and Ranković the latter. Detailed plans were worked out for the push into Serbia, and orders were issued to the units in eastern Bosnia to support Dapčević's breakthrough.

We knew and rejoiced at something which the British refused to see: in Serbia an uprising was spreading which was bigger and stronger than in 1941, especially in the southeastern regions. Every day new units were created—unequipped and inexperienced, but full of spirit. Contacts with Serbia grew. A group of Serbian functionaries came to Vis, headed by Blagoje Nešković, the party secretary. Koča Popović was sent into Serbia as commander: from the underground and out of the shadow of gallows, the Communists of Serbia emerged as a government and an army.

The uprising grew and gained strength in Macedonia as well. Svetozar Vukmanović-Tempo arrived at Vis from Macedonia, overjoyed at the self-confident militancy of the Macedonians. He had ample personal reason for this, because it was he who had effected the transformation to armed struggle in Macedonia. Somewhat maliciously, we in the leadership saw special significance in the fact that it was the Macedonians themselves who would cure the Bulgars, both of the right and of the left, of their Great Power megalomania. We weren't at all surprised when Tito stated, concerning the "unclear stand" of the leftist Fatherland Front of Bulgaria with respect to Macedonia, "Even if Stalin asked us to back down regarding Macedonia, I would not agree!" At the time—it seemed strange and incomprehensible that Stalin could have even contemplated something of the sort.

On July 20, 1944, von Stauffenberg made his attempt on Hitler's life. The incident made no deep impression on us. We took it as an attempt by a nationalist, probably with Western ties, to save Germany from destruction and occupation. And later, when in the West public attention turned to Hitler's plans for a guerrilla war on German soil, in an article in *New Yugoslavia* I pointed out that no such resistance could be organized because it lacked an idea, had no external front, and would not meet with popular support.

The British were too directly and unaccountably enmeshed in Yugoslav affairs to permit us to theorize, to devise schemes. The Central Committee decided that Tito should not accept an invitation to visit General Henry Maitland Wilson, the Allied Commander of the Mediterranean Theater, to which Tito had at first agreed. Tito so informed the British on July 22, explaining that as president of the National Committee, which had received the prerogatives of a government at Jajce, he could accept invitations only from governments and not from individual commanders. In addition to this formal, though for us essential, reason, there were others no less important for canceling that visit. Šubašić was also to have been with Wilson. We weren't happy to get involved with him in British headquarters, as he hadn't yet carried out the agreement concluded on Vis. Besides, King Peter II was in Italy at the time, and we had reason to believe that General Wilson intended to arrange a surprise meeting between Tito and the King. Such a meeting could only have created confusion among our radicalized followers, who were united by the decisions at Jajce. The British didn't even conceal their efforts at the time "to water down Tito's wine." We knew that such stratagems would involve us in explanations to our followers, perhaps even in a precipitous radicalization. Besides, since at that time Italy was full of Chetniks and political émigrés of all shades, we were afraid for Tito's life, all the more so since relations with the British aroused our extreme wariness.

General Wilson soon repeated the invitation in the name of the British government. This time Tito made the visit. He returned with promises of weapons and the delivery of our fleet, as well as with pleasant impressions of an American air force general!

At that time disputes arose between us and the British over the bombing of our cities, specifically of Split and Belgrade, which were still in the hands of the Germans. We might have swallowed the bombing of

Split: it was a small city with a harbor, and not every bomb could be expected to hit a military target. But when reports reached us from Serbia describing the bombing of Belgrade on Easter Sunday, 1944, we could no longer sit still. Our informers in Belgrade had sent us detailed and precise information about garrisons, supply dumps, and depots. We sent along this information, with drawings, to the British. The Allied air force carried out a blanket bombing of the city, which was overjoyed at the sight of the Allied squadrons. Not one military target was hit, except for the haphazard destruction of a group of Germans here and there. The devastation was no less, and the despair even deeper, than that caused by the German attack on April 6, 1941. This is how it was told to us on Vis, and how it lies buried in the memories of the people of Belgrade to this day. That bombing aroused a double bitterness in us: emotional, because we pitied a city of legendary suffering which Hitler had turned into a ruin and a place of torment; political, for we suspected—and at times believed—that the Allies were carrying out such bombings in order to make postwar rehabilitation and administration harder for us Communists.

Tito summoned a representative of the British mission and poured out his bitterness at him, along with the categorical demand that in the future not one of our cities be bombed without our consent. Moreover, Tito threatened to spread the word that the Allies were carrying out bombings of places which had no military targets. Thereafter the Allies bombed several cities, but in consultation with us and only when it was justified from a military standpoint. Thus key cities and industrial installations were spared. It has stuck in my memory that certainly military targets in Zagreb had been pinpointed, but the experience with the bombing of Belgrade and Split spared Zagreb from a similar fate.

Tito received a new invitation, this time to visit Field Marshal Harold Alexander, commander of the front in Italy. Tito was accompanied by Sreten Žujović and Arso Jovanović.

In talks with the British commander Tito consented to a landing in Istria, on condition that Allied units collaborate with our army alone and respect our government. Though only a small region was involved, which was not even under Yugoslavia but Italy, Tito wasn't happy about this arrangement. Yet he couldn't refute the military argument: by opening up a front in Istria and around Trieste, the Allies would threaten the German rear in Italy and get closer to Austria. Such a landing never did take place, either because operations took a different turn or because the British never went beyond the planning stage.

In the course of that visit, on August 12 Tito met with Churchill in Naples. Tito returned to Vis on August 15, impressed by Churchill and pleased by his talks with him. Churchill and the commanders had promised reinforcements and a greater regularity of shipments. Indeed, our units within reach of the Allies now received heavy weapons as well, and acquired a full capacity and the capability of liberating the islands and the entire coast. Churchill agreed to our demands concerning Istria, but not Trieste. He asked Tito if he would meet with the King, but did not insist on it when he noticed Tito's critical attitude and reluctance.

Šubašić had come from Italy to work out with Tito the details of the agreement, thus substantiating Tito's impression from his talks with Churchill that the British had finally abandoned the idea of operations in Yugoslavia and thus also any intervention. Indeed, Churchill did insist on legality and on the King, but he accepted Tito and the new regime as a reality. The meeting with Churchill strengthened Tito's self-confidence and authority.

Tito had other reasons for his confidence as well: our units had smashed the Germans and the Chetniks on the Ibar River and were successfully penetrating into Serbia. All speculation regarding Serbia and Chetnik hegemony in it no longer made any sense. There was nothing for Churchill to do but accept a new Yugoslavia or else get entangled in an unpopular and drawn-out conflict—a conflict incomparably more difficult than the British intervention in Greece, given our strength and difficult terrain. The excuse and self-justification by some Serbian nationalists that Churchill had betrayed them could hardly be more foolish and undignified. Was it up to him to wage for them a war which their leaders had long since lost through their own self-interest and absurdity? In politics no one can be responsible for someone else's defeat, for someone else's fate.

On his return, Tito could not contain his respect and enthusiasm for Churchill, even though he saw ironic skepticism on our faces. He flared up: "It isn't as simple as you think! Yes, Churchill is an imperialist, an anti-Communist! But you won't believe it, his eyes were filled with tears when he met me. He almost sobbed, 'You're the first person from enslaved Europe I have met!' Churchill even told me that he had wanted to parachute into Yugoslavia, but he was too old!"

Žujović was the most skeptical, and was again displeased with Tito's attitude. He told me, as if in passing, "The English are clever: an escort of warships, and naval maneuvers in honor of the Old Man, and I see that it's had its effect on him!"

I was aware that Tito was suddenly giving in to new and powerful impressions, but also that such impressions didn't last and didn't go deep, if they ran counter to Tito's objectives and prestige. Yet these weren't the reasons why I passed over Žujović's comments in silence: I avoided repeating criticisms, and wouldn't mention Žujović's now, had not his later defection to Stalin struck me as calculated and gross.

At that time Turkey had severed diplomatic relations with Germany. I note this because the question as to when Turkey would enter the war had become a joke among us. Even the most simple-minded soldiers and illiterate peasants sensed that Turkey's entry into the war would ease the pressure along our fronts and reduce our afflictions. Yet now, when the Red Army was already penetrating into the Balkans and our own army was rushing to meet it, Turkey's militancy no longer attracted any attention, much less excitement. Our people generally remember and esteem the Turks as a brave, independent people. But in war, soldiers admire only those who shed their blood for ideals. As for the Swedes, to the north of the European battleground, they attracted the attention only of the party leaders; Sweden's business deals with Germany, the conversion of Swedish ore into Nazi weapons—all this was confirmation of our theories concerning opportunism and betrayal by the Social Democrats. After our run-in with Stalin in 1948, we began to think differently of Sweden's neutrality. In a conversation with me Kardelj called it a wise policy, and I agreed with him. Moreover Kardelj affirmed, "The policy of old Yugoslavia to stay out of the war wouldn't have been bad, had there not been weakness and fascism behind it." However, I couldn't agree with such thinking. There were too many "ifs," and too few prospects for the Communists.

We appeared to be entering into a more fruitful and open collaboration with the Western Allies. During Tito's stay in Italy, Šubašić also got going; he promised the fulfillment of the agreement, and recognized the government authorities that AVNOJ and the National Committee had established. The King finally abandoned the Chetniks and recognized Tito as commander over all the forces of resistance. We made fun of the King, not for his reckless intention of saving his crown, but because we knew that it was too late for the Chetniks to obey their King.

However, a new dispute arose, so deep that it stirred up doubts in us and a bitterness reserved only for the enemy. In Montenegro the Germans and Chetniks had forced our units into devastating engagements for the protection of our wounded. They simply went after our wounded, whom our units dared not and could not abandon, even though they thereby lost the initiative and dwindled in numbers from battle to battle. Though there were convenient landing places, our pleas to the Allies to send planes for the evacuation of the wounded went unanswered. The situation finally became so critical that our units had no choice but to perish along with the wounded or else abandon them to SS executioners and Chetnik cutthroats. Our anger was intensified by the fresh memories of the horrors and desperation of the Fifth Offensive. Tito decided on extreme measures: he called in Maclean and told him that he could not see in the attitude of the Allies either carelessness or good will, that he would not pass it over in silence, that there would be consequences. I remember Maclean when he left Tito: this reserved, cold-blooded Scot was confused and downcast as never before. And he acted energetically: on August 22 Allied planes evacuated over a thousand wounded from Bajovo Polje to Italy in the course of a single day. There was an instant sense of relief on Vis. And rejoicing. And good will toward the Allies: they weren't so bad after all, maybe it was just a misunderstanding on the part of some of their services. Our units carried out a breakthrough that same evening. The night following, they stole up to the Chetniks' position and killed off three to four hundred of them—all that fell into their hands. Two days before that, Mihailović had reached an agreement with Nedić on obtaining arms from the Germans in a common defense of Serbia. As our victory became obvious, our domestic opponents began to dissolve into reckless, desperate gangs, and we began to feel the need for forgiveness.

It was then, on August 30, in the spirit of the talks with Šubašić, that Tito directed an invitation to the Croatian Home Guards, the Chetniks, and the Slovenian Home Guards to join us by September 15, with the guarantee that their collaboration with the occupation would be forgiven. Though terse and general in its wording, the invitation met with opposition and grumbling in our own ranks. No one was against amnesty and the subsequent absorption of the misled and the forcibly mobilized—the ignorant. Nor did the grumbling come from those who took this to be a maneuver, nor from those who accepted automatically everything that came from above. But the flame of revolution still blazed, and people were still waging a war of survival and vision. The revolution could not be anyone's monopoly, for there were still plenty of revolutionaries who were truly and totally revolutionary, who pondered every turn of events with their own minds. They were the ones who asked themselves and their friends: how can our soldiers take in those who killed and tortured their comrades? Our army was shaped not only by heroic suffering and death, but also by an arduous, self-sacrificing transformation of consciousness and the reception of new ideals. How could such an army accept something the very rejection of which gave it life and made it a world in itself? I sensed this resistance not only in the ironic remarks made by the intellectuals I worked with, but murkily in Ranković as well. He felt that the offer applied to peasants, while concrete cases could be decided as they came up. I was not against Tito's offer, at least not its essence. But I lacked the conviction necessary for my propaganda work. Kardelj showered me with enthusiasm: "Come on, Djido, it's a wonderful thing! That will demobilize them! Of course it doesn't apply to murderers and ringleaders. Don't worry, that kind won't come forward anyway."

I got up the nerve to bring it up with Tito as well, though I expected him to be angry and impatient with me. "Well, you know, this isn't time for sectarianism," Tito replied, as if I had caught him in a compromising situation. "We need the Allies to recognize us. We're just giving the quislings a last chance—showing our generosity. Then we shall see." The response to our invitation by the quislings, so far as I know, was hardly significant—not even in Croatia where the offer had its greatest reverberation. As for the resentment inside our own ranks, that died down in the fray of battle and the hostility of the warring sides. It was almost as if nothing had happened. No one could stem the tide of a blood-drenched ideology, much less reconcile anyone.

On September 6 the Red Army reached the border of Yugoslavia, and on September 9 it broke into Bulgaria, where a coup took place and a government under Communist domination was set up. Hitler's Balkan armies found themselves in a maze of uprisings and a blind alley. Meanwhile on September 11 the American army reached the border of Germany.

On the occasion of the Red Army's arrival on our border, Tito issued an order of the day. The order was written in stock wartime phrases, without inspiration. There would have been more poetry, had Tito entrusted the writing of that historic order to me or Zogović. Yet not even the most poetic order could have evoked the passion we had felt while the Red Army was bleeding on distant battlefields for the survival of Russia and the Slavs, and for the liberation of Europe. Though still seen in the glowing light of our common ideals and sacrifice, once it came into our own territory the Red Army became a part of our own realities, outsmarting the British, outmaneuvering the Americans, and outfoxing the Bulgarians. Politics and life called for measured words and reason, regardless of ideology and our former beliefs and sentiments.

The Soviet mission insisted that Tito visit the Soviet Union. Tito wanted to go, not only to obtain the heavy weapons without which our army couldn't take Belgrade or wage a frontal war, but to put an end to pressures and schemes for the partition and restructuring of Yugoslavia.

Tito's immediate entourage also favored Tito's leaving the island, if not for the Soviet Union then for the liberated territory on the mainland. True, the British gave no cause for suspicion, they had even stopped reminding us that we were under their protection. Yet there were disputes, such as the breaking off of negotiations with UNRRA in Bari because of the insistence that its own personnel distribute aid in Yugoslavia. We saw in this attempted move the setting up of intelligence and political bases, and the belittling of our authority, but the Allies, primarily the Americans, remained adamant. The rupture with UNRRA was not the crucial cause of our mistrust, which was constant and organic, but it deepened it. Now we were more suspicious than ever, perhaps all the more so because of British cordiality and candor. What if the British surrounded the island and suddenly landed a superior force? We were also mindful—far too often—of Stalin's warnings, transmitted through me. There were even comical touches to all this. Our soldiers plotted to kill a British dog called "Tito," and they probably would have done it, had not the British assured us that in their country dogs were given the names of popular persons.

Our fears were especially aroused by the Warsaw uprising, as were our conflicting feelings over the sufferings of the Polish people under the very noses of the indifferent Soviet troops: if the Western powers had set such a trap for the Russians, what then were they thinking of doing to us? We saw events in the light of our own situation. Is this not a condition for victory?

Tito's departure was carried out in the greatest secrecy, at midnight of September 18–19, in a Soviet plane. Ranković arranged the departure, and I helped him with some of the details. Korneyev, chief of the Soviet mission, and Ivan Milutinović were traveling with Tito. There was some hesitation over Tito's dog: what if he should bark? A solution was found: Tito would hold the dog and pet him. So contingencies were taken care of.

The secrecy of Tito's departure was so total that not even high party officials who worked in the Central Committee and Supreme Staff knew about it. Vladimir Bakarić recalled recently in an interview that he, as well as other high officials, had been informed by me three or four days *after* Tito's disappearance. During those three or four days we avoided the British, then finally admitted that Tito had left for some liberated area in Yugoslavia. The British were angry, but we ignored their protests. I had suggested to Ranković before that we announce over Radio Free Yugoslavia that Tito had landed in western Serbia, where he had first begun the struggle, and had been greeted by enthusiastic crowds. Shortly after, I heard that a group of British officers in search of Tito had fallen for our misinformation and had parachuted into western Serbia.

Tito landed in Rumania. From there he flew to Moscow, where in the name of the National Committee he signed an agreement with the Soviet government regarding the operation of Soviet troops on Yugoslav territory bordering on Hungary. The agreement called for the withdrawal of the Soviet troops after the operations, and for the jurisdiction of the National Committee on liberated territory. At the same time Tito was promised armaments for twelve infantry divisions, planes for two air force divisions, and the acceptance of officers for training.

No one on Vis, not even Tito and Korneyev, had known that such an agreement would be concluded, though it had been anticipated that the Red Army would have to fight on Yugoslav soil. It had not been Tito's intention, so far as I knew, to seek the help of the Red Army, except for armaments, particularly tanks, without which we couldn't liberate Belgrade. For us Belgrade was an important center, a decisive one for the establishment of the National Committee under the new conditions in Europe and the Balkans.

Tito recounted later that he had asked Stalin for eighty to a hundred tanks, to which Stalin replied, "We shall move a corps." (Tito quoted this in Russian: *Dvinem korpus.*) Looking back, I maintain that at the time our forces would not have been able, even with those hundred or so tanks, to liberate Serbia and take Belgrade. For the Germans, the Balkans had become the right flank of the Eastern front, and the German forces in Serbia were to be joined by the forces from Greece. Probably Tito sensed this or realized it while in Moscow, and agreed—though reluctantly—to the participation of Bulgarian troops in the operations on Yugoslav soil. Military reasons also played a part in this agreement, and not just Moscow's effort to enable the new Bulgaria to atone for its old sins. Yet this gives no reason to the Soviets, and particularly none to the Bulgars, to boast of their decisive role in liberating this or that part of Yugoslavia, still less the whole of it. We had the forces with which to smash the Germans and to enlarge our liberated territory, and we could finally have taken Belgrade, just as we had Zagreb, Sarajevo, and Ljubljana. With respect to our domestic enemies, particularly the Chetniks, we were already incomparably stronger. Draža Mihailović's general mobilization of September 1, 1944, was a failure. His corps of three hundred men each scattered, and he withdrew into Bosnia with the remnants before the entry of the Red Army.

I recall also some colorful details of Tito's meeting with Stalin that time. Stalin made fun of his marshals on the telephone. To Malinovsky: "You're sleeping over there, sleeping! You say you have no tank divisions? Well my grandmother could have waged war with the help of tanks!" To Konyev: "You talk of the possibility of reduced fighting. If it's reduced fighting you want, go to the English. That's how they fight!" Stalin regarded the prohibition against the King's return as premature. "You don't have to take him back forever," he remarked. "Just temporarily, and then at the right moment—a knife in the back!" Stalin was enigmatically silent when Tito told him that we would resolutely resist any British landing for purposes of intervention. Tito had two meetings with Stalin: the first in the Kremlin, which was almost cool, perhaps under the impact of tsarist buildings and offices, and the other at a party in Stalin's villa where Tito, not used to drinking, withdrew to vomit, accompanied by Beria's cynical comment: "Never mind, never mind, it happens."

Vis had become deserted. Officials and clerks were being flown into the liberated towns of Serbia, in preparation for the entry into Belgrade. Kardelj and I were among the last to leave. In Croatia there again was muted talk of setting up shop on their own, causing Kardelj, in agreement with Ranković and myself, to propose to Tito that Hebrang be replaced. Bakarić had been on Vis the whole time, co-ordinating the sectors of the National Committee. Everyone liked him because of his vast intelligence and balanced tolerance, and we agreed to make him the new secretary in Croatia. Kardelj and I were assigned to effect the replacement.

We also had to reach some agreement of views with the Communist party of Italy concerning Yugoslav territories which Italy had appropriated, Trieste in particular. The Italian party was undergoing regeneration; in the light of its traditionally nationalistic campaigns, these questions were important and sensitive. Kardelj and I had the job of finding common ground with the Italians.

We flew to Bari and from there we left for Rome by car, accompanied by our intelligence officers. We met Palmiro Togliatti in the evening, in a clandestine house. The conversations lasted far into the night; we were exchanging information rather than working out agreements. At the time Count Sforza was not only claiming Trieste, but seeking to ensure an Italian presence in Istria and the neighboring islands through the internationalization of Rijeka (Fiume). In a speech delivered on Vis, Tito had reacted sharply against any such proposals, giving life to the old slogan: what belongs to another we don't want, what belongs to us we won't give up! But we and the Italian Communists had no reason to quarrel, though we differed in some respects. Especially sensitive for the Italian Communists was the question of Trieste, given the Italian majority in that city, and Italian nationalist aims that dated from the struggles against Austria. It was Togliatti's idea to have a Yugoslav-Italian condominium over Trieste and its surroundings. But neither Togliatti nor anyone in the Italian Central Committee ever agreed to Yugoslavia's annexation of Trieste.

At the time it was easy to reach an agreement with Togliatti; it just flowed, so to speak, out of Togliatti's analysis. Stocky, with a high forehead, deliberate in his movements and expression, Togliatti looked like a parliamentary leader or a professor of humanities. The smoothness and precision of his sentences revealed that his mind worked much faster than his tongue. Even as one of the leaders of the Comintern, under the pseudonym Ercoli, Togliatti had enjoyed the reputation among our leaders of being a brilliant and wise speaker. In conversation he was unaggressive, yet stubborn. Slowly and methodically, he dealt with a problem from all sides until he got to the heart of it. He came to Belgrade several times after the war, and each time was attentively heard and cordially treated. Trieste was always the reason for his visits. My impression was that Togliatti came with the desire to find a compromise, but not with a ready-made formula. And the final agreement was always a combination of Togliatti's formulation and Tito's additions. The Italian Communists gained practical advantages for their elections, countering rightist charges and fascist rantings. For us the benefit lay in the easing of tensions between the two countries and in avoiding a quarrel between two Communist parties.

From our meetings with Togliatti I concluded that he was one of the wisest Communist leaders. Today I would add that he gave less evidence than any other Communist leader of arbitrariness and coercion. His strength lay chiefly in his verbal, logical dialectics, and not in the impulse for survival and control. When the Yugoslav party clashed with the Soviet Union in 1948, Togliatti sided with Stalin. But unlike the French

Communist leader Maurice Thorez, he did so discreetly and unaggressively, with expressions that were not brutal. We surmised that Togliatti had guessed the import of the Soviet attack and was biding his time. Consequently, we too were restrained in our criticism of Togliatti. I wonder to this day how Togliatti felt, what had gone through his refined Machiavellian mind at the time of the Moscow purges. Everything was undoubtedly clear to Togliatti; he took part in everything. He convinced himself, if he was not already convinced, that this was that Marxist-Hegelian historical necessity in its purest, most total form and expression. Undoubtedly he also learned something: that the institution is more important than the idea, and Italy more important than Muscovite internationalism.

In Rome we visited Saint Peter's, the Vatican, and other famous buildings. I was intoxicated by their beauty. To my surprise they looked more like luxurious palaces than places of worship. We returned via Naples, through a devastated, hungry, frantic land. We stopped off at Pompeii. The war had driven the tourists away, so we strolled all alone through streets where people walked two thousand years ago, in humble joy that a moment of existence in eternity had also befallen us.

At the airfield in Bari, we met with Dr. Šubašić and Sava Kosanović. Belgrade had been liberated the day before: October 20, 1944. They were waiting for a plane to Belgrade, while Kardelj and I were off to liberated territory in Croatia. We all four understood how crucial the liberation of Belgrade was: they hoped to establish a coalition government there; we, to entrench our authority and implement our ideals. In the humid dusk, pierced by the lights of the airfield and the shouts of the crews, we chattered on. Šubašić continuously brought up the conditions for a coalition government, while Kardelj stressed the legality and authority of AVNOJ. Kosanović would jump in, first on one side, then the other. He had been with Šubašić on Vis during our recent negotiations, and left the impression of being nervous and candid. We thought of him as being "American," though not to the degree that Šubašić was "English." At one point Kosanović and Kardelj went off on their own tangent in the discussion, while Šubašić assured me that he was keeping the Soviets informed of our negotiations, and that he had always opposed the persecution of the Communists. I couldn't resist telling him, even if in a restrained way, "But it was while you were viceroy in Croatia that camps were built for Communists! And when you left the country, you didn't order the release of imprisoned Communists, so the first act of the Ustashi was to execute around a hundred of them, mostly leaders!" For a moment

Šubašić was petrified. Then he tried to assure me, to swear to me, that he had ordered their release, but that lower officials had failed to carry the order out.

Kardelj and I flew off first, to Croatia. On the plane I told Kardelj what I had thrown up to Šubašić. "You shouldn't have," Kardelj said in agitation. "We're still negotiating."

I replied, "And he shouldn't have boasted of shielding Communists. If he doesn't know where he stands, we'll have to tell him!"

When I later recounted that exchange with Šubašić, Tito smiled, shaking his head, while Ranković roared with satisfaction.

As the plane was approaching an improvised airfield near Glina, I thought: What if the Germans and the Ustashi in Glina have discovered our signals and placed them on one of their own fields? In that case, we would defend ourselves with pistols until their machine guns cut us down. I suspect that Kardelj was thinking the same thoughts as he sat there mute and withdrawn. The highest functionaries, practically all of them, met us at the airfield. We talked about the importance of the liberation of Belgrade and our relations with Šubašić, then they showed us to rooms with prewar comforts. The food and the service too were on a prewar bourgeois level.

That morning we held a meeting with the Central Committee members. Kardelj had the floor. He reviewed the work of the Committee, without being either sharp or too general. He presented Hebrang's replacement as a necessity, to make it possible for Hebrang to take charge of the economy in the National Committee, that is, the future coalition government. Hebrang minimized his errors, portraying them as isolated incidents and reflex responses to momentary circumstances. Even though the discussion was reasonable and calm, a dark discontent hardened within him. The economy and rehabilitation of the country, in which he was to play a significant role, was an extraordinarily prestigious task, especially given the ravages of the war, but this did not even approach the status of an authoritarian leader of a people. Hebrang could only be such a leader, or a malcontent.

The discussions continued into the afternoon and focused on current problems. The most entertaining part was the conversation, over dinner, with old Dr. Ante Mandić regarding spiritualism. Like most champions of South Slav unification who were disappointed in Yugoslavia as a government and a social system, Mandić was also intrigued by the full gamut of spiritualism, from the scientifically researched phenomena to the conjuring of illiterate peasant women. He recounted with fervor all sorts of incredible cases, and was charmed by the supposedly naive curiosity of us Communists. We looked upon this phantasmagoria as a poor substitute for an ideal. Nevertheless, the charming old gentleman relieved our tensions amid the tangle of wartime politics.

On October 22 Kardelj and I flew back by night to Bari, and then—that same night or the next—to Serbia, to an improvised airfield near Valjevo. Regular frontal warfare was in progress there, as well as a more secure, more comfortable life for the leadership; but the worries and duties did not seem to have lessened.

And That Was Freedom

1

The thought nagged at me that we were being received in Valjevo the way the Serbian generals were received on their way back from the Salonika front at the end of World War I: a master bedroom in a prominent burgher's house, tiptoeing so as not to disturb our sleep, and when we awoke, jam preserves and brandy—the traditional Serbian prebreakfast offering—followed by a plentiful breakfast. That is how our whole army was welcomed by Serbia—militant, suffering, and politically creative Serbia.

The next day I continued to Arandjelovac, where I picked up a Willys jeep which had been assigned to me on Vis, and learned to drive it. That afternoon, with an escort, I proceeded by jeep for Belgrade, where we arrived before dusk. Everywhere there were scenes much like those in Valjevo: young peasants with colorful bags, in traditional army caps and braided leather sandals with upturned toes, being escorted to command posts by local dignitaries and veterans of the Salonika front; high school boys and apprentices standing patiently in front of recruiting stations to be given assignments; caravans of carts with food and herds of swine and cattle, all for the army; and our slogans everywhere, in bright colors, with bigger and thicker letters than ever.

The headquarters of Dapčević's corps had moved out of the villa formerly occupied by Neuhausen, the German economic governor of Serbia, at 15 Rumunska Street, so that Tito could move in. I spent the night in the temporary headquarters on Andra Nikolić Street, where Dr. Ribar later resided. I was immediately overwhelmed by detailed accounts of the violent behavior of Soviet soldiers, their grabbing of revolvers and watches from our officers, and their looting and raping. Dapčević was also stunned by all this, and perhaps even more by the Soviet officers' efforts to belittle our contribution in the battle for Belgrade. Late that night, long after I lay down to sleep, these violent scenes and distorted views of Soviet men kept me from falling asleep. Everything that was worth admiring and remembering—the youth meetings in the stadium during the battle for Belgrade; the publication of a newspaper with the first printing press that we took over; the carpets spread in front of Soviet tanks; the fighting by bare-handed citizens; the steadfastness of the Soviet soldiers and the daring of our own in the attacks—all this receded before our anguish and doubts that the Red Army was not all that we had imagined, not all that it should be.

One day I came across a Soviet command leaflet for soldiers on entering Yugoslavia—a superficial bureaucratic explanation to the effect that Yugoslavia was an allied country in which there were patriots who had risen against the Germans. But Stalin had drummed one single thought into his army; the struggle for victory. He had drummed it into everyone, from top to bottom. One day I went to Novi Sad on business. On my way back, I was stopped by a Soviet soldier who was driving a horse-drawn cart piled high with little sacks, pots, and bedspreads. "Is this the road to Berlin?" he asked. That was the only direction he knew about ever since the Russian army had turned around at Stalingrad and Kharkov, and the only one in which he was sure of finding his regiment. I later heard that Red Army men were hardly more considerate toward the population in their own country.

As for the liberation of Belgrade, to this day I remember Mitra's touching reaction as the battle was going on: "I felt," she told me that same night, "that my heart would burst as I watched Belgrade, stretched out on the hill like a giant, disappear in the smoke and din."

On the next day, October 24, I set out for the Hotel Majestic, where the National Committee, high officials, and members of AVNOJ were located. There was much talk there, and good food. The following day I moved into a little room and began work on agitation and propaganda, which at that particular moment meant the acquisition and mobilization of the means required for such activities. I had my previous team—Zogović, Dedijer, and others, and the army aided me in every way: revolutionary armies are insatiable in their demand for ideas and news, so they take tender care of the people and instruments that fulfill this need.

That same day Vladislav Ribnikar told me that *Politika* could start up in two or three days, but that there was a problem about *Borba,* which as the party newspaper ought to have priority, though the presses of the pro-Nazi *Novo Vreme* (New Times), which had been assigned to it, were partially damaged. I told him that *Politika* did not have to wait, because it was an important newspaper—in its own way as important as *Borba.* I record this because some people attribute the decision for *Politika*'s first issue to Tito. There probably was some talk of this on Vis, if not in Jajce as well. But it was not Tito who made the immediate decision, because he was still in Romania. Nor did I decide by myself, but in consultation with other comrades. Be that as it may, *Politika* appeared on October 27, and *Borba* on November 15, while Radio Belgrade began to broadcast on November 10.

That afternoon, I stopped by on business in the corps headquarters on Andra Nikolić Street. Everyone was busy there, so I sat in the waiting room. Suddenly the stairs creaked under a heavy-set, rough-looking Soviet general. I thought to myself: this must be the liberator of Belgrade, Zhdanov. I had already heard how irascible he was, and how at a banquet he had been attracted to a pretty Partisan girl; they say she went with him—prompted a little by our people—so the victor wouldn't remain unsatisfied amid the general celebration. I decided not to get up, even if it caused an incident. General Zhdanov stopped and demanded, with the muscles of his jutting jaws twitching, "Why don't you salute?"

Drawing myself up from the armchair, I replied in Russian, "And you should be able to recognize the ranks of allied armies!"

Zhdanov turned angrily and stalked out. At that, Dapčević and several staff members came downstairs. They calmed my anger, but were immensely pleased.

It was that evening, or the next morning, that we learned of Milutinović's disappearance. He had set out from Romania, where he had been with Tito, for Belgrade, to take over the finance post to which he had been appointed on the National Committee. He never got there. Some soldiers reported that they had been on a small boat which, on the night of October 23, came upon some mines as it sailed up the Danube. On the boat were some Soviet officers and a Yugoslav general, mustachioed and thin, who said, "I don't know how to swim—I'll drown!" Everything pointed to the worst—that Milutinović had perished.

Misfortunes of my own descended on me as well. I had been told earlier that my father had been killed by Albanian fascists in Kosovo.

And then the former inmates of the camp at Banjica told me that they had seen my younger brother Milivoje, battered and out of his head, as the police dragged him off to be tortured. He was brought to the camp at Banjica, but they didn't know who he was until a gendarme recognized him. Then Vujković informed the Special Police and undertook to win him over in a conversation. They tortured Milivoje some two months, then hauled him off to the execution ground. All that stayed behind him was a brief official record: "Same replied to all questions with 'I won't talk' and 'I wish to die honestly.'" My older brother Aleksa, who had been killed in 1941, was still appearing to me when Milivoje's torments became a part of my past and my hopes. I was often to feel the almost physical presence of my two dead brothers in my prison cell, and always took heart after my encounters with them in memory. It was then that I also learned about my middle sister Dobrana, whom the Chetniks had killed. She was pregnant at the time. I hope that her death alongside her beloved husband was not as hard as it was for Milijove. My grieving mother understood that to lose one's life was necessary, as it always has been. When I met her later in Belgrade, I tried to console her: "You've lost so much in this war!"

She had already made room for both the dead and the survivors in her life, so she answered, "I have given a lot of life, so I have enough left."

Tito returned to Belgrade from Romania on October 27, 1944, about noon. He came by boat, passing over the skeleton of the wrecked Pančevo Bridge, to land on a muddy shore where only we members of the Central Committee and top commanders awaited him. He had hardly exchanged greetings when he began to explode: why was nothing being done in Belgrade—the people were suffering, the fifth column was active, Milutinović hadn't been found, the front in Srem was disorganized! I detected in this an imitation of the Russians, probably of Stalin himself. But this didn't last long; by that afternoon, when we Central Committee members gathered around him, he had reverted to his own style of quick, almost abrupt decision-making, with personal openness and solicitude. At Tito's insistence the search for Milutinović continued, rather more methodically and persistently. After a few days he was found in a channel of the Danube, and recognized only by the Soviet medal and state money that he kept on his person. He was buried with honors. And so his wife, an ordinary secretary who had waited for him through nine years of prison and three and a half years of war, now received his unrecognizable body.

Tito's first act, on arriving from Moscow, was to create the Guard on November 1. I told Ranković that I didn't like this restoration of royal forms. He replied, "It's necessary, that's tradition!" A strong unit was created, under Tito's personal command and with a name such as the kings gave. Perhaps this was an inspiration from Moscow, carried out in our own fashion; a Soviet general had come to organize security for Tito.

In other ways more had been accomplished in Belgrade than could be expected, considering that we had the services only of the army, which was engaged in the front, and the secret police, which had not yet established itself. When we got into Belgrade we encountered not one—literally not one—member of the party. There were thousands of sympathizers, even wildcat nonparty groups, but the party members had been wiped out in camps, in gas extermination trucks, and on execution grounds. At the execution ground in Jajinici night after night—every night in the course of three and a half years—hundreds of hostages and patriots, mostly Communists and their sympathizers, were executed. Lack of resourcefulness—the failure to accommodate organizational forms to changing conditions—also contributed to this state of affairs. The party leadership was brave and persevering, but rigid and uncreative. The terror by the Gestapo and Nedić's Special Police, uninhibited by human considerations or legal regulation, had snuffed out not the resistance but the organization; an urban guerrilla cannot hold out against an army and terror.

Even before our entry into Belgrade we had established the criteria according to which the adherents of Nedić and Ljotić would be killed on the spot. This had already been announced, though most followers of Nedić and Ljotić were withdrawing with the Germans. To be sure, among the executed there were also those whom the worst, most unjust court would have spared. But wars, and especially revolutions and counterrevolutions, are waged according to dogmatic and ideological criteria—criteria which in the course of extermination become a passion and a practice, a custom and a virtue: a person is guilty not necessarily of having done something, but simply of having belonged to something. Rage and hatred were important ingredients in such decisions, but not decisive. To such men the judicial process seemed senseless, except in spectacular cases involving leaders: a court could not hand down hundreds, thousands of death sentences. The bitterness of our followers, and our assessment of the matter, combined in a decision: three years of fighting on the side of the Germans, or participation in punitive expeditions and in burning and massacring one's own people, were too bloody a record to allow the participant to be treated as someone who had simply been misled.

In Belgrade there were still spies, groups of Mihailović's men, all kinds of informers, provocateurs, as well as those who hadn't succeeded in getting away or who didn't consider themselves guilty of a capital offense. But ONZA, the secret police, was methodically tracking them down, spreading its dragnet, conducting arrests and executions. Even the highest officials were moving about securely; special measures were taken only for Tito, while the others went around with only a single bodyguard. Public utilities were also restored—water, electricity, even the supply system; the worst problem was the children—those without parents, without milk and bread, without homes. Military units grew rapidly; the influx of new recruits diluted the experience and stamina of the veterans. Luckily, however, the Germans were also having trouble recovering, so that the front in Srem had at least become stabilized. Meanwhile Kardelj, in his negotiations with Šubašić, had cleared the way for a new agreement, and the National Committee was infusing new life into its activities.

Thus Tito had no reason to be dissatisfied, nor was he for long—except regarding the behavior of the Soviet troops and the attitude of Soviet commanders, who were resentful of our admonitions or else oblivious to them. All of this went on simmering, and the simmering grew worse with each new and more drastic case. The city committee informed us that at Čukarica, on the outskirts of Belgrade, Soviet soldiers had raped and slashed open a pharmacist, to whose funeral five thousand people came as a demonstration. The behavior of the Soviet soldiers became more crucial for our influence and our stabilization. This was all the more the case as illusions about the Red Army, and consequently about the Communists themselves, were being destroyed, while we, still organizationally undeveloped, operated in a fluid, partly hostile, uncontrolled environment.

Finally, since no end was in sight, our feelings boiled over, and the Politburo resolved to speak with the chief of the Soviet military mission, General Korneyev. To lend authority, Tito felt it necessary for all four members of the Politburo to attend the talks, as well as the two most prominent commanders, Peko Dapčević and Koča Popović. Soviet representatives had awarded the Order of Kutuzov to our highest officials and commanders, while Tito received the Order of Suvorov. I was away at the time, therefore Tito presented me with the medal later. So all of us—except Kardelj, who later received the Order of Lenin—were wearing these medals when General Korneyev came to Tito's villa for the talks.

We were lined up, so that Korneyev immediately sensed something serious and unpleasant in the offing, and assumed a stiff, offensive stance. He barely listened to Tito, insisting that these were isolated cases exaggerated by reactionaries. We stared at each other. Tito went on with restraint, while Korneyev kept interrupting him with mounting anger. Finally I blurted out, "The problem is that our opponents are taking advantage of this: they keep comparing the assaults by Red Army soldiers with the behavior of English officers, who don't engage in such assaults!" At this Korneyev reddened and got up: "I protest most emphatically against the insults being leveled at the Red Army by comparing it with the armies of the capitalist countries!"

This virtually ended our conversation with Korneyev. Nevertheless, the Soviet command began to establish order and the attacks decreased in number—partly also because the Soviet troops were transferred to Hungary, an enemy state which did not idealize them, and which they didn't have to regard as an ally.

As soon as Korneyev had left, Kardelj reprimanded me, "You shouldn't have said that!"

Ranković added, "It really is awkward."

Tito didn't chide me, though he seemed ill at ease. Only Dapčević was overjoyed.

It never occurred to any of them, though, to alter their attitude to me or to diminish my role because of that incident, out of regard for the Soviet government. Indeed, the Politburo made me the speaker at the celebration of the October Revolution, held in the recently restored National Theater. Soviet agents began to spread reports that I was a Troskyite. However, Ranković was able to expose the source of that first attempt to discredit a member of the Central Committee. Having run into the stone-wall resistance of our party, the Soviet agents stopped the campaign. But I really suffered over the misunderstanding, over the scorn which these new idealized Soviet men had for me.

The agreement with Šubašić would probably have been reached before November 2, had he not had to check with the Western Allies. We agreed to a coalition government, on condition that the Allies promise beforehand to recognize it, and the King be replaced by a regency nominated by Šubašić and the National Committee.

Kardelj and Šubašić thereupon flew to Moscow, to consult with the Soviet government on the agreement.

The talks between Kardelj, Šubašić, and the Soviet government progressed amid mutual satisfaction that a legal way had been found to affect the change in Yugoslavia. Stalin had told Kardelj that, during recent talks with Churchill, he had agreed to a fifty-fifty division of influence in Yugoslavia. No one in our leadership saw anything inappropriate or even unpleasant in that at the time. We understood it as the neutralizing of British intervention, that is, as leaving Yugoslavia to its own inner forces; we ourselves were already not only the dominant but almost the only force in the country. As for the fifty-fifty formula, presumably the Soviet Union was merely dealing with the realities of the capitalist Great Powers, and so was adopting their ways for the sake of greater ultimate socialist aims. After the clash with Stalin in 1948 that almost forgotten "fifty-fifty" was remembered in Yugoslavia with bitterness and horror. But it was held against only Stalin. It was taken for granted that for Churchill, an imperialist, such ways were congenital. When Churchill received me in 1951, I asked him, "Now you too regard Yugoslavia as useful?"

He replied, "I have always done so."

I persisted: "But you made the fifty-fifty agreement with Stalin."

"Yes, but that agreement had to do not with territory but with influence," Churchill concluded.

Kardelj took a trip to Sofia as well, for talks regarding unification with Bulgaria. This was an old idea of the Balkan socialists which had been revived in Moscow, and tossed in during Tito's meeting with the Bulgarian leaders during his sojourn in Romania. In any event, it was a constant topic of conversation. Kardelj returned from Sofia greatly disappointed. "Incredible!" he recounted. "The Bulgarian Communists are simply fakers! They're glad to see you, they laugh with you—and then they trick you!" The Bulgarian leaders envisioned unification with Yugoslavia as a dual Bulgarian-Yugoslav state, which would have offended and belittled the Croats and Slovenes, to say nothing of the Serbs. According to their plan, the capital was to be Sofia. Tito was to be president of the government, and Georgi Dimitrov secretary of the party; in other words, power was to be in Bulgarian hands. Kardelj, as well as our leadership, looked upon unification with the Bulgarians as a process in which national rivalries would disappear.

Vojvodina, the region to the north of the Danube, was liberated at about the same time as Belgrade. With this, problems arose regarding the German and Hungarian minorities in the area. In actual fact the Politburo was saddled only with the problem of the Hungarian minority, because the fate of the German minority—the Volksdeutscher—was foreordained. Roughly speaking, there were 500,000 Germans and as many Hungarians. A sizable number of the German population—I heard about one-half—withdrew with the German Army, while the Hungarians stayed, perhaps because the Red Army was already waging successful though very severe battles in Hungary.

Our Germans had so embittered our army and people that their expulsion from our soil was mentioned many times in the Central Committee. Yet perhaps we might have changed our minds had not the Russians, Poles, and Czechs already decided for, and partially carried out, the expulsion of their Germans. We adopted this stand without any discussion, as something which the German atrocities made understandable and justifiable.

The remaining German population had been encamped in two or three villages. Before long the furniture of these German homes began to make its way into the offices and houses of our officials. German maids—excellent housekeepers, according to the wives of officials—were used as unpaid labor. The Central Committee later prohibited this practice as shameful. A large portion of the land covered by the postwar agrarian reform was in fact German. From these German camps, in which hunger and disease raged, and prostitution with the guards flourished, groups were systematically taken away and expelled across the Austrian border. Ranković later admitted that conditions in the camps were awful, and Kardelj pointed out that we thereby lost our most productive inhabitants. I believe that Tito kept silent because he felt it had to be so. I too drowned out my own consciousness and conscience in inevitabilities and the memory of German atrocities.

Early in September 1948—that is, after the clash with Stalin—Tito took Ranković and me along to hunt deer at Belje. As we rode along in coaches like counts, I noticed a long row of barracks with laundry hanging outside and poorly clad women and children milling around. The gamekeeper told me that they were "Schwaben," as the German minority is called in Vojvodina. I went into the barracks: about fifteen families lived there, separated by blankets and tent cloth. The women told me that they and their husbands were working without pay, that they had taken for themselves a patch of land behind the barracks and had made a garden out of it, that they had no rights and no medical care. Slaves. I brought this up when we all gathered in the hunting lodge. Ranković said that there were some thirty thousand such cases; they had stayed behind largely because of the labor shortage. We were horrified—

as if it had nothing to do with us, as if it hadn't happened under our own roof! Tito snapped, "They must be given civil rights like everyone else!" What do I think of this today? One becomes a conqueror by serving and sacrificing for one's legions; it is the fatal, unforgivable error of conquerors to ordain the destinies of men and nations according to wartime views and circumstances.

On the other hand, German prisoners of war were well treated by any standards. During the battle for Serbia and Belgrade, the killing of prisoners had stopped. True, commanders here and there tolerated the killing of Gestapo and SS men, but not army personnel. By the end of the war there were well over 100,000 German prisoners in Yugoslavia. They expressed the desire to work, and received more or less the same wages as our workers. They were reliable and methodical, and in great demand wherever there was a quick and complicated construction job to be done. They gained respect, and were even liked. We kept them too long—until 1949–1950—but we parted with them in a friendly manner. Here was still more proof that peoples should be judged by what they are like not just in war but in peace as well.

The problem of the Hungarians, on the other hand, was discussed and decided at a meeting at Tito's villa. Specifically, the party leadership of Vojvodina had adopted the view—a view that was undoubtedly based on the mood of the Serbian population—that the Hungarians should also be expelled from Yugoslavia. It was a reaction and a judgment by analogy: the Hungarian occupation forces and Hungarian agents had vied with the Nazis; they did not shoot Serbs in mass executions as in Kragujevac and Kraljevo, but only cast them under the ice in the Danube, or drove them from their lands, which they then gave to Hungarians. The party leadership of Serbia was in a quandary.

The Central Committee argued strongly against the expulsion of the Hungarians. The chief reason for our stand was the Soviet government's opposition. Soviet representatives had already made it known to us that Hungary would be socialist. Nor had the Hungarian army or local units pursued us as furiously as the Germans. Hungary was not so large that we had to fear it in the future, as we feared Germany. Among the Hungarians there was a resistance movement, however weak, and even fighting units, however few. Finally—though no one could explain why this didn't apply to the Germans as well—women and children could not be held accountable for the crimes of their rulers.

I stayed at the Hotel Majestic for a couple of weeks, and then moved to *Borba*—into the apartment of the former owner of *Novo Vreme*. Since the editorial board was made up of the leading comrades in Agitprop, *Borba* was the main newspaper from which our soldiers and activists in the rear got their news. Even then Tito didn't want *Borba* to be the organ of the Central Committee, but rather of the Communist party: that way there was less responsibility for the Central Committee, if it made a mistake. However, Tito accepted my arguments—in which Kardelj ably supported me—that we should not introduce censorship, though censorship existed in the Soviet Union. Kardelj and I stressed that censorship was unpopular, and that we Communists had long struggled for freedom of the press. I also explained that our staff was made up almost entirely of party members, so the line would be followed anyway. Our decision not to introduce censorship had more consequences than I could foresee at the time: the Yugoslav press has been less dull and the journalists more independent than anywhere else in Eastern Europe.

Following his return from Moscow, Šubašić flew off to London to convince the King to accept the regency and the agreement with Tito. The King resisted, under the influence of his entourage. He even threatened to withdraw Šubašić's mandate—and to repudiate the agreement with Tito! All this lasted much too long—for a king who was so weak and opponents who were so strong. Tito demanded of Šubašić that he stop haggling and return to the country. It never came to that: Churchill and the Soviet government exerted pressure on the King, who then accepted the regency, all of whose members were favorable to Tito. This hopeless and confused resistance by the King gave us good reason to launch a campaign against him; "We want Tito, we don't want the King," was our slogan, and that slogan didn't die down until the Constituent Assembly abolished the monarchy on November 29, 1945.

Friction with the British over the landing of commandos in Dubrovnik in larger numbers than had been contemplated would not have aroused our mistrust and ideological ire, had it not been for the fact that an American mission was still stationed with Draža Mihailović, and that the British had begun their intervention in Greece against ELAS.* Because of the efforts of German units from Greece, whose retreat across Serbia had been cut off, to push through from Montenegro toward Mostar and Sarajevo, we had asked the British for help in artillery. They gave us this help readily, and these batteries played an important if not decisive role in crushing the German penetration. But the British also insisted that their own commandos protect the batteries. The landing of

* The Communist-led Greek resistance army, which was battled by British forces following the German retreat from Greece.

some six hundred commandos heartened our opponents and awakened our own latent suspicions. We ordered our units in Dalmatia on the alert, and asked the British to stop further disembarkations.

Meanwhile the U.S. mission under Lieutenant Colonel Robert H. McDowell had gone beyond gathering information and evacuating American pilots from Chetnik territories, which also provoked our irritation and mistrust. In our top echelons ironic comments were made: the Americans were picking up whatever the English had left; the Americans didn't know what was going on; and the Americans were playing with the devil.

The British intervention in Greece at the beginning of December 1944 saddled me and my section with a delicate and special role. Our propaganda was an expression of a consistent leftist line. Though military and revolutionary necessities called for fiery zeal and implacability, the latter were certainly heightened by the intellectually more doctrinaire cadres of Agitprop. Tito tolerated this "leftism": militancy was a vital necessity. But this "leftism" could also spoil profitable and promising negotiations with the West and with Šubašić, and provoke criticism by the Soviet government. When British troops under General Ronald Scobie launched the intervention, Tito was unwilling to come out against it, particularly since the Soviet press had ignored it. Nevertheless, Tito gave us his consent to cover the story—but with care, not aggressively. The tone and line of argument grew sharper, along with the sharpening of our relations with the Western powers over Trieste, and over the return of our gold and our ships. At the beginning of 1945, at a closed lecture for leading Communists, I analyzed the significance of General Scobie's intervention. Ranković told me that the British mission had gotten wind of that lecture: caution and circumspection had already become necessary even among the elect.

Even if we hadn't been engaged in a war, and in the process of transforming ourselves into a state, we Communists were so few in number that we couldn't be exclusive in our approach. This was also required by our collaboration with the democratic parties, even though they had shrunk into small groups which we didn't allow to organize and spread; they were to stay put in a "united front" under a joint program, and eventually wither away with the establishment of socialism. Latitude and patriotism in word and in deed, were not contrary to the strengthening and monopolistic proliferation of the party.

I participated in negotiations with the democratic groups in Serbia, in the name of the Central Committee. We had long discussions and often tussled over details, but always arrived at an agreement: the heterogeneity of the participants made the transition easier for the Communists and cushioned collisions, while those groups had nowhere to go except with us. Those who stand out most distinctly in my memory are Jaša Prodanović, the leader of the Republicans, and Dragoljub Jovanović, the leader of the left-wing Agrarians.

Uncle Jaša, as everyone called him, was already in his eighties, a relic of the old democratic Serbia with its political party system. We Communists also had blood ties with Prodanović: his son had perished in the war. He never mentioned it. He kept his grief to himself, and this bond didn't seem to figure in his views; antimonarchism was for him not only the highest principle, but apparently his one political passion. With him honesty was a creed, and education an end in itself; everything about him emanated from the study, isolated from the currents of time and the storms of life. Once when he and I were engaged in a sharp exchange, he left to go to the toilet, and I went after him to make sure that he was all right. Later he said of me, "Yes, he's tough, but he's decent, too; he followed me in case I needed help." And once he shouted at me, "For me it's the form that counts, not the content! You can proclaim Communism tomorrow, provided you've got the majority of votes!" I later repeated this to the comrades, who all laughed: for us, only the content was important. Prodanović also became vice-president of the Yugoslav government—a symbolic post in his eyes and our own. His only job was to receive complaints over arrests and expropriations; those he considered valid he would pass on to Ranković, who went through them patiently and carefully. Perhaps that is why the business community called Jaša "the Wailing Wall of Serbia."

Dragoljub Jovanović was the only prominent ally of the Communists who steadily—and stealthily—guarded both his party's independence and his own. He was incorruptible and fearless, but without verve or any great ideas, utopian, an amalgam of visions and pettiness. It was easy to reach agreement with him so long as it didn't threaten his independence. I had no clashes with him in the course of negotiations. He simply repressed any flexibility in his ideas and aims—even in his personality. We leaders, particularly from Serbia, secretly nursed a prewar hatred of him, and lamented our error in not having liquidated him during "that mess" at the time of Belgrade's liberation. Jovanović had hidden out during the war and didn't reappear in public until after the liberation of Serbia.

The Orthodox Church, however conservative and secretly monarchist, was known for its patriotism. It ignored for the moment the godlessness of the new regime, and invited us to take part officially in the thanksgiving service in the cathedral on the occasion of the liberation of Belgrade. At a meeting of the Central Committee we discussed Tito's going, but decided against him because he was a Croat, a "Catholic," and the leader of the party. We designated instead the chief of the Supreme Staff, Arso Jovanović, and also Moša Pijade. We then remembered, with laughter, that Pijade was Jewish, but that didn't matter, since he wasn't representing a religion but rather the state.

The day of the Serbian Saint Sava, the medieval statesman and educator, was a traditional school holiday, and we accepted the proposal by the Serbian leadership that we continue to observe it. But how were we to inject our own new element into this traditional observance honoring the most eminent Serbian saint? At a meeting I held with the Central Committee of Serbia, it was Zogović who showed himself the most inventive in updating and in "Marxifying" Saint Sava's image. He suggested that we focus on the Saint Sava of folk legends, the wise doer of good deeds who taught children to plow, to weave, and the like. A few years later we abolished the celebration of Saint Sava's day in our schools. Certainly I too was for that at the time, though today I maintain that the destruction of the past can only destroy the future of the destroyer.

Early in 1945 a delegation of the National Committee was sent to Moscow to seek economic and military aid. The delegation was headed by Andrija Hebrang and also included Arso Jovanović, the chief of the general staff, and my wife Mitra. Both as a group and individually, they were subjected to criticisms of conditions in Yugoslavia and of particular leaders. These criticisms consisted most often of half-truths, directed mainly against Tito, namely that he was more concerned with the restoration of royal palaces than with military operations; he had little understanding of military, and economic, matters; too much was being made of him personally. At the same time Hebrang was writing report after report for the Soviet leadership. He didn't do this in secret, nor would we have taken it for a deadly sin, had he not undermined our leadership.

Stalin took for himself the final tragic role in demoralizing our delegation. He received the delegation at a party in the Kremlin, and while Molotov coldly provoked them, Stalin fell into a tragic pathos. He claimed that it wasn't Tito he had attacked, but the Yugoslav Army and its organization—and me. He shed tears as he spoke of my "attack" on the Red Army, "an army which pushed its way across thousands of kilometers of wasteland, which didn't spare its blood even for you, yet this very army was attacked by none other than Djilas! Djilas, whom I received so well! Does Djilas, who is himself a writer, know what human suffering and the human heart are? Can't he understand the soldier who has gone through blood and fire and death, if he has fun with a woman or takes some trifle?" He proposed frequent toasts, joked, teased, wept, kept kissing my wife because she was a Serb, but with the taunt, "I'll kiss you even if the Yugoslavs and Djilas accuse me of having raped you." Arso Jovanović defended our army and wept, until Hebrang chided him for contradicting Stalin. Mitra cried, too. "How could you keep from crying," she told me on her return to Belgrade, "when you see Stalin in tears?"

Mitra's account shook and stunned me. Yet not so much that I couldn't see in all this an attempt to weaken, if not even depose, Tito and Tito's leadership. Mitra quickly snapped out of her depression, as the other leading comrades were also told of the dramatic supper with Stalin.

The building owned by the Madeira Restaurant was remodeled for occupancy by the Central Committee. Agitprop got some space there, too, and I ended up spending more time there than at *Borba*. Ranković's office was on the same floor, which facilitated our close contact. Even at that time the Soviet intelligence service was recruiting people in our top agencies, including veteran party members in sensitive posts. The Soviet agents justified it as follows: "Yes, our relations are good now, your leadership is truly devoted to the Soviet Union, but one must play it safe. We've bad experience with Trotskyites and other enemies and foreign agents in the party." In all this, to be sure, methods were also employed that were hardly ideological—women were seduced, Soviet actresses used as bait, gifts passed around, and the like. They even tried to recruit a woman code clerk in the Central Committee who came running to Ranković in great agitation: she couldn't understand why any "security" was required outside our own Central Committee and without its knowledge. In such cases Ranković would invite over the chief of the Soviet intelligence service, Colonel Timofeyev—if indeed that was his name and if indeed he was a colonel—and pedantically cite the facts at him. The embarrassed Colonel Timofeyev would promise, "This won't happen again! Some overzealous people are doing it on their own!"

Tito reacted to this recruiting with anger and resolution: "This must be stopped! What's the meaning of all this? Aren't we also Communists?

Today someone plants spies among us, tomorrow our heads may be in danger. This demoralizes our people and kills their confidence in the leadership."

In reaction to these Soviet "misunderstandings," we made it a policy to put Tito forward as our leader, particularly in our propaganda, and always mentioned Tito's name along with Stalin's in anything that concerned us. Tito was magnified: he grew out of our own need for autonomy and identity, and not simply out of a domestic power struggle.

The uprising by the Albanians in Macedonia and Kosovo didn't disturb our top echelon, though it dragged on from the end of 1944 till March 1945. For us this was largely a military problem. Our regime was as foreign to the Albanian peasant masses as that of the Serbian kings, and our effort to send Albanians to the front for their "re-education" was used to good advantage by the profascist Albanian leaders and the Germans. Despair and hopelessness: Albania was also in the hands of Communists, and units of Albanians fought on our side. The public knew hardly anything about this broad uprising. Later I inquired of General Savo Drljević, who quelled the uprising, if there had been any killing of children and raping of women: "No, we were strict in preventing that. But several shepherd boys were killed here and there, and there was looting. By that time our army was based on general conscription, and we had some criminals and chauvinists." There was some regret about this, but only after the rebellion had been brought under control.

Soon after Mitra's return from Moscow, I moved from *Borba* into the Rosche Villa in Dedinje, an exclusive suburb of Belgrade. The scramble to get villas in Dedinje began with the liberation of Belgrade—a scramble first for the abandoned ones and then, because there were so many aspirants, for those that had been expropriated for both legitimate reasons and reasons not so legitimate. I didn't succumb to that scramble—not until Zogović's wife was told about the modest villa of the Rosche family, whereupon they took the downstairs with one room, while Mitra and I took the upstairs with two. As it was, this modesty of mine didn't last long; the separation from my own high circle was enough to smack of discontent and deviation. In the fall of 1945, when my mother, brother, and sister moved in with me, I had to move also because of cramped quarters.

In the acquisition of Dedinje villas there was social climbing, jealousy, and backbiting. But the expropriation of factories, workshops, and stores belonging to collaborators was accepted with guarded enthusiasm by our top echelons, and with open enthusiasm by the Communist rank and file. Among the latter, not only deserving Communists but recent converts had come into managerial posts and were handling large funds.

To be sure, there were people who had gotten rich by working closely with the Germans. On the whole, the wealthy had behaved cravenly and venally under the occupation, so when the expropriation got under way, every prominent capitalist could be accused of collaboration. Factories, workshops, estates, hotels, and apartment houses had been utilized by the enemy regardless of the attitude of the owners. One day the woman who owned the Majestic Hotel, and whose son I knew to be a sympathizer, came to see me at the Central Committee. They had arrested her husband, a well-known architect, as a follower of Ljotić. She said to me, "My husband and Ljotić have known each other since their youth, and Ljotić used to stay at the hotel. But my husband is not a follower of his. He was arrested because of the hotel. Take the hotel, but let my husband go. He'll always earn enough for us to live on." I intervened. The husband was immediately released, but the hotel was expropriated on the grounds that German officers had lived in it.

That is how the nationalization—that is, the expropriation—of the large estates and the property of the well-to-do was carried out. This became the pattern for the whole country as each region was liberated. The top echelons had no qualms over what they viewed as nationalization—a nationalization with special "popular" motivations. It was an extension of a war which was national in its inspirations and slogans, and Communist in its core and objectives.

At the same time special shops with token prices were set up for high officials, their supplies commensurate with the rank of their privileged clientele. The best was the "diplomatic store," which supplied diplomats and the highest officials. Immediately there was abuse—buying for relatives—so that limitations were imposed, though not effectively. The shops were organized on the Soviet model. Though necessary at first, for reasons of security and administrative functioning, they were a luxury and a waste from the start, particularly in a time of deprivation. They were maintained far too long—until 1951. Privilege begins as a necessity and then becomes established as a right.

Following the Yalta agreement, the British and Soviet governments applied pressure on King Peter to accept the proposed regency. On March 7, 1945, the Tito-Šubašić government was announced; we Communists became a royal government, though only formally and without a king. I was included in the cabinet. It was Kardelj's idea that the federal units ought to have representatives in the cabinet, so I became the minister for Montenegro. Later I became a minister without portfolio. I didn't have any particular assignment in the cabinet except on a temporary basis. In effect, I had a title and a salary. My real work was in the Central Committee, until my fall in January 1954.

Problems developed with the Soviet representatives over the composition of the new government as well. The first was over the appointment of Milan Grol, the leader of the Democratic party, as vice-premier. Grol had impressed Tito and Kardelj by his levelheadedness and culture, and they decided to include him in the government—without consulting the Soviet representatives. The Soviets looked upon this as infiltrating from within by the British, even though it could not be said that Grol worked for anyone except himself and his fellow party members.

The second problem was the government declaration, which Kardelj and I edited. It listed the three Allies on an equal footing—the U.S.S.R., Great Britain, and the United States. The Soviet Union was not singled out, and that was "proof" enough of deviationism and a bias toward the West.

And finally, there was Tito's dinner for the newly arrived ambassadors of the three Great Powers: he gave the British ambassador, and not the Russian, the place of honor. In actual fact, Tito was acting according to protocol: the British ambassador was the doyen. To his inner circle Tito complained, "Really, one doesn't know what these Russians want!"

It was with my arrival in Belgrade in October 1944 that my migraine headaches began, incurable and unexplained to this day. Headaches aside—if there is no trouble, man will invent it—for no apparent reason, and amid power and plenty, there developed within me an emotional alienation from and conflict with Mitra. We would have an argument and then not speak for days, for weeks. At that time, breaking up with wartime and "party" loves was quite common—probably that was why I didn't do so at that time. Or perhaps the time had not come yet; I was not yet my own man, my ideas and emotions had not yet sorted themselves out.

2

The Soviet government lost no time in underscoring its dominant role in Yugoslavia through the mutual aid treaty with the newly formed government. The minister for foreign affairs, Šubašić, took all the credit, even though everything had already been arranged behind his back. On the surface the agreement made no sense: an alliance against a Germany which was on the verge of collapse, by countries which had been fighting together against it for nearly four years. But that was just a form, part of a pattern by which the Soviet government was establishing and reinforcing its influence.

Tito headed the delegation; without him, it wouldn't have had the necessary character. Besides Šubašić, the delegation also included two economic ministers. And myself—not only because Tito wished to take along someone close to him, but even more to iron out through direct contact my "insult" against the Red Army. Tito told me with an enigmatic but friendly smile, just two or three days before our departure, that I too was to go to Moscow.

The delegation left on April 5, 1945, in a Soviet plane. Tito was so airsick that he vomited. As for me, the closer we got to Moscow—under a somber, opaque sky over a gloomy and ravaged land—the more keenly I felt an already familiar and bitter loneliness; for months prior to this trip, I had been made to feel like a transgressor who had little chance of redeeming himself. Both the plane and my faith swayed painfully, and yet it all seemed quite unreal. The only support that I felt came from

Tito. He had said nothing about the whole affair, but there was no change in his relationship of cordial patronage toward me.

Though he had vomited again just before our plane touched down in Moscow, he endured the ceremonial welcome by sheer will power, as if he hadn't been ill throughout the entire trip. Resistance hardened within me as Molotov extended his hand without even batting an eye to show that we knew each other. Tito was lodged separately in a villa, while they put us up in the Hotel Metropole. There, dubious Katiushas and Natashas offered themselves to us Communist ministers by phone. I didn't ask Šubašić whether he had been exposed to such enticing propositions, because as a Communist I was ashamed of such methods in the homeland of socialism.

The economic ministers had practical affairs to attend to, while Tito and Šubašić were involved in protocol. The treaty had been prepared in advance; the only problem was the comparison of translations, which I took care of in a moment. I made the rounds of museums and old bookstores, but unfortunately not of the theaters, because evenings were taken up with banquets and receptions.

The treaty was signed on April 11, in the Kremlin, with Stalin drinking toasts to Yugoslav-Soviet friendship—the only charming episode. Stalin was sulky toward me. After supper, the toasts, and the inevitable films, he shook hands with me without a word. But I was by now more relaxed, whether because of the tolerable atmosphere or because I had grown tougher even toward Stalin. Then, at a dinner in the Kremlin, Stalin relaxed even toward me, at least in that he took notice of me with his inscrutable eyes.

At that dinner Stalin—and only he could have had the boldness—broke up the stiff, artificial atmosphere. As a rule we and the Soviet representatives addressed each other as "comrade," but at banquets and receptions, because of the presence of Šubašić or Western representatives, we used "mister." Stalin, however, raised his glass and addressed Tito as "comrade," adding that he would not call him "mister." Shouting across the table, he also made jokes, and flustered the senile and half-blind President Kalinin while guffawing loudly. But he never made jokes at his own expense: a divinity is a divinity only if he acts like one. The apotheosis of Stalin and his domination were felt more completely and directly in that circle than at the suppers in his villa.

I too was invited to an intimate supper in Stalin's villa. I was pleased about that, though I had a presentiment of an encounter with Stalin over my insult to the Red Army. The supper was attended on the Soviet side by Stalin, Molotov, Malenkov, Beria, Bulganin, and General Antonov; on our side, by Tito and us three Communist ministers. As for Šubašić, who didn't even know of our banquet with Stalin, our hosts put on a special supper for him about which he boasted the next day.

Tito was offered the place at Stalin's right. I sat across, to the right of Molotov. Face to face with Stalin, I suddenly acquired confidence.

As soon as the toasts and jests had warmed up, the conversation took on a comradely directness, and Stalin "remembered" to liquidate the dispute with me. He did it half jestingly: he poured a small glass of vodka for me to drink to the Red Army. At first—perhaps because of the great strain—I didn't understand what Stalin intended. Hard liquor gave me headaches, and so—in order not to spoil the party—I drank beer. I couldn't refuse the glass which Stalin offered me. I wanted to drink it to his health.

"No, no," Stalin insisted with pointed banter—as it dawned on me what he intended—"but just for the Red Army. What, you won't drink to the Red Army?"

I drank to the last drop, as if making a confession, but without shame or embarrassment, for it was to Stalin.

I briefly explained to Stalin the reasons behind my remark about the Red Army. All this was obviously long since clear to him: in the Soviet displeasure over my so-called attack, there prevailed a Great Power sensitivity for which accuracy and good intentions are irrelevant. Interrupting me, Stalin began to hold forth: "Yes, you have, of course, read Dostoevsky? Do you see what a complicated thing is man's soul, man's psyche? Well then, imagine a man who has fought from Stalingrad to Belgrade—over thousands of kilometers of his own devastated land, across the dead bodies of his comrades and dearest ones! How can such a man react normally? And what is so awful in his having fun with a woman, after such horrors? You have imagined the Red Army to be ideal. And it is not ideal, nor can it be, even if it did not contain a certain percentage of criminals—we opened up our penitentiaries and stuck everybody into the army. The Red Army is not ideal. The important thing is that it fights Germans—and it is fighting them well, while the rest doesn't matter."

The quarrel over the behavior of the Red Army was thus patched up—or so we thought. At the time my first impression of Stalin became even stronger: an excellent, still lively memory, and a vivid, concrete imagination. Today I would add that Stalin seemed to possess a special intuitive power to see into men. I felt divested in front of him, and even glad that it was so. It was in no way apparent—nor did we know of it—that he had lost his son Yasha in the war. He had grieved over him for two or three days, then accepted it as a necessity and gone on as if nothing had happened.

Stalin teased Tito with good-humored mischief. Most often it was with unflattering comments about the Yugoslav army, and praise for the Bulgarians: "The Bulgars had their weaknesses and enemies in their ranks. But they executed a few score—and now everything is in order. The Bulgarian army is very good—drilled and disciplined. And yours, the Yugoslav—they are still Partisans, unfit for serious front-line fighting. Last winter one German regiment broke up a whole division of yours. A regiment beat a division!"

Stalin's praise of the Bulgarian army was not simply to goad us Yugoslavs: the Soviet leadership, like the czarist, was more favorably inclined to the Bulgarians. The Panslavic Congress in Sofia which had taken place just before our visit to Moscow, was held on Soviet initiative, to build up the prestige of the Bulgarians—the only Slavs to have fought on the side of Germany and Italy. We in Belgrade accepted this without the least bit of envy. I had headed the Yugoslav delegation to that congress. The Bulgarians were extremely hospitable—indeed, extravagantly so. The most unpleasant, the most dissolute impression was made by the Bulgarian Exarch; there was nothing priestly even about his appearance—bloodshot and bursting with fat. Though he had only yesterday served their own tsar and blessed the Bulgarian army that had spread devastation through Greece and Macedonia, he got it into his head that he would outsmart the Communists by imposing his patronage on this assemblage, about which the only thing Slavic was the hackneyed phraseology of Russian imperialism, since the motives and aims were Communist. At a certain point during the Exarch's thanksgiving service in the Sofia cathedral—which was an endowment of the Russian tsar—the Exarch, as required by the liturgy, went from dignitary to dignitary with cross in hand. In a fit of nationalist and Slavic intoxication, they all kissed the cross—members of the Bulgarian Central Committee, Soviet generals, the proud and the embarrassed Poles, and the meek and mild Czechs. When my turn came, I said that I was an atheist, and the Exarch drew back in silence. Our delegation followed my lead. In it was a professor of medicine, Jevrem Nedeljković, a non-Communist from Serbia. I asked him later,

"Why didn't you kiss the cross? You're religious."

He replied, "I can't kiss a Bulgarian cross."

Over the years, Stalin had embraced all sorts of increasingly irrational ideas and impulses with an ever more autocratic and total power. And we were led by him, out of necessity, just like our forebears in the enslaved and powerless South Slavic lands.

Stalin then proposed a toast to the Yugoslav army, adding, "But which will yet fight well on level ground!"

Tito restrained himself. Whenever Stalin made some witty remark, even though at our expense, Tito looked at me with a smile, and I would return his look with sympathy. But when Stalin said that the Bulgarian army was better than ours, Tito couldn't take it anymore. He shouted that the Yugoslav army would soon show that it had gotten rid of its weaknesses.

It seemed as if Stalin and Tito held a grudge against each other. Without striking at Tito personally, Stalin took sideswiping jabs at conditions in Yugoslavia. Tito treated Stalin as his senior, but without humbling himself, and even countered criticisms of Yugoslavia.

Tito said that socialism today was advancing in ways different from those of the past. This gave Stalin an opportunity to say, "Today socialism is possible even under the English monarchy. Revolution is no longer necessary everywhere. Just recently a delegation of British Laborites was here, and we talked about this in particular. Yes, socialism is possible even under an English king."

I interjected that in essence Yugoslavia had a Soviet government because the Communist party held all the key positions, and there were no serious opposition parties. Stalin retorted, "No, your government is not Soviet—you have something in between de Gaulle's France and the Soviet Union." Deep down I didn't agree with this assessment of Stalin's, particularly since it seemed to me that he not only didn't know, but didn't wish to recognize, the essence of the change in Yugoslavia.

Stalin didn't tolerate monologues, not even his own, though he did most of the talking. Only Tito and Molotov participated in the conversation. I did some talking, the rest almost none at all.

"This war is not as in the past," Stalin said. "Whoever occupies a territory also imposes on it his own social system. Everyone imposes his own system as far as his army can reach."

He also expounded on the reasons for his "Panslavic" policy: "If the Slavs keep united and maintain solidarity, no one in the future will be able to move a finger."

Someone remarked that it would take the Germans more than fifty years to recover. Stalin observed, "No, they will recover, and very quickly.

Give them twelve to fifteen years and they'll be on their feet again. And this is why the unity of the Slavs is important."

At one point he hitched up his pants and cried out, "The war shall soon be over. We shall recover in fifteen or twenty years, and then we'll have another go at it."

One could tell, especially from Molotov's remarks, that the Soviet leaders respected Churchill as someone farsighted and dangerous.

As always there were many toasts, and much drinking and eating. On our way back, Tito, who also couldn't take large quantities of liquor, remarked, "I don't know what the devil is wrong with these Russians that they drink so much—plain decadence!"

Even Tito could not understand why the Soviet top echelons drank so desperately. In any event I made a tally: the controversial questions had been settled—for political reasons of state, and not out of comradeship and cordiality.

We also had dinner at Dimitrov's, in a villa near Moscow. Tito and Dimitrov reminisced about the Comintern. There was talk of uniting Bulgaria and Yugoslavia, and some excellent Soviet performers who bored us just a little. Otherwise, the evening was pleasant and a bit sad. One could sense Dimitrov's apathy: the Russians were keeping him in Moscow while the rest of the émigrés had long since returned to Bulgaria and were assuming the highest positions. Russia is inscrutable even to the Russians, even when everything is clear and in order. Is that not its strength, its misfortune, and its magic?

While in Moscow, we received an invitation to visit the Ukraine. The Soviet leaders had established commissariats for foreign affairs in the Ukraine and Byelorussia as a means of getting these republics into the United Nations. These commissariats had no staffs, and certainly no independent foreign policies. Having been put together hastily, they had not yet gotten around to the simplest formalities of organization. The suggestion had probably come from Kiev—from Nikita Khrushchev, who was then secretary of the Ukrainian party and premier of the government—that on our way back we visit the Ukraine.

Khrushchev, his commissar for foreign affairs Manuilsky, and other Ukrainian officials met us all the more cordially, since they regarded our visit as a confirmation of their own political status. Everything was less formal than in Moscow, less suspicious and tense, more lively and direct. Even the Ukrainian language sounded more friendly to us.

The Ukrainian leadership was dominated by Khrushchev, by virtue not only of his function but also his personality. He held everything in his hands, more directly than Stalin, but without inspiring such awe. With Stalin, one detected no tie with the people, except in a strange ideological sense. Stalin was interested in goals, but very little or not at all in people and how they lived. Khrushchev tried in every respect to be a man of the people: even in a general's uniform, he was casual and at ease. He was certainly popular in his speech and manners, but in his thinking—in his mode of thought—he was the Marxist of party schools, with an inherited peasant practicality. And his sense of humor was roguish and rather vulgar. He seemed to be the sort that holds grudges: despite Manuilsky's cajoling and chiding, he refused to applaud a singer at the opera who had entertained the Germans. But he didn't exhibit any vindictiveness. Even his behavior toward the singer was not revenge, but rather a childish, dogmatic grudge. He was extremely talkative and delved into petty details. One could have arguments with him without straining relations. Actually, he was a populist democrat, an authoritarian without formalism and self-adulation. Short and stocky, but agile, he ate and drank well, like a man of the people who had suddenly struck it rich. Unlike Stalin, for whom ideas were a means, Khrushchev used ideas readily to justify practical reversals in policy.

Situated on hills beside a large river, Kiev reminded us of Belgrade. But Kiev was more beautiful and better preserved—as we would have wished Belgrade to be. It was not until April 20 that we flew back. The Ukraine, with its meager, half-Russianized public life, receded into facelessness and hopelessness. Doubt about the official portrayals of Soviet reality took root in my consciousness as a reality of its own.

3

The delegation of the Yugoslav government was still in Moscow when on April 12, 1945, our army broke through the Srem front. The front was defended mostly by the Germans, who in the course of five months had built up its defenses with their superior experience with this type of warfare. And our army broke through that front using methods which were almost worthy of the armies of the Great Powers. In February, in connection with my army propaganda work, I had visited the staff of the First Army, the one that was holding the Srem front. The roads were muddy. Desultory firing echoed nearby. The headquarters teemed with activity; messengers were coming and going. I gathered then from my conversations with the commander, Dapčević, that without Soviet instructors and arms we wouldn't have been able to break through the front. But it is also true that our officers learned quickly and perfected their otherwise rich fighting experience.

The breakthrough strategy had been worked out by the headquarters at the front and the general staff, of course with Tito's approval. But the order for the breakthrough was signed by Ranković, as the most prominent member of the Supreme Staff, because Tito was in Moscow as head of the government delegation. The penetration of the Srem front was effected by units of all the Yugoslav peoples except the Slovenes, who were prevented from being there by geographical and other reasons. But the greatest losses were suffered by the Serbs, who had the largest number of new units, and the largest number of new recruits in their old units. This was the greatest and bloodiest battle our army ever fought. By its generous self-sacrifice and unstinting losses, Serbia once again confirmed its defiance and political awareness. By the time he reached Zagreb, Dapčević's army had lost around 36,000 dead. The German units, not fully manned and certainly less confident than before, nevertheless successfully resisted our inexperienced army. Even Pavelić's Home Guards fought better when, at the end of 1944, Ustashi formations were poured into them to give them a core.

Yet it was that very battle, that breakthrough, which has been most frequently criticized. The criticism doesn't come only from opponents—some of whom claim, out of frenzied hatred, that all that vast slaughter was calculated in order to exterminate either the Serbian youth or else the Croatian state. I have heard well-intentioned criticisms even from participants: essentially, that all that fighting was superfluous because the Germans were collapsing, and the new government was already being established in Belgrade.

This view seems superficial to me, though I see its rationale. A war, particularly a revolutionary war, is always waged by forces with definite goals. Every war and every battle is begun by certain concrete forces, taking into account external and internal circumstances, for the purpose of carrying out definite tasks. Hence Clausewitz's most famous dictum: war is the extension of politics by other means, that is, by weapons. On the Srem front we had against us not only the Germans but also domestic enemies who were ready, at the fall of Germany, to place themselves under the protection of the Western powers, which were suspicious of and inimical to Communism. Had that front not been held by Communists, it might have been wise to put off the breakthrough.

In the spring of 1945, reports reached Belgrade of preparations within Pavelić's regime—involving leaders of the Croatian Peasant party—to declare an independent Croatia, and place it under the protection of the Western powers. These reports were confirmed, in part, when Croatian Foreign Minister Mladen Lorković's plot to overthrow Pavelić was uncovered. Draža Mihailović's main force, though shrunken and demoralized, had withdrawn into northern Bosnia, under the sheltering wing of the Germans. While the Chetnik forces of Djurišić and Djujić were making their way toward the Western Allies, those under Draža's direct command were heading toward Serbia.

Emotional and moral impulses for the breakthrough on the Srem front grew and pressed continually, particularly after January 1945, when despite their inferior forces the Germans scored successes against three of our armies: in Srem they made a breach, however shallow and brief, into the First Army; in northern Bosnia, they were successful against the Second Army in the defense of Bijeljina; and on the Drava River they forced the Third Army across to the left bank, into Hungary. This stung the rank and file, let alone the officers. The glory of the Partisans dimmed swiftly alongside the experienced and well-equipped Red Army, in battles along the level front. I will never forget Dapčević's despair, when he came from the front to Belgrade for a council, over some forty artillery pieces that the Germans took from him in that advance: "One German mortar is more effective than two of my batteries!" Or the criticism that the commander of the Second Army, Koča Popović, leveled against his own men: "Untrained peasant youths and schoolboys! Most of the trained men of Serbia are prisoners of war." And when I remarked to Popović that the Germans might capture his artillery, too, he retorted—though there was no rivalry between Dapčević and himself—"Oh, no! I'm keeping mine on the right bank of the Drina!"

Filled with zeal, the army yearned for battle, for a breakthrough, to show that they were ready for such fighting. Nations, movements, and armies justify their existence through their peculiar character, their distinctiveness. Yet moral impulses, though at times decisive in politics, could have been set aside had it not been for the reality, which was so unfavorable and ill-omened, and for our assessment of that reality, which was weighed down with concern and ideology.

While the Zagreb government was conspiring and plotting, and the Chetniks were heading west to join the Allies or returning to Serbia, the Western Allies hastened to force us out of the Julian March, and especially out of Trieste. Anticipating this attitude by the Allies, Tito rushed in the Fourth Army and on May 2, 1945, occupied Trieste—except for a few bunkers, which the British held onto. That same day, Field Marshal Harold Alexander demanded that we withdraw from Trieste, and during negotiations in Belgrade on May 5 General William Morgan, Alexander's chief of staff, insisted on our withdrawal behind a line which he drew on a map. Our units around Trieste were exposed to the frightening roar of hundreds of Allied planes, while Churchill menacingly refused to dismiss captured German divisions. We were rather afraid of a conflict with the West. Even if there were no direct conflict, we feared that groups which had but yesterday received German support would receive fresh and even stronger support now from the West. For this reason our armies were spurred on, not only to smash the Germans, but also to close off the frontier with Austria and Italy.

The German forces in Austria had capitulated, and the German government had accepted surrender in negotiations with the Western Allies, yet we were still quarreling with the West over Trieste, and fighting the Germans and the Home Guards, and spilling blood around Zagreb, which we did not enter until May 8.

On May 9 we leaders greeted the unconditional surrender of Germany—Victory Day—in bitter loneliness. It was as if that joy were not meant for us. No one invited us to the feast, even though—both as a movement and as a people—we had helped prepare it through the most terrible sufferings and losses. It is certain that 305,000 Partisans lost their lives; that for the better part of the war this new Yugoslav army tied down on its territory over twenty German divisions; that according to official statistics, Yugoslavia lost between 1.7 and 1.8 million inhabitants. This includes deaths in camps, massacres, and bombings, and both those who died fighting against Communism and those whom we Communists executed. One always gives round figures, as if human beings were not involved. It was not even possible to establish an exact figure: hundreds of thousands were killed anonymously, uncounted. That number may even be larger, though it seems to me—perhaps this is wishful thinking—that the total is exaggerated, unless it also includes the German national minority which fled or was expelled. But even without the minorities, the losses were horrible, particularly for a country of sixteen million inhabitants. At that time these losses were felt wherever one went; they were present in everything we touched or thought about.

The world celebrated peace and victory, while we were still waging war on a grand scale. Revolutions have their own peculiarities; they create special situations and mental states. Not even the Germans dared surrender to us, to say nothing of the Chetniks, the Ustashi, and other bitter mortal enemies. This may explain why I don't remember how and where I spent Victory Day on May 9. It had an impact of my consciousness, but not on my mood. Joylessness and bitterness overcame most of the leading comrades. Tito was obstinate: for several years Yugoslavia did not celebrate May 9 as its Victory Day but May 15, when our enemies laid down their arms. It was concern over how this looked to the outside world—and in time, a better perspective on our own importance—that finally caused us to adopt May 9 as our Victory Day as well.

The end of our war—May 15, 1945—found me in Montenegro, which, as its representative in the Yugoslav government, it was my duty to visit. The hospitality there would have been too much even for a rich, unravaged land. There were also undeserved, quite flattering panegyrics in my honor. I was showered with flowers, though Montenegro hardly abounds in them: Montenegrins grow flowers but little, and are scarcely able to distinguish the various kinds. Some episodes might have been amusing, if they hadn't portended disappointments—a reality different from the one we had hoped for and bled for.

Before the walls of Kotor only local officials and a column of skinny, shabbily dressed schoolchildren greeted us. The officials made excuses—the telephone system had broken down, as usual—and persuaded us to wait until the people had gathered: party members and Communist Youth had been sent out to round them up. As we exchanged impressions and reminisced about the war, some three or four hundred citizens showed up in their finest dress. The people of the Bay of Kotor are cultivated like the Italians, and traditionally look upon Montenegrins as savages given to plundering and feuding. Perhaps that is a little how they looked upon us as well, with our uniforms and weapons. When the people had gathered, the secretary of the district committee or the district president decided that the time was now ripe to make a welcoming address, climbed up on a counter in the outdoor market place, and delivered it. After him, I climbed up too—as the main attraction. I began approximately as follows: "In the course of the National Liberation Struggle . . ."

I was interrupted by the shrill shouting of a rawboned girl: "Long live the National Liberation Struggle!"

The people, with a trained unanimity, shouted, "Long live!"

I: ". . . under the leadership of the heroic Communist party . . ."

The girl: "Long live the heroic Communist party!"

The people: "Long live!"

I: ". . . and the leader and teacher of our peoples, Comrade Tito . . ."

The girl: "Long live the leader and teacher of our peoples, Comrade Tito!"

The people: "Long live!"

I: ". . . in alliance and with our great allies . . ."

The girl and the people repeated this, too.

Unsettled already, because all of this was but one single sentence, I continued: ". . . allies, primarily the Soviet Union. . . ."

I never finished that sentence. It was clear to me by this time that nobody was paying the least attention to the meaning; they were merely echoing hallowed clichés, repeating ritually the word symbols that it was the girl's job to signal to them. This, then, I thought to myself, is how my agitation and propaganda look out there. This is the extent to which the spontaneity of the people has been organized—it's grotesque! Even though my speech is unimportant, a matter of protocol, because of all this organization I won't even be able to deliver it. As if in jest, I blurted out, "Listen here, girl! Let me unload the heavy burden I assumed on my way down to Kotor, and I give you my word of honor that I'll mention all those slogans at the end!" The people of Kotor laughed uneasily and the officials sourly, while the girl withdrew, abashed.

It seemed as if all of Yugoslavia was synthesized in Montenegro, in the boundless confidence of the victors and the silence and shame of the vanquished. There were scorched walls along torn-up, demolished roads; rivers without bridges; railroad tracks with splintered, ripped-out ties. In the forests, outlaws—four to five hundred in Montenegro. We were escorted by truckloads of soldiers and jeeps full of OZNA agents. Titograd was so devastated by Allied bombings—they say there were over twenty—that it resembled an archaeological excavation through which only one path had been cleared. The people of Podgorica had scattered to the villages or to the caves around the Morača River. From these caves there still came smoke and the cry of children.

Aid from UNRRA was beginning to arrive. With it came jokes about Montenegrin laziness: when UNRRA sent pickaxes and shovels, the Montenegrins allegedly returned them: "That must be for some other fraternal republic—not for us!" And we leaders were amazed at our own generosity: UNRRA distributed equally to the Chetnik and the Partisan needy, though the latter grumbled: in Montenegro they didn't seem to know that they were bound to nondiscrimination by UNRRA's rules, and by our government's agreement with that organization.

The Montenegrin leaders were a reproduction of the central administration in Belgrade, only here, because of the poverty and familiarity, everything was made more obvious in its grotesqueness and profligacy. Blažo Jovanović, president of the government and secretary of the party, got the villa of the British legation to the Montenegrin court, while the rest got something less sumptuous, according to rank. I was also put up in Blažo's villa. In the evening we sat and talked in the *Locanda*—"inn" in Italian—just as in the days of Prince Nicholas. I slept badly and took a walk around the courtyard in the early morning, so as to get a view of myself and my course of action under the new conditions; I reasoned with myself that all these limitations and all this hierarchy was temporary and incidental, that our teachings and sacrifices would never allow us to become merely a new face for the old regime. I heard some young people singing, but their song quickly vanished into the mist-covered crags. Along the Lim, the Tara, and the Morača, everywhere on our journey we had met people: faceless crowds gathered at bridges, roadsides, public buildings. Out of the ruins and the hatred, life was unfolding with an irresistible momentum. Could not art, literature, be my path, my role in that renewal and creation?

In Montenegro, during its withdrawal at the end of the war, the German Army left a trail of heroism, though the domination by Nazism over that army and over Germany has suppressed in the world's mind even the thought of such a thing. The German 21st Corps had to withdraw from Greece across Albania and Montenegro; to its aid came the 91st Corps, which had made its way out of Kosovo and into the valley of the Lim. With the 21st Corps there also retreated the Montenegrin Chetniks, desperate men with fighting capacity. On the long and only way from Greece to the Drina River, the Germans were confronted by a devastated land, demolished bridges, enemy units, and the vengeance of those who were deceived, if only for a moment, by their reputed invincibility. Hungry and half-naked, they cleared mountain landslides, stormed the rocky peaks, carved out bypasses. Allied planes spotted them easily and used them for leisurely target practice on bare mountain slopes, at river fords, in deserted towns and villages. Their fuel ran out, their motorized equipment gave out and was destroyed; everywhere along the road there were charred and overturned trucks. We were told that the Germans killed their gravely wounded, whom they couldn't get out. They seized farm animals—anything they could find to eat. They took worn-out, shabby peasant clothing. No one begrudged them that, because they didn't molest civilians or burn dwellings. In the end they got through, leaving a memory of their martial manhood—albeit a fleeting and unrecorded memory. Apparently the German Army could wage war—and far more successfully at that—without massacres and gas chambers.

It was in Montenegro that I welcomed the end of our own fighting: on May 15, 1945, the German forces—estimated at 130,000—finally recognized us as a legal Allied army and laid down their arms. There were individual cases of retribution in the course of the disarming. The captured Germans were sent to camps and put to work. Along with the Germans, our enemies who collaborated with the invaders or bound their destiny to the fascist powers—the Chetniks, the Ustashi, Home Guards, and the Slovenian Home Guards—also laid down their arms. Some of these groups got through to the British in Austria, who turned them over to us. All were killed, except for women and young people who were under eighteen years of age—so we were told at the time in Montenegro, and so I later heard from those who had taken part in these senseless acts of wrathful retribution.

These killings were sheer frenzy. How many victims were there? I believe that no one knows exactly, or will ever know. According to what I heard in passing from a few officials involved in that settling of scores, the number exceeds twenty thousand—though it must certainly be under thirty thousand, including all three of the groups just cited. They were killed separately, each group on the territory where they had been taken prisoner. A year or two later, there was grumbling in the Slovenian Central Committee that they had trouble with the peasants from those areas, because underground rivers were casting up bodies. They also said that piles of corpses were heaving up as they rotted in shallow mass graves, so that the very earth seemed to breathe. In Zagreb, too, purges were conducted according to Belgrade standards. Serbian and Croatian nationalists each echo the other in claiming that their own side was treated more harshly than the other side. The same men, the same aims, the same means—they had to operate more or less in the same way everywhere.

The fate of Draža Mihailović's forces was even more terrible, because of the uncertainty and duration of their calamity. Along with refugees and soldiers from scattered units, Draža spent the winter of 1944–1945 on Vučjak Mountain in northern Bosnia. Typhus raged while discord and recriminations tore them apart. It was here that Draža was joined by the Chetniks and refugees from Montenegro, led by the self-willed Pavle Djurišić. Disputes arose out of Montenegrin-Serbian differences: Djurišić was for pushing to the northwest, toward the British, while Draža was for returning to Serbia. OZNA had previously gotten its hands on one of Draža's radio operators, who gave away the code and agreed to send spurious messages: the link between Draža and OZNA in Belgrade functioned well. Ranković and his helpers thought up the idea of enticing Draža into Serbia. In fact, I helped to compose the counterfeit messages which depicted the situation in Serbia as favorable to Draža. Draža also sent out saboteurs, in German planes and with German equipment, whom OZNA was waiting for.

Draža had no strength or resoluteness to deal with Djurišić, so he ceremoniously saw him off on his uncertain journey. Djurišić tried to make his way through the Ustashi army and lost much blood in the attempt. There in an alien region, for days on the marshy flatlands

Montenegrin women were heard weeping over their fallen brothers, fathers, and sons. It was then that Djurišić, under the influence of the Montenegrin separatist Sekula Drljević, entered into negotiations with the Ustashi. The Ustashi induced a large group of Chetnik officers—and with them the writer Dragisa Vasić—to surrender, then disarmed and slaughtered them.

Draža had no prospect of getting through to the Western Allies; we controlled the bulk of the territory and our units were numerous, and to go through the Ustashi territory, he would be courting the same fate that had befallen Djurišić. Laying waste to Moslem Fojnica on the way, Draža set out for Serbia. It is not clear to what extent that decision had been influenced by OZNA's bogus messages; certainly they encouraged him in his despair and illusions. On that march the Chetniks fought with great bitterness, and simply stomped over one of our brigades. But our forces methodically closed in on them and caught them in an impenetrable ring—in that same area where the Partisans had to break out of the German encirclement during the Fifth Offensive in May and June of 1943. In the Partisan memorial by the Sutjeska there must be more than one Chetnik bone. No prisoners were taken: they were shot down by those same peasants who had also attacked our field hospitals. Some seven thousand were killed; around four hundred escaped, among them Draža with a small group—only to be caught later. Our aviation pounded the Chetniks for days on end. Our commanders would say, "Let the Chetniks see what it was like for us!" And to the inquiries of Soviet instructors as to where the planes were flying off to, the answer came, "They're going after some rebels against the state"—an answer which sounded strange, incomprehensible, to a new Soviet generation which had forgotten what a revolution was.

Draža's force was destroyed at about the same time as the one in Slovenia. Tales of horror reached Montenegro along with tiny groups of crushed Chetniks. No one liked to speak of this particular experience—not even those who made a show of their revolutionary spirit—as if it were only a horrible dream. I too was stunned when I heard about it from the Montenegrin leaders, for a punishment is just only if meted out individually. But my astonishment lasted only a moment. After all, side by side with the invader they had waged war for many years against the children of their own people; they had run to the new masters; they had burned, tortured, slaughtered; they did not change even on the long road of retreat with the invader; they took no prisoners. The penetrating and decisive thought in these reflections was that the comrades, the party, thought it necessary to put an end to our hesitation and to Chetnik hopes.

I tried to persuade my comrades and myself: "Their treason has no end. The people and the homeland have no need of those who sullied their name and betrayed them in their most fateful hour. To kill collectively, without investigation, is senseless and contrary to our teaching, our faith in people—but in this case we've had enough! They all had a chance to see, to understand, and to leave." I don't claim I spoke these very words, but I did react in that spirit, defending our ideology and decision from doubts, momentary and eternal, about collective retribution, violence, and death. And everyone agreed, with bitter conviction.

Who issued the order for this extermination? Who signed it? I don't know. It is my belief that a written order didn't exist. Given the power structure and the chain of command, no one could have carried out such a major undertaking without approval from the top. An atmosphere of revenge prevailed. The Central Committee did not decide that. And what if it had? Doubtlessly the Central Committee would have gone along with those in power. There was never any voting anyway. And I would have agreed—perhaps with some reservation which would not have threatened my revolutionary resolve, my adherence to the party, and my solidarity with the leadership. As if there were no justice, truth, and mercy outside the ideology, the party, and an aroused people, and outside us leaders as their essence! We never spoke of it either in the Central Committee or privately among ourselves. Once in a rambling conversation—after the clash with the Soviet leadership in 1948, of course—I mentioned that we had gone too far then, because among the executed also were some fleeing for ideological reasons alone.

Tito retorted immediately, as if he had long since come to a final, though hardly comforting, conclusion: "We put an end to it once and for all! Anyway, given the kind of courts that we had . . ."

And so it was, somehow—"once and for all." Yet OZNA continued to carry out executions, according to its own often local and inconsistent criteria, until late in 1945, when at a meeting of the Central Committee Tito cried out in disgust, "Enough of all these death sentences and all this killing! The death sentence no longer has any effect! No one fears death anymore!"

The war and the revolution were at an end. But the hatreds and divisions continued to bring destruction and death, both inside and outside the country. The Allies threw us out of Trieste and its environs after the Soviet Central Committee informed us that, after such a terrible war, the U.S.S.R. could not embark upon another. The world seemed to turn against us Yugoslavs and we were alone, growing lonelier all the while. The revolutionary war was still going on inside us—we who were so fresh and self-confident. It was as if all this had not been a calamity, terrible and majestic in its totality.

Revolutions begin new epochs, whose direction no one can foresee, let alone determine. Would life be life if it had to conform to hypothesis? Revolutions must take place when the political forms are unable to develop reasonable and just solutions. Revolutions are justified as acts of life, acts of living. Their idealization is a cover-up for the egotism and love of power of the new revolutionary masters. But efforts to restore prerevolutionary forms are even more meaningless and unrealistic. I sensed all of this even then. But choice does not depend only on one's personal outlook but also on reality. With my present outlook, I would not have been able to do what I had done then.

Let us hope that in the end monolithic ideological revolutions will cease, even though they have roots in idealism and idealists.

Biographical Notes

Vladimir Bakarić (1912–)
A leading Croatian Communist who helped organize Partisan resistance in Croatia during World War II. He has since held high government and party posts in Croatia.

Mitar Bakić (1908–)
A leading Montenegrin Communist who helped organize the Partisan uprising in Serbia and Montenegro during the war. He held high positions in the government during and after the war. Proclaimed a National Hero.

Savo Brković (1906–)
Montenegrin Communist who held various military and political posts among the Partisans during the war. He was a member of the Montenegrin government and of the Yugoslav Assembly after the war. Proclaimed a National Hero.

Rifat-Trso Burdžoić (1915–1942)
Montenegrin Communist of Moslem background who helped organize the Partisan uprising in the Sandžak during the war. He was killed by Chetniks. Proclaimed a National Hero.

Radoje Dakić (1915–)
Montenegrin Communist who held various party positions in Montenegro during the war. He held high posts in the labor movement after the war.

Peko Dapčević (1913–)
Montenegrin Communist who fought in the Spanish Civil War, joined the Partisan uprising in Montenegro, and became commander of the First Army. In

1953 he was named chief of the Yugoslav General Staff, but was demoted as a result of being indirectly involved in Djilas's troubles with the party.

Vladimir Dedijer (1914–)

An editor of the Communist party newspaper *Borba* and a member of the agitprop section during the war. He later became a member of the party's Central Committee. He wrote two important accounts of Partisan history: *Diary* and *Tito,* both of which have been published in English. He broke with the party in 1954, and has since devoted himself to writing history and teaching.

Georgi Dimitrov (1882–1949)

Prominent Bulgarian Communist and a high-ranking official of the Comintern who lived in the Soviet Union for many years. He returned to Bulgaria at the end of World War II to lead the Communist party there, and became premier in 1946.

Mitra Mitrović Djilas (1912–)

Serbian-born first wife of Milovan Djilas. She joined the Partisans in 1941 and did party organization work. After the war, she held important education posts in Serbia.

Andrija Hebrang (1899–1948)

Prominent Croatian Communist and leader of the Partisan movement in Croatia during the war. In 1946 he was found guilty of wartime cowardice and collaboration with the Ustashi, and was relieved of all his posts. After being arrested while allegedly fleeing to Romania in 1948, he committed suicide; some sources claim he was murdered.

Duane T. Hudson (–)

Captain in the British intelligence service who headed the first British mission to Yugoslavia in 1941. He went first to Tito's headquarters, then to Mihailović's.

Arso Jovanović (–1948)

Officer in the prewar Royal Army who joined the Partisans in 1941 and helped organize their army, serving as chief of the General Staff through 1946. When Tito broke with Moscow in 1948, he sided openly with the Soviet Union. He was killed by border guards while trying to escape to Romania.

Blažo Jovanović (1907–)

One of the organizers of the Partisan uprising in Montenegro in 1941. He held high Communist political posts during and after the war. Proclaimed a National Hero.

Edvard Kardelj (1910–)

A leading Slovenian Communist who received prewar training in Moscow and was an organizer of the Partisan uprising in Slovenia in 1941. He later became a member of the party's Central Committee and in 1945 vice-premier of the new Yugoslav government. He had been, and is still, Tito's second-in-command and a leading party ideologist.

Boris Kidrič (1912–1953)

A leading Slovenian Communist who, with Kardelj, organized the Partisan uprising in Slovenia in 1941. He held high political posts in Slovenia during and after the war, became a Politburo member in 1948, and was in charge of the Yugoslav economy from 1946 until his death.

Sava Kovačević (1905–1943)

Montenegrin Communist who helped organize the Partisan uprising in Montenegro in 1941. Noted for personal bravery, he was killed in the battle at Sutjeska in 1943. Proclaimed a National Hero.

Dimitrije Ljotić (–1945)

Serbian fascist leader and head of the Serbian Volunteer Corps, which served under the Germans during the war and fought both the Chetniks and the Partisans. He died in an automobile accident while fleeing Yugoslavia in 1945.

Vlatko Maček (1879–1964)

Prewar leader of the Croatian Peasant party. Opposed to both the fascists and the Communists, he spent the war under house arrest in the fascist puppet state of Croatia, left Yugoslavia when the Germans retreated, and finally settled in the United States.

Fitzroy H. R. Maclean (1911–)

Brigadier who led a British mission to the Partisans, 1943–1945. After the war, he was a member of Parliament and was made a baronet. He is the author of *Escape to Freedom*.

Dmitri Zakharovich Manuilsky (1883–)

Soviet Communist party official and diplomat. He held important posts in his native Ukraine and in the Comintern.

Veselin Masleša (1906–1943)

Leading Communist publicist and journalist of Bosnian origin who joined the Partisan uprising in Montenegro in 1941. He was a propagandist and editor of the Communist party newspaper *Borba* during the war. He was killed in the battle at Sutjeska in 1943. Proclaimed a National Hero.

BIOGRAPHICAL NOTES

Draža Mihailović (1893–1946)

Colonel in the prewar Royal Army who organized the Chetnik resistance to the German occupation in 1941. He was promoted to general and named minister of war by the royal government-in-exile. When fighting broke out between the Chetniks and the Partisans, he collaborated with the Italians and later with the Germans. The British supported him until 1944. Reluctantly dismissed by King Peter, he was tracked down by the Partisans, captured in 1946, tried as a traitor, and executed.

Ivan Milutinović (1901–1944)

A leading Montenegrin Communist and member of the Central Committee who helped organize the Partisan uprising in Montenegro in 1941. He held high political and military posts during the war. In 1944 he was drowned when his boat hit a mine in the Danube. Proclaimed a National Hero.

Vladimir Nazor (1876–1949)

Noted Croatian author and teacher who joined the Partisans in 1942, at the age of sixty-six, shared in their wartime ordeal, and held various posts.

Milan Nedić (1877–1946)

General in the prewar Royal Army and minister of war. In 1941, under the German occupation, he became prime minister of a puppet government in Serbia. He was captured by the Partisans at the end of the war and is said to have committed suicide during the pretrial investigation.

Ante Pavelić (1889–1959)

Croatian fascist leader who in 1941, under Axis sponsorship, became head of the puppet Independent State of Croatia. His special troops, the Ustashi, massacred hundreds of thousands of Serbs, Jews, and Gypsies, and fought both Partisans and Chetniks. At the end of the war he escaped from Yugoslavia and took refuge in Argentina and Spain. Allegedly, he died in Madrid in 1959.

Peter II Karadjordević (1923–1971)

King of Yugoslavia after his father, King Alexander, was assassinated in 1934. His cousin Prince Paul ruled as regent until March 27, 1941, when the regency was overthrown by an officers' coup opposed to collaboration with the Axis. When the Axis Powers invaded Yugoslavia, Peter fled and established a government-in-exile in London. As the Partisans grew stronger, he was forced to accept a coalition government, was forbidden to return to Yugoslavia, and in 1945 was deposed when Yugoslavia was declared a republic. He spent the rest of his life in exile and died in the United States in 1971.

Moša Pijade (1890–1957)

Prominent Yugoslav Communist of Serbian Jewish origin. With Djilas, he led the Partisan uprising in Montenegro in 1941. He held high political posts during and after the war and was a member of the Central Committee and the Politburo. Proclaimed a National Hero.

Krsto Popivoda (1910–)

Prominent in the Partisan uprising in Montenegro from 1941. He was a member of the Central Committee and the Supreme Staff during the war and held high positions in the postwar government. Proclaimed a National Hero.

Koča Popović (1908–)

Communist volunteer in the Spanish Civil War who was interned in France. He joined the Partisans in 1941, commanded various units, and was chief of the General Staff from 1946–1953. He became foreign minister of Yugoslavia in 1946. Proclaimed a National Hero.

Alexander-Leka Ranković (1909–)

A leading Yugoslav Communist of Serbian origin who was a member of the Politburo from 1940. Captured and tortured by the Gestapo in 1941, he was rescued by a daring Communist raid. He served on the Supreme Staff throughout the war. After the war, he was minister of the interior and head of the military and secret police. He fell from power in 1964, ostensibly for abusing his authority, and was expelled from the party in 1966.

Ivan Ribar (1881–)

Prominent Croatian leftist before World War II. He joined the Partisans in 1941, became a Communist, and helped organize both sessions of AVNOJ (Antifascist Council of National Liberation of Yugoslavia). He was president of the Constituent Assembly in 1945.

Ivo-Lola Ribar (1916–1943)

Son of Ivan Ribar. He was an active Communist before and during the war, a member of the Central Committee, and head of the Communist Youth. He was killed by a German bomb while taking off on a mission to the Allied Middle East Command in 1943. Proclaimed a National Hero.

Vladislav Ribnikar (1900–1955)

Liberal prewar editor of the Belgrade newspaper *Politika*. He joined the Partisans in 1941, became a director of their news agency, TANJUG. He was minister of education in the postwar government.

BIOGRAPHICAL NOTES

Ivan Šubašić (1892–1955)

A prewar Croatian political leader who was a member of the Croatian Peasant party and became governor of Croatia in 1939. He fled Yugoslavia in 1941, served the royal government-in-exile, and in June 1944 became its prime minister. In this capacity he concluded two agreements with Tito, which led to a coalition government, with Tito as premier and himself as foreign minister. He resigned in 1945 for political reasons and reasons of health.

Velimir Terzić (1908–)

Captain in the prewar Royal Army. He joined the Partisans and the Communist party in 1941, became a general, and held high military posts during and after the war. He went with Djilas on the first Partisan mission to Moscow in 1944.

(Josip Broz) Tito (1892–)

Wartime and postwar leader of Yugoslavia. Born in Croatia, he was a locksmith and metalworker. Arrested for antiwar propaganda during World War I, he was sent to the front with the Austrian army, was wounded and captured by the Russians. He joined the Red International Guard during the October Revolution in 1917. Back in Yugoslavia, he joined the Communist party and rose in its ranks. He became secretary-general of the Yugoslav party and reorganized it. He led the Partisan movement from 1941. In 1945 he became premier of a coalition government, then head of the new People's Republic of Yugoslavia. He has been head of the state and the party ever since, surviving a quarrel with Stalin and a complete break with the Soviet Union in 1948.

Vlatko Velebit (1907–)

Croatian Communist who joined the Partisans in 1941. He served on the Supreme Staff. In 1943 he headed a military mission to Great Britain. After the war, he served as ambassador to Italy and to Great Britain. He was a delegate to the United Nations in the 1960's.

Radovan Vukanović (1906–)

Montenegrin Communist who commanded various Partisan units in the war. He became a general and held high military posts after the war. Proclaimed a National Hero.

Svetozar Vukmanović-Tempo (1912–)

A leading Montenegrin Communist and member of the Central Committee. During the war he served on the Supreme Staff, went on missions to Bulgaria, Greece, and Albania, and became Tito's personal representative in Macedonia. He held high positions in the postwar government. Proclaimed a National Hero.

Radovan Zogović (1907–)

Montenegrin Communist, journalist, and author. He did propaganda work for the party during and after the war.

Sreten Žujović (1899–)

Serbian veteran of World War I and long-time Communist. He was a member of the Central Committee and the Politburo before World War II. He helped organize the Partisan uprising in Serbia in 1941 and became a member of the Supreme Staff. Finance minister in the postwar government, he lost his party membership and high office when he sided with Stalin against Tito in 1948.

Index

Adžija (Communist), 10
Albahari, Mika, 116, 136, 211
Alexander (King of Yugoslavia), 129
Alexander, Harold, 400, 442
Alevandrov (officer), 383
Andjelić, Djukan, 87
Andjelić, Milovan, 66, 84, 87
Andrejević-Kun, Djordje, 6
Antonov, General, 435
Atanasov, Shterin, 357
Atherton, Terence, 144–45
Atrocities (executions, reprisals, massacres)
  Chetnik, 83, 98, 139–41, 149, 250–51, 304, 352, 402–3, 446–48
  German, 93–94, 113, 114, 283, 285, 291, 394, 402–3, 422–23
  against Gypsies, Jews and Serbs, 197–98
  Italian, 124–25, 182, 220, 268
  in Jajinci, 419
  prisoner execution, as policy, 227
  Russian, 420
  against spies, 61–62, 75, 84–85, 124–27
  in spring 1942, 155–56, 168–69
  Ustashi, 139–41, 192–93, 211–12, 291–94, 320–21, 326–27, 333, 409, 446–48
  against White Guards, 338
Auersperg, Prince Karl, 336, 337
Avgustinčić, Antun, 346, 379

Babić, Radomir, 73
Babović, Cana, 125, 191–92
Bailey, S. W., 228
Bakarić, Vladimir, 98, 312–14, 305–7, 451
Bakić (police official), 30–31
Bakić, Mitar, 56, 57, 142, 451
  Chetniks and, 148, 169
  in Fifth Offensive, 303
  S. Kovačević and, 202
  in new leadership (spring 1942), 157
  in offensive (spring 1942), 146, 152, 155
  sent to Dalmatia, 210
  sent to Serbia, 72, 82
Bašić, Ljubo, 160–61
Bećković, Matija, 161
Bećković, Vuk, 161
Belinić, Marko, 328
Beneš, E., 389
Beria, Lavrentia, 390, 407, 435
Berus, Anka, 318, 319, 329
Biroli, Pirzio, 26, 74
Bojović (lawyer), 351
Bojović, Archpriest, 29
Bojović, Ljubica, 39
Bošković, Lazar, 168
Bošković, Momir, 120
Bošković, Sekula, 168
Božović (painter), 65
Božović, Boro, 181
Božović, Saša, 181
Brkić, Duško, 324, 328, 329
Brković (mother of Savo), 14–15
Brković, Savo, 15–17, 24, 78–79, 152, 451
Broz, Herta, 240, 242, 244, 256, 311

INDEX

Bukharin, Nikolai, 389
Bulganin, Nikolai, 435
Buljan, Vice, 210
Bultović (brothers), 281
Burdžović, Rifat-Trso, 17–18, 62, 51, 71, 451
  in Chetnik-Partisan clash (October 1941), 88
  in Italian offensive (July 1941), 41
  perverts and, 127
Burić, Savo, 279

Čagorović, Djuro, 30, 76
Čalić, Dušan, 325
Cankar (man of letters), 340
Cavallero, Ugo, 215
Čengić, Ferid, 287
Cerović, Komnen, 71
Cerović, Stojan, 161–62
Četković, Mihailo, 87
Četković, Pero, 170
Chetniks, 28, 33, 93, 123, 205
  atrocities involving, 83, 98, 139–41, 149, 250–51, 304, 352, 402–3, 446–48
  Bihać operation against, 207
  Borač and, 158
  clash with Partisans in Serbia (October 1951), 88
  defeat of, in Serbia, 351–52
  defeating (1944–45), 40–3, 407, 443, 446–49
  disintegration of, 327
  Djurašinović killed by, 53
  driven from Boža, 72
  failures in Slavonia, 325
  in Fifth Offensive, 246–48, 270, 292, 296–98
  in Fourth Offensive, 215–45
  and German atrocities, 94
  German view of, 235
  hostility of (after January 1942), 141
  Kruščić killed by, 53
  Kuč killed by, 53
  Moscow and collaboration of, with invaders, 143–44
  and myth of Djilas responsibility for reprisals, 83
  oppose occupation, 96–97
  Partisans compared with, 124
  people abandoning, 250–51, 402
  Ružica killed by, 54
  and B. Selić, 46
  spread of movement of, 116, 117
  in spring 1942 attacks, 141–42, 146–48, 152–56, 159–61, 166–71, 174–79, 182–83
  use of term, instead of guerrilla, 71
  weaknesses of, 148–50
  Western support for, 350, 376
  in winter 1942 attacks, 131–35
Churchill, Randolph, 369
Churchill, Sir Winston, 187, 228, 248, 369
  and Allied influence in Yugoslavia, 422
  and captured German divisions, 442
  Deakin and, 252
  message of, on invasion of France, 390
  messages of, to Tito, 375–76
  and Peter II, 425
  Russian respect for, 438
  and Soviet-Yugoslav clash, 349
  Stalin on, 388–89
  Tito meets, 401
Ciano, Count Galeazzo, 215
Clausewitz, Karl von, 441
Čolaković, Rodoljub, 4–5
Čopić, Brando, 206–7
Čosić, Dobrica, 252
Čuković, Mirko, 287
Cvetković (Jereza leader), 44

Dakić, Radoje, 16, 17, 24, 156, 163–64, 451
Damjanović, Mališa, 51
Dangić, Jezdimir, 141
Dapčević, Peko, 57, 76, 77, 415, 451–52
  and Ceklin atrocities, 125
  Chetniks and, 148
  command of, 182
  and entering Serbia, 398
  and execution of spies, 84
  in Fifth Offensive, 303
  in Fourth Offensive, 221
  and Garibaldi Division, 351
  and incident with Zhdanov, 417
  in Italian offensive (July 1941), 31
  Kovačević and, 202
  in march to Bosnia (June 1942), 190
  in new leadership (spring 1942), 157
  in offensive (spring 1942), 159–61, 167–68
  on opening guerrilla war, 19
  and Red Army behavior, 416, 420, 421
  on Strem front, 440–42
Davičo (writer), 128
Davidović, Ljuba, 129

Deakin, F. W., 253, 301
Dedijer, Olga, 199, 210
Dedijer, Vladimir, 83, 103, 142, 144, 192, 452
in agitprop, 416
in Fourth Offensive, 217, 219, 231, 234–35
in German offensive (November 1941), 108–13
military mission to Russia and diary of, 381
press requisitioned by, 196
and publication of *Borba,* 310
report on meeting Stalin, 393
in spring 1942 fighting, 177–78
wife of, exchanged, 199
wounded, 311, 354
De Gaulle, Charles, 355, 437
Demonja (soldier), 325
Dimotrov, Georgi, 355, 422, 438, 452
and military mission to Russia, 380–84
report on meeting, 393
Dippold, Benignus, 233
Djilas (father), 417
Djilas (mother), 418
Djilas, Aleksa, 44, 71, 418
death of, 78, 87–88
first ambush by, 22–23
in Italian offensive (July 1941), 37, 46, 47, 51–54
joins guerrilla war, 6, 8–11, 15
Djilas, Dobrana, 243, 418
Djilas, Milovoje, 98, 418
Djilas, Mitra, 62, 77, 83, 174, 240, 452
arrested, 5–7
changes in, 91
in Chetnik attack (January 1942), 132
in conference of women, 210
conflict with, 432
in delegation to Moscow, 428–30
Djordjević and, 125
estranged, 195–96
in Fifth Offensive, 269, 286–89
in German offensive (November 1941), 105
on government ministers, 93
O. Jojić and, 98
on liberation of Belgrade, 416
named to party leadership, 120
in new regime, 371, 372
occupying a room with, 206
and offensive of spring 1942, 146, 152
Pijade and, 175
and publication of *Borba,* 310
in raid on Drvar, 394
and raid on munitions factory, 101, 102
report on meeting Stalin, 393
Djilas, Štefica, 52
Djilas, Stevan, 167
Djilas, Stoja, 53
Djilas, Tofil, 53
Djokić, Lieutenant, 298–99
Djordejević, Mika, 120, 125–26
Djujić (Chetnik), 441
Djukanović, Vojo, 13
Djurašinović, Milosav, 23, 52
Djurašinović, Unčic, 52–53
Djurić, Željko, 113
Djuričković, Boško, 30, 31, 191
Djurišić, Pavle, 244, 297, 441
fate of, 252–53
against Italians, 29
as leader, 150–51
Ustashi and, 447–48
Vasojević clan and, 160
Djurović, Milinko, 148
Dostoevsky, Feodor, 435
Dragićević, Veljko, 71
Draguna, Aunt, 86, 87
Drapšin, Petar, 322–29
Drinka (secretary to Pijade), 225
Drenović (Chetnik), 205
Drljević, Savo, 168, 170, 430
Drljević, Sekula, 66, 76, 448
Dudić, Dragojlo, 113
Dugalić, Duško, 191, 192

Eden, Sir Anthony, 219, 348
Ehrenburg, Elya, 202
Eileen (mistress of Nedeljković), 50, 51
Engels, Friedrich, 247
Executions, *see* Atrocities

Franco, Francisco, 223
Francetić, Jure, 192, 199
Franz Ferdinand (Archduke), 207
Frol, Ivo, 243, 244

Gani-beg, 69
Gaulle, Charles de, 355, 437
Gažević, Nikola, 289, 290
Goethe, Johann von, 230

INDEX

German offensives
November 1941, 102–14
January 1942, 141
September 1942, 194–99, 202
January 1943 (Operation Weiss; Fourth Offensive), 215–45
May 1943 (Operation Schwarz; Fifth Offensive), 245–305
August 1943, 325, 332–33, 343–44
1944, 327, 368, 376
Goebbels, Joseph, 241
Gorkić (secretary of Central Committee), 69
Gorshkov, General, 373
Gošnjak, Ivan, 319–20
Gottwald, 249, 381, 382
Gregorić, Pavle, 314, 317, 324
Grigoryev, Major, 373
Grol, Milan, 431–32
Grujić (school friend), 171
Gutman, Dr., 297

Hamović, Rade, 190
Haris, Ilja, 326
Hebrang, Andrija, 209–10, 452
in delegation to Moscow, 428, 429
policies of, 315–17
replacing, 407, 410
Hebrang, Olga, 209, 316
Hegel, G. W. F., 50
Hitler, Adolf, 5, 118
attempt on life of, 399
Belgrade and, 400
and crushing Yugoslav resistance, 215, 216
German officers and, 234, 236
local feuds and, 198
Napoleon's campaign compared with that of, 206
Ott on, 241
retreating armies of, 404
Serbians and, 94
truce negotiations and, 245
Holjevac, Veco, 319–20
Hudson, Duane T., 68–72, 105, 108, 112, 114, 227, 452
Humo, Olga, 375
Huot, Louis, 349

Ilić, Pavle, 246
Italian offensives
fighting capacities in, 21
first attacks on Italians, 22–26
July–August 1941, 29–48, 51–58
December 1941, 115–17, 119, 120
January 1942, 132–37
spring 1942, 141–42, 146, 165–66, 180–83
January 1943, 219–21, 223–45
April 1943, 246–48
May 1943, 252, 259
Ivanović, Dragiša, 282, 284, 291
Ivo (groom to Tito), 195

Jakšić (commander), 263
Jančić, Olga, 244
Jančić, Štefica, 244
Janković, Danilo, 269
Jevdjević (Chetnik), 215
Jojić, Olga, 98
Joseph II (King of Yugoslavia), 337
Jovančević (Communist), 168–69
Jovanović, Arso, 50, 56, 67, 347, 452
and Allied front in Balkans, 212
and attack on Plevlje, 116
in attack on Ustashi (August 1942), 199
bitterness of, over surrender of government, 70
in Borač operation, 158
as chest player, 370
as chief of Supreme Staff, 311
in delegation to Moscow, 428, 429
in Fifth Offensive, 248
functions of, 398
and Istria landing, 400
in Italian offensive (July 1941), 35, 37–39, 43, 44, 48
in march to Bosnia (June 1942), 188–90, 193
as military chief, 29
in offensive in spring 1942, 146, 173, 177–78
and opening of war, 19–20
Orthodox Church, liberation of Belgrade and, 428
Plevlja defeat and, 116, 117
in raid on Drvar, 394
sent to Serbia, 72, 77, 82
sent to Slovania, 207
and Stanišić, 27, 28
Tito and, 183, 222
on war on Russian front, 73
work habits of, 393

Jovanović, Blažo, 18, 27, 67, 76, 150, 452
  and Chetnik surrenders, 351
  and death of Aleksa Djilas, 78
  in Fifth Offensive, 270
  in Italian offensive (July 1941), 31–33, 56
  leads attack on Italians, 73, 74
  in negotiations with Lašić, 117
  in new leadership (spring 1942), 156
  and opening of guerrilla war, 14–19, 25
  as political commissar, 29
  in reorganization of party, 62, 63
  residence for, 445
  sent to Albania (August 1942), 200
  and Tadićes, 165
Jovanović, Dragoljub, 69, 427
Jovanović, Lidija, 33, 200
Jovanović, Nikola, 84
Jovanović, Rusa, 62
Jovanović, Žikica, 100
Jovanović-Jarac, Djordje, 127–28
Jurišević, Lenka, 243

Kalinin, Mikhail, 434
Kant, Immanuel, 230
Kapicić, Jovo, 170, 171
Karadžić, Vuk, 156
Kardelj, Edvard, 91, 92, 118, 201, 249, 319, 453
  and Allied influence in Yugoslavia, 421–22
  characteristics of, 93, 344
  censorship opposed by, 425
  and Chetnik attack, 97
  at December 7, 1941 meeting, 118–19
  and definition of struggle in Serbia (1941), 100
  on Djilas, 83, 345
  on errors in Serbia in 1941, 99
  and explusion of German minority, 423
  in Fifth Offensive, 293
  in formalizations in new regime, 371
  and formation of First Proletarian Brigade, 122, 123
  in forming new regime, 346, 353–59
  functions of, 352
  in German offensive (November 1941), 105, 108, 114
  Hebrang and, 316–17
  Kocbek and, 336
  and Ljubljana, 338–39
  and making Tito a marshal, 360
  on massacres, 338
  military rank of, 371
  and new government, 431, 432
  in new regime, 372, 409–11
  and offer to enemies to join Partisans, 404
  Pijade and, 38
  on policy of neutrality, 402
  in raid on Drvar, 394
  and Red Army behavior, 421
  relations with, 95
  and return to Bosnia (1943), 343
  Russian medal received by, 421
  and Slovenia, 141, 332, 334, 340, 341
  and Soteska Manor, 337
  and spies, 10
  and Šubašić, 396, 420
  Tito and, 362, 363
  Vis and, 407
Keitel, Wilhelm, 215
Keršovani (wife), 374
Keršovani, Otokar, 10, 374
Keserović (Chetnik), 151
Khrushchev, Nikita, 438–39
Kidrić, Boris, 334, 453
  Kocbek and, 336
  and Slovenia, 340
  and Soteska Manor, 337
Kidrić, Zelenka, 339
Kladarina (commissar), 263
Kljajić, Filip, 108
Kocbek, Edvard, 334, 336, 343
Kolarov (Bulgarian leader), 382–83
Končar, Rade, 10
Konyev, I. S., 384, 385, 407
Korneyev, General, 373, 405, 406, 421
Korotayev, General, 384, 385
Kosanović, Sava, 409
Kovačević, Boriša, 190
Kovačević, Sava, 453
  characteristics of, 153–55
  command of, 182, 183
  death of, 279–82, 295
  and disarming of peasants, 165, 166
  in Fifth Offensive, 257, 260–63, 267, 273–80
  in Fourth Offensive, 220, 223, 239
  inspection tour by, 165–66
  in new leadership (spring 1942), 157
  Orović and, 165
  reaches Bosnia (1942), 201–2
  and Tadić family, 164

INDEX

Kovačić, Goran, 218, 294–95
Kovijanić, Stanija, 50, 87
Kovijanić, Vuko, 87
Krleža, Miroslav, 253, 303
Kruščić, Vukman, 53–54
Kuč, Milan, 53–54
Kulenović, Skender, 206–7
Kun, Djordje, 361
Kušić, Milinko, 133–39, 191, 211

Lakićević, Milutin, 54, 171–72
Lalaković, Petar, 178–79
Lalaković (officer), 71, 72
Lalić, Mihailo, 41
Lašić, Djordje, 39, 117
Lawrence, T. E., 69
Lazarev, Boško, 168
Lehman, Herbert, 381
Lekić, Danilo, 224, 300
Leković, Voja, 88, 120
Lenin, V. I., 59, 127, 143, 285
Leskovšek, Luka, 334
Ljotić, Dimitrije, 48, 93, 95–96, 99, 419,
    431, 453
Ljumović, Božo, 14, 16, 18, 75–76
Löhr, Alexander, 215, 237
Lorković, Mladen, 441

McDowell, Robert H., 425–26
Maček, Vlatko, 315, 371, 453
Maclean, Fitzroy H. R., 348–49, 403, 453
Maksimović, Desanka, 52
Malčick (official), 369
Malenkov, Georgi, 435
Malinovsky, Rodion Ya., 407
Mandić, Ante, 410–11
Mandić, Gligo, 220
    in Fifth Offensive, 273–76, 283, 300
Manuilsky, Dmitri Z., 367, 381, 383, 438,
    439, 453
Maria Theresa (Empress of Austria), 337
Marinović, Dika, 77
Marković, Miloš, 234
Martinović, Dr., 76
Marušić, Slobodan, 66, 84
Marx, Karl, 50, 59, 127, 247, 285
Masleša, Veselin, 163, 453
    in Bihać operation, 208
    death of, 293–94
    in Fifth Offensive, 286, 287
Massacres, *see* Atrocities
Matetić, Vlado, 328
Matić (writer), 128
Mayakovsky, Vladimir, 371
Mazolini, Sefafin, 13–14
Medenica, Miloš Dragišin, 36
Medenica, Novo, 66
Medenica Radun, 36
Mesić (commander), 384
Meštović, Julka, 131
Mićanović, Veljko, 17–18, 156–57
Mihailović, Draža, 141, 317, 454
    abandoned, 251–52
    agreement reached with Tito, 88
    Allies and support for, 347–50
    attacks Communist, 97, 98
    at Ba congress, 377–78
    denounced (spring 1942), 182
    fate of forces of, 447–49
    fighting (1944), 377, 441
    followers of, in Belgrade, 420
    in Fourth Offensive, 219, 226–28
    in German offensive (November
        1941), 112
    killing followers of, 98–99
    and Lašić, 117
    moderation opposed by, 352
    Moscow and, 144
    Nedić and, 403
    on political spectrum, 96
    Todorović and, 144, 145
    trial of, 126
    U.S. mission with, 425
    Western and royal government relations with, 68, 71, 72, 114, 248, 253,
        368, 375–76
Mika, Aunt, 170
Mikailovich, Sergei, 139
Miletić, Bozo, 281
Miletić, Petko, 156, 281
Milosavljević, Ljubinka, 131–32, 135
Milosević, Simo, 278, 294–95
Milutinović (wife of Ivan), 418
Milutinović, Ivan, 20, 24, 146, 454
    and Andjelić, 87
    and attack on Plevlja, 116
    criticized (December 1941), 122–23
    death of, 393–94, 417, 418
    in Fifth Offensive, 257, 261–64, 266,
        269, 275, 279–83, 286–93
    letter of, to Central Committee, 85–86
    in offensive in spring 1942, 154–56, 174,
        181–82
    Pijade and, 17

in Plevlja defeat, 116, 117
replaces Djilas, 79–81
role in reprisals, 84, 85
Sekulić and, 150
and Tadić family, 164
and Tito's visit to Russia, 405
Vasojević Chetniks and, 169
working with, 151–52
Zogović and, 163
Milutinović, Todor, 20, 24
Minić, Ljubo, 45–47, 66
Mitrović, Dr., 194, 196
Mitrović, Golub, 304
Mitrović, Stefan, 31–33, 76
Molotov, V., 119, 348, 381
in conservations with Stalin, 437, 438
and delegation to Moscow, 434–35
military misson and, 386–91
provoking Yugoslav delegation, 428
Morgan, William, 442
Mugoša, Dušan, 200
Mussolini, Benito, 88, 304

Napoleon I (Emperor of the French), 57, 206
Natlačen (viceroy), 339
Nazor, Vladimir, 212, 268–69, 454
Croatian leadership and, 317
in Fifth Offensive, 263–64, 303
in Fourth Offensive, 221–22
on return to Croatia, 310–13
Nedić, Milan, 99, 111, 123, 141, 205, 454
administration of, broken, 100
attack by (January 1942), 131, 132
Djurišić and, 252–53
fighting (1944), 377
and German atrocities, 93, 94
in German offensive (November 1941), 114
Mihailović and, 403
in political spectrum, 96
reign of terror of, 352, 419
Nedeljković, Dušan, 50–51, 82
Nedeljković, Jevrem, 436
Nedeljković, Raja, 283, 286, 288–89, 295
Nešković, Blagoje, 351, 398
Nešović, Slobodan, 17
Neuhausen (German governor), 415
Nicholas, Prince, 158
Nikić, Nikota, 60–61
Nikolis, Gojko, 201
Ninčić (émigré minister), 375

Njegoš (writer), 150
Novaković, General, 145
Novović, Bogdan, 38
Nušić (playwright), 346

Obradović, Branko, 220, 225
Opačić, Čanica, 320
Oreščanin, Bogdan, 319
Orović, Savo, 157, 165
Ostojić, Zaharije, 71, 72, 227
Ott, Hans, 198–99, 241–42

Pantić, Miša, 136
Pap, Pavle, 10
Paunovic, Davorjanka (Zdenka)
fears and rages of, 80, 175–77, 240, 311
in German offensive (November 1941), 109
in march to Bosnia (January 1942), 188
in raid on Drvar, 394
and raid on munitions factory, 102
Ranković breaks with, 370
run-in with, 254
Pauker, Ana, 357
Pavelić, Ante, 12, 141, 205, 330, 454
fighting by forces of (1944), 441
Hebrang and, 209
Hitler and, 215
massacre under, 198
royal government and, 226, 396
"state" of, 207, 329
units of forces of, in Russia, 384
Pavićević, Aleksa, 66
Pavlović (lawyer), 108
Pavlović, Dara (Vera Blandie), 111, 114
Pećanac, Kosta, 71, 205
Pekić (wife of Pavle), 289, 290
Pejačević, Count, 328
Pekić, Pavle, 156, 178, 289
Penezić, Slobodan, 101–2, 136
Perović, Puniša, 30
Peter II (King of Yugoslavia), 187, 298, 375–76, 454
abandons Chetniks, 402
possibility of Tito meeting, 399, 401
regency and, 425, 431
replacing, 421
representative of, 395–403
Stalin on, 407
Petronijević, Andra, 129–30

INDEX

Pijade, Moša, 44, 47, 50, 51, 201, 252, 455
  and animal farms, 175
  Borba and, 203
  and Central Committee, 16, 118
  characteristics of, 37–38
  and charge of opportunism, 80–81
  and dissolution of Comintern, 249
  Djilos opposed by, 16–17, 25, 181
  and expulsions of Popivoda and L. Jovanović, 200
  in Fifth Offensive, 254, 256, 257
  in Fourth Offensive, 225, 229
  in forming new regime, 354, 356, 357, 361
  functions of, 352
  in Italian offensive (July 1941), 37, 56
  and "Letter from Comrade Veljko," 63
  military rank of, 371
  and Minić, 47, 48
  and Orthodox Church celebration of liberation of Belgrade, 428
  and people's committees, 141
  and Plevlja defeat, 117, 123
  and L. Ribar, 363
  sent to Durmitor, 57
  sets fire to Zabljak, 178
  and unjustified executions, 156
  Vasojević Chetniks and, 168–70
Plamenac, Petar, 76
Planojević, Mileva, 247
Plećević, Čeda, 129
Popivoda, Krsto, 56, 157, 200, 247, 455
  in Provisional Committee, 16, 17
  and scaling movement to guerrilla proportions, 24
  and spies, 57–58
Popović, Koča, 174, 194, 347, 455
  characteristics of, 370
  in Fifth Offensive, 300–1
  in formation of new regime, 360
  in Fourth Offensive, 230–36, 241
  in German offensive (November 1941), 104
  and Red Army behavior, 420
  sent to command in Serbia (1944), 398
  on Strem front, 442
Popović, Major, 67
Popović, Milorad, 292
Popović, Vladimir, 209
Poptomov (delegate), 357
Posderac, Hamdija, 268
Pozderac, Nurija, 264, 268, 272
Predavec (engineer), 378
Prica (Communist), 10
Princip, Gavrilo, 207
Prodanović (son of Jaša), 427
Prodanović, Joša, 426–27
Pucar, Djuro, 206–7
Purić, Bozidar, 317

Radić (Chetnik leader), 205
Radonjić, Miljan, 69
Radović, Marko, 16
Ranković, Andja, 179–80
Ranković, Alexander-Leka, 10, 91, 115–16, 212, 455
  and British intervention in Greece, 426
  characteristics of, 93, 370
  Chetniks characterized by, 97–98
  and collaboration with the King, 375–76
  and creation of the Guard, 419
  on dating uprising in Serbia, 100
  and death of Andja, 179
  at December 7, 1941 meeting, 118, 119
  and defeat of Chetniks, 351
  and dissolution of Comintern, 249
  Djilas punishment and (spring 1942), 181, 182
  education of cadres under, 201
  on expulsion of German minority, 423
  and expulsions of Popivoda and L. Jovanović, 200
  and fall of Mihailović, 447
  In Fifth Offensive, 248, 257, 262, 293, 294, 299–302
  and First Session of AVNOJ, 209
  in formalizations of new regime, 371–72
  in forming new regime, 346, 354–56, 358, 361–62
  in Fourth Offensive, 218–19, 225, 229, 230, 237–39, 244
  functions of, 352, 398, 429
  in German offensive (November 1941), 105–8, 112
  Gošnjak compared with, 320
  Hebrang and, 210, 407
  interrogation of Ustashi by, 196–97
  Kovačević and, 202
  Kusić and, 211
  and liberation of Slovenia, 339
  and Livno attack, 194

and military mission to Russia, 379, 380
and negotiations with Chetniks, 94–95, 97, 98
in new regime, 370, 375
in offensive (spring 1942), 146
and offer to enemies to join Partisans, 404
organizational work under, 141
Ott and, 242
Pijade and, 17, 38
political sections created by, 195
and prison matters, 101
in raid on Drvar, 394
and raid on munitions factory, 102, 103
and Red Army behavior, 421
and report on meeting Stalin, 393
and Sandžak region, 174
and secret police service, 250
speaking with, 94–95
and Strem front, 440
Šubašić and, 395
Tito and, 188, 362, 363
and Tito's visit to Russia, 405, 406
and Tito-Zdenka relationship, 176
and Uncle Jaša, 427
Žujović and, 301
Ranković, Novica, 76
Reprisals, *see* Atrocities
Ribar, Ivan, 129, 200, 310, 455
and death of his sons, 358–63
in Fifth Offensive, 263–64
in new regime, 371
residence of, 415
and return to Bosnia (1943), 343
sent to Otočac, 332
and Slovenia, 342
Ribar, Ivo-Lola, 17, 65, 115–16, 129, 200, 455
death of, 358–63
at December 7, 1941 meeting, 119
in German offensive (November 1941), 103–5
in offensive of spring 1942, 181
as orator, 210
pamphlet written with (June 1942), 187–88
relationship with, 95
sent of Zagreb, 141
Ribar, Jurica, 358–59
Ribbentrop, Joachim von, 215
Ribnikar, Vladislav, 5, 7, 416–17, 455
Richtman (Communist), 10
Rip, Ružica, 54
Ristić (writer), 128
Roatta, Mario, 215
Roberts, Walter R., 83, 145, 234
Robespierre, Maximilien de, 134
Rokvić, Mane, 312
Roosevelt, Franklin D., 187, 349–50, 389
Rukavina, Ivo, 313, 314
Rus, Josip, 334–36

Saint-Just, Louis de, 134
Sakharov, Major, 373
Saranović, Milovan, 165
Šašić, Jevto, 324–25
Sava, Saint, 377, 428
Savić, Pavle, 379
Šćepanović, Jovo, 51
Šćepanović, Mileva, 51–52, 88
Šćepanović, Tadiša, 51, 52
Scobie, Ronald, 426
Šegrt, Vlado, 324
Sekulić, Bajo, 77, 117–18, 150
Selić, Blagota, 44–45
Selić, Boba, 43, 54
Selić, Vladeta, 43, 54
Selić, Vojo, 54
Sentjurc, Lidija, 334
Sforza, Count, 408
Sikorski, W., 389
Šilović (physician), 374
Simonov, Konstantin, 386
Simonović, Jagoš, 44–45
Škerović, Slobodan, 6
Šlander (Communist), 334
Smirnov, Vladimir, 176, 224
Smodlaka, Josip, 395
Smodlaka, Slaven, 396
Sočica (governor), 175
Šolaja, Simo, 193
Srna, Hasan, 136–37
Stalin, Joseph, 52, 109, 116, 117, 204
and Allied influenced in Yugoslavia, 422
anger of, over independent Yugoslavia, 367
characteristics of, 434–36
conversations with, 385–91, 434–39
demoralize Yugoslav delegation, 428–29
and formation of First Proletarian Brigade, 120

Stalin, Joseph (*cont.*)
gift for, 379–80
and leadership of Tito, 429–30
and military mission to Russia, 379–93
publically mentions partisans, 162
quarrel with (1948), 118–21, 144, 210, 301, 402, 422, 423
and Red Army struggle for victory, 416
Tito imitates, 418
Tito on Macedonia and, 398–99
Tito meets, 406, 407
Tito receives gift from, 391
Togliatti and quarrel with, 408–9
use of works of, 247
warnings on safety given by, 405
Stalin, Yasha, 436
Stana (cousin to M. Djilas), 189
Stanišić, Bajo, 27–28, 351, 352
Stanojlović, Moma, 280–81
Stauffenberg, Claus, Graf von, 399
Stefanović, Svetislav, 146, 152
Stevenson, Ralph, 395
Stilinović, Marijan, 109
Stoecker, Major, 229, 237, 239–40
Stojadinović (Jereza leader), 44
Šubašić, Ivan, 389, 390, 409, 456
in coalition government, 431
confronting, 409–10
in delegation to Moscow, 433–35
as King's representative, 395–403
negotiations with, 420–422, 426
and regency, 425

Tadić, Ljubo, 164
Tadić, Spasoje, 164
Tadić, Tadija, 164–65
Terzić, Velimir, 168
as chief of staff of Croatia, 310–11
crossing the Drina, 246
in Fourth Offensive, 217–19, 223, 229
in Kolašin attack (spring 1942), 170
in military mission to Russia, 379, 385, 386
Tito and, 183
Thorez, Maurice, 409
Tiljak (police agent), 10
Timofeyev, Colonel, 429
Tito (Josip Broz), 72, 114, 137, 408
agreement between Mihailović and, 88
article by (December 1942), 205
Atherton and, 145
and attack on Plevljo, 116–17
becomes marshal, 359–62
biographical note on, 456
birth of, 370
and bombing of Belgrade, 400
and British intervention in Greece, 426
and British mission (1943), 253
censorship and, 425
charateristics of, 95, 176–79, 183–84, 188, 201, 222, 232, 361, 394–95, 402
Churchill meets, 401
Churchill messages to, 375–76
in coalition government, 431–32
crossing the Drina, 246–47
at December 7, 1941 meeting, 118–19
decisions of (spring 1942), 172–74, 183
and decisions of Central Committee, 122
defeating Chetniks, 97, 449
in delegation to Moscow, 433–39
diet of, 116
discussions on future with, 344
Djilas dismissed by (1941), 79–83, 85–86
errors criticized by (spring 1942), 181–82
and expulsion of German minority, 423–424
in Fifth Offensive, 248–49, 254–60, 264, 265, 293, 294, 299–304
and First Session of AVNOJ, 208–9
in formalizations in new regime, 371, 372
forming new regime, 346–59
forms First Proletarian Brigade, 120–23
forms Second Proletarian Brigade, 138
in Fourth Offensive, 216, 219–32, 240, 244
in German offensive (November 1941), 103–14
Gorkić and, 69
Gošnjak and, 320
groom of, 195
Hebrang and, 209–10, 315–17
and Herta, 241, 242
illusions of, 141
and Istria landing, 400
A. Jovanović and, 207
Kolašin attack and (spring 1942), 171
Kovačević and, 202
as leader, 362–63
and liberation of Slovenia, 339–40

in march to Bosnia (June 1942), 188–93
meeting with (October 1941), 91, 93
meeting of leadership (October 1943), 343
as military commander, 92
and military mission to Russia (1944), 379, 380, 383, 385, 392, 393
and mobile shock battalions, 151
Moscow and (1943), 143–44
and Moscow recognition of Yugoslav émigré government, 199
Nazor and, 212
near capture of, 387*n*
and new regime, 367–70
and offer to enemies to join Partisans, 403, 404
and opening of guerrilla war, 4, 5, 7–8
in operations in Serbia (1944), 377
Orović and, 165
and Orthodox Church, 427–28
parting with, 7
and People's Youth, 65
picks leadership, 37–38
Pijade and, 17, 203
and Plevlja defeat, 116, 117
poem on, 346, 371
popular uprising and instructions of, 23–25
and publication of *Politika,* 417
punishment ordered by (June 1942), 187
in raid on Drvar, 394
and raid on munitions factory, 102
and Red Army behavior, 420–21
reputation of, 310
residence of, 415
in return to Croatia, 313
returns to Belgrade (October 1944), 418–21
Russian attempts to weaken leadership of, 429–30
and Russian mission (1944), 373–75
safety of (1944), 388, 389
and secret police, 250
and sectarian errors in Serbia (1941), 99
and situation in spring 1942, 142–43, 146
Stalin and leadership of, 429–30
Stalin imitated by, 418
Stalin and Macedonian question, 398–99
Stalin meets, 406, 407
Stalin's gift to, 391
in Strem front, 440
and struggle in Albania, 200
Šubašić and, 395–97
and terms partisan and guerrilla, 57
Trieste and, 442
and unification with Bulgaria, 422
Vasojević Chetniks and, 169
and victory celebration day, 443, 444
and victory in Serbia, 352
view of peasants held by (1941), 99–100
visits Russia, 405–6
weight gained by, 346
wounded, 265
Zdenka and, *see* Paunović, Davorjanka
Todorović, Boska, 144, 145
Todorović, Miljalko, 196, 304
Togliatti, Palmiro, 407–9
Tomić, Mirko, 110–12
Tomović (worker), 6
Tomović, Budo, 16, 62, 64, 65, 150, 180
Tomović, Lazar, 168
Topalović, Živko, 253, 378
Trotsky, Leon, 389

Usković, Vido, 16
Ustashi, 9, 205, 243–44
activities of, in Croatia, 322–25
atrocities involving, 139–41, 192–93, 211–12, 291–94, 320–21, 326–27, 333, 409, 446–48
authority of, 10
Bihać operation against, 207–8
and Borač operation, 158
Chetnik opposition to, 96
defeat of (1945), 443, 446
and eastern Bosnia, 141
in Fifth Offensive, 259, 291, 292, 295–97, 302
fighting (August 1942), 194–99, 202
fighting, in Croatia, 314
followers of, 10–11
in Fourth Offensive, 216–18, 225, 232, 233, 242
German view of, 235–37
ideology of, 11–12
Livno taken from, 211

Ustashi (*cont.*)
Partisans compared with, 124
raid by (September 1943), 330
strength of, 198

Vasić, Dragiša, 251–52, 448
Velebit, Vlatko, 199, 456
in Fourth Offensive, 229, 230, 234–35, 240–44
and military mission to British, 358
Veljko (Djilas pseudonym), 63
Veselin, Vojo (Stranjo), 281
Vidmar, Josip, 334, 335, 343, 360
Vlahov (delegate), 357
Vlahović, Veljko, 249, 367, 381, 382
Vučinić, Drago, 76
Vučinić, Dušan, 66, 67
Vučković, Ljubo, 190
Vučo (writer), 128
Vujačić, Marko, 165, 264
Vujošević, Periša, 16
Vujović, Djuro, 176, 418
Vukanović, Radovan, 456
attacks Italians, 73
in Fifth Offensive, 257, 262–64, 276–77, 283, 286–91, 298, 300
Vukmanović-Tempo, Svetozar, 4, 122, 190, 456
in beginning armed struggle, 5–6
and relations with Allies, 350
sent to Belgrade, 207
and uprising in Macedonia (1944), 398
Vukošević, Pariša, 14

Wilson, Henry Maitland, 399

Zarubica, Milorad, 52
Zdenka, *see* Paunović, Davorjanka
Zečević, Father Vlada, 108
Zečević, Milutin, 6
Zečević, Vlado, 114, 201
Zhdanov, General, 416
Zhukov, Georgi, 386
Zidanšek, Tomšić, 334
Zogović (wife of Radovan), 430
Zogović, Radovan, 38, 207, 457
in agitprop work, 416
and arrival of Red Army on border, 404
characteristics of, 372
and First Session of AVNOJ, 208
and image of Saint Sava, 428
in new regime, 371
poem by, 346, 371
purge and, 163–64
in raid on Drvar, 394
report on meeting Stalin, 393
and secret police, 250
Zora (Mitra's sister), 7
Žujović, Sreten, 118, 119, 457
and Allies landing in Istria, 400
ambition of, 395
on the British maneuvering of Tito, 401–2
characteristics of, 93
Chetniks opposed by, 98
on Djilas, 345
Djilas punishment and (spring 1942), 181
in Fifth Offensive, 257, 300–1
in First Session of AVNOJ, 208
forming new regime, 246, 354
in German offensive (November 1941), 108
Gošnjak and, 319–20
Hebrang and, 315, 316
as lecturer, 247
and meeting of leadership (October 1943), 343
military talent of, 184
on Mitrović, 304
and offensive (spring 1942), 146
as orator, 210
organizational work under, 141
in raid on Drvar, 394
in return to Croatia, 313, 314
sent to Otočac, 332
in Slavonia (September 1943), 333–34
tours Croatia, 310
visiting, 95